SPSS® Statistics Workbook

by Jesus Salcedo, PhD
and Keith McCormick

SPSS® Statistics Workbook For Dummies®

Published by: John Wiley & Sons, Inc., 111 River Street, Hoboken, NJ 07030-5774, www.wiley.com

Published simultaneously in Canada

For general information on our other products and services, please contact our Customer Care Department within the U.S. at 877-762-2974, outside the U.S. at 317-572-3993, or fax 317-572-4002. For technical support, please visit https://hub.wiley.com/community/support/dummies.

Wiley publishes in a variety of print and electronic formats and by print-on-demand. Some material included with standard print versions of this book may not be included in e-books or in print-on-demand. If this book refers to media such as a CD or DVD that is not included in the version you purchased, you may download this material at http://booksupport.wiley.com. For more information about Wiley products, visit www.wiley.com.

ISBN 978-1-394-15630-6 (pbk); ISBN 978-1-394-15632-0 (ebk); ISBN 978-1-394-15631-3 (ebk)

SKY10048251_052223

Contents at a Glance

Table of Contents

Introduction

Do you need to use IBM SPSS Statistics for school or work and you feel that you need more practice to become proficient? Maybe you eventually get the results you need when you work in SPSS, but it feels like you aren't doing so efficiently. Maybe you're about to embark on an important assignment or research project, and you feel like some more practice (with carefully presented solutions) is just what you need.

Not to worry. Help has arrived. *SPSS Statistics Workbook For Dummies* is a collection of exercises and solutions to move you towards mastery with SPSS. It is a companion book to *SPSS Statistics For Dummies*, 4th Edition, but you don't need to read both. (You see more about how the books work together later in this introduction.)

We (Keith and Jesus) have trained tens of thousands of students during in-person workshops for SPSS Inc. (before IBM acquired SPSS). That experience, over many years, means we've looked over the shoulders of attendees as they've attempted exercises just like these. We can anticipate possible mistakes before you make them — and if you practice the examples in this workbook, you won't make them at all. We've used that experience, along with the topics that IBM has identified as important for certification, to keep you focused on the critical skills for using SPSS efficiently.

About This Book

This workbook is designed to develop your ability to perform the most common and important tasks in the efficient use of SPSS Statistics. Like all books in the *Dummies* series, it features easy-access organization and doesn't bog you down in unnecessary details. We carefully listed, discussed, debated, and narrowed the topics to maximize an efficient investment of your time.

We understand and value academic pursuits, but this is not an academic book. This is an eminently practical book written by two authors who have spent decades leading software training workshops. However, if you're a student, a researcher, or an academic, we have you covered. We explain critical statistical concepts, choosing the right technique, and interpreting your results. All that and more is explained from a practical point of view.

So why choose this workbook?

- Dozens of practical examples with step by step solutions
- Advice from two lifelong users of SPSS Statistics

- Topics carefully chosen and prioritized for the tasks you will encounter most often
- Tips and tricks to avoid the most common mistakes from two authors who have observed (and made) every mistake that is possible to make in SPSS

We follow a few conventions in this workbook:

- Most chapters in this workbook will have the following repeating structure: explanation, example, questions, and answers. The answers are found at the end of each chapter. While skipping around chapters and sections to find what you need is encouraged, you'll probably have the best luck reading each section, as indicated with a major heading, in its entirety.
- Most of the exercises use the menus and dialogs, so when we show you a series of mouse clicks to access a specific area of the menus, we list the steps like this: choose File ⇨ Open ⇨ Data and load the GSS2018.sav file.
- Two chapters, 15 and 16, focus on using SPSS Syntax. SPSS Syntax programming code appears in `monofont` to help differentiate it from other text.
- You are encouraged to write directly in this workbook when trying interpretive questions. When the questions prompt for an interpretation like "Describe your result," space is provided for you to write your answer.

Foolish Assumptions

This book is for you if you have IBM SPSS Statistics and want to practice. Because it's a workbook, you get dozens of examples to help you prepare data, create variables, produce statistical analyses, and even write SPSS Syntax. You can even check out practice questions for IBM's certification exam in the final section of the book.

This workbook has been written as a companion to the book *SPSS Statistics For Dummies*, 4th Edition. That book, which is written for first-time users of SPSS, has more theory, overview, and initial setup. You aren't required to have both books, but they make a perfect pair. *SPSS Statistics For Dummies* is the best place to start if you are at ground zero. You can start there, and then try the exercises in this one. Or, and we like this option even better, use them side by side.

This book is about IBM SPSS Statistics. We sometimes call it SPSS Statistics, or even just SPSS. You may encounter something called IBM SPSS Modeler. That is different software, and this book will not discuss SPSS Modeler at all (except just now).

Icons Used in This Book

TIP

The tip icon marks tips and shortcuts to make working in SPSS Statistics quicker and easier.

REMEMBER

Remember icons highlight information that's especially important to know. These reminders represent best practices in using SPSS.

TECHNICAL STUFF

The technical stuff icon indicates information of a highly technical nature that provides helpful context. You can skip this info to complete a practice example, but it will help you more fully understand SPSS.

WARNING

The warning icon tells you to watch out! It notifies you of important information that may save you headaches when using SPSS.

EXAMPLE

The example icon presents a fully worked-out solution as a reference to assist you with the practice problems.

Beyond the Book

This workbook is just one part of the support we provide for mastering SPSS. For details about significant updates or changes that occur between editions of this workbook, go to `www.dummies.com`, type **SPSS Statistics Workbook For Dummies** in the Search box, and open the Download tab on this book's dedicated page.

That page is also where you'll find the book's cheat sheet, which contains advice for exploring a new data set, identifying which menu to use to perform various important data preparation operations, and finding the correct graph type for your data by utilizing level of measurement.

You can find a wealth of related information at the companion book's Dummies page by going to `www.dummies.com` and typing **SPSS Statistics For Dummies** in the Search box. Videos describing the topics covered in *SPSS Statistics For Dummies*, 4th Edition and ten SPSS gotchas can be found at `https://keithmccormick.com/SPSS4Dummies`.

Where to Go from Here

We wrote this book in a nonlinear way. You can skip around. It's okay. We skipped around when we wrote it, too. But it does have a structure. If you have a deep interest in a broad topic, you might want to complete an entire part of the book.

If you need practice getting your data in and set up, check out Part 1. Part 2 is all about creating variables with the Compute dialog and by using other transformations. If you need to apply a formula or modify a variable, check out the dozens of examples in this part.

If you need practice producing statistical analyses and interpreting them, you need Part 3. Part 4 is all about graphing, providing lots of practice making nearly every graph type that SPSS Statistics supports. You also learn about editing output.

If you are intrigued with Syntax but have been afraid to dive in, see Part 5. You get a chance to try Syntax and then check your work. Many of the tasks are similar to those in Part 2, except you work in the Syntax window in Part 5.

In the final section, Part 6, we warn you about skills that are particularly tricky and then offer some practice questions to help you decide if the IBM SPSS Statistics certification exam is something you'd like to try.

1 Getting Data into and out of SPSS

IN THIS PART . . .

Bring data into SPSS and show results in other applications.

Define data properties.

IN THIS CHAPTER

» Resolving trouble with delimiters on import

» Exporting output

Chapter 1

Working Through Import and Export Challenges

In this chapter you learn about ways to transfer data into and out of SPSS. Getting data into SPSS is the first step before any analysis can be done. If the data is available in an SPSS data file (.sav file extension), bringing that data into SPSS is easy. If your data comes from another program such as Excel or is in the txt, CSV, or SAS format, you can import that data into SPSS with just a little more work. The examples in this chapter demonstrate some complications that arise when importing these types of files.

After SPSS analyzes your data and displays results in easy-to-understand tables and graphs, you might want to use the results in another application to share your findings with others. This might involve

- Formatting tables as cleanly as possible for clarity of presentation and ease of viewing
- Exporting output to a file format that can be read by any user, such as Portable Document Format (pdf)
- Doing post-processing on pivot tables in another application, such as Microsoft Excel

Applications such as Microsoft's PowerPoint or Word can display the results of your analyses as plain text, as rich text, or as a metafile, which is a graphical representation of the output. Pivot tables can be pasted or exported to Microsoft Excel with each cell of the pivot table in a separate Excel cell.

In this chapter, you both copy and paste as well as export your SPSS output to another application.

Importing Data

This section contains an example of a procedure you can follow to read data from Excel files into SPSS. Along the way, SPSS keeps you informed about what's going on so there won't be any big surprises at the end. Here are a few things to consider to make the import process a little easier:

- When reading the Excel file into SPSS, the file must not be open in Excel.
- Variable names are read from the first row of the Excel spreadsheet. However, the blank spaces in the variable names are removed because blanks are not allowed in SPSS variable names.
- The measurement levels of variables are assigned based on the criterion defined in Data options. Variables with a small number of values are set to nominal. Variables with many values are set to scale.
- SPSS reads only the values in the cells of the spreadsheet. Formulas in the spreadsheet will be computed and these computed values imported to SPSS.
- The formulas and other spreadsheet characteristics associated with the cells are not imported to SPSS.

TIP

We strongly recommend opening the original file in the software where it is stored so you can see how the file is structured.

In the following example, the Excel workbook has two worksheets: The first is the title page and the second contains the data. Do the following to read this data into SPSS:

1. **Choose File ⇨ Import Data ⇨ Excel.**
2. **Select the GSS2018 Title.xlsx file and then click Open.**

   You can download the file from the book's companion website at `www.dummies.com/go/spssstatisticsworkbookfd`.
3. **In the Worksheet drop-down list, select the GSS2018 worksheet.**

   An Excel file can contain more than one worksheet, and you can choose the worksheet you want from the drop-down list, as shown in Figure 1-1.

Also, if you've elected to read only part of the data, use the Range drop-down list to specify the range of Excel cells that you want to import. Use the five check boxes, as needed, to specify whether the names of variables appear in the first row, the percentage of data to use to determine the variable type, and how to handle leading or trailing spaces.

TIP

Inspect the data preview to make sure the variables and data will be read properly.

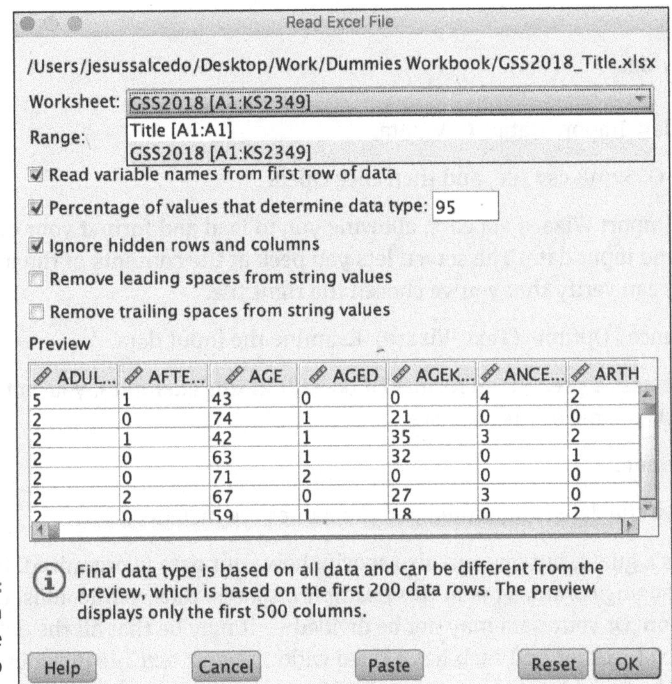

FIGURE 1-1: Select which data in the spreadsheet to include.

4. **Click the Read Variable Names from the First Row of Data check box.**
5. **Click OK.**

   Your data appears in the SPSS window.

6. **Switch to the Variable View tab to examine the variable definitions and make any changes.**

   SPSS makes a bunch of assumptions about your data, and some of those assumptions are probably wrong.

7. **Save the file using your chosen SPSS name, and you're off and running.**

See the following for an example of importing data into SPSS.

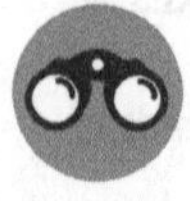

EXAMPLE

Q. Text files are another common source of data. Many spreadsheet programs and databases can save their contents in a text file format. Commonly used delimiter characters are tabs or commas.

Import the file GSS2018.csv to SPSS. Note that the comma-separated file has variable names in the first row. The variable names are in the first row and the data begins in row 2.

A. Do the following:

a. Choose File⇨Import Data⇨CSV Data.

b. Select the GSS2018.csv file, and then click Open.

The Text Import Wizard appears, allowing you to load and format your data. Examine the input data. The screen lets you peek at the contents of the input file so you can verify that you've chosen the right file.

c. Click Advanced Options (Text Wizard). Examine the input data.

If your file uses a predefined format (it doesn't in this example), you can select it here and skip some of the later steps.

d. Click Continue.

e. Specify that the data is delimited and the names are included.

SPSS takes a guess, but you can also specify how your data is organized. It can be divided using commas (as in this example), spaces, tabs, semicolons, or some combination. Or your data may not be divided — it may be that all the data items are jammed together and each has a fixed width. If your text file includes the names of the variables, you need to tell SPSS.

f. Click Continue.

g. Specify how SPSS should interpret the text.

You can tell SPSS something about the file and which data you want to read.

Perhaps some lines at the top of the file should be ignored — this happens when you're reading data from text intended for printing and header information is at the top. By telling SPSS about it, those first lines can be skipped.

Also, you can have one line of text represent one case (one row of data in SPSS), or you can have SPSS count the variables to determine where each row starts.

And you don't have to read the entire file — you can select a maximum number of lines to read starting at the beginning of the file, or you can select a percentage of the total and have lines of text randomly selected throughout the file.

h. Click Continue.

i. Specify tab as the delimiter.

SPSS can use commas, spaces, tabs, and semicolons as delimiting characters. You can even use some other character as a delimiter by selecting Other and then typing the character. You can also specify whether your text is formatted with quotes (which is common) and whether you use single or double quotes.

REMEMBER

Strings must be surrounded in quotes if they contain any of the characters being used as delimiters.

j. Click Continue.

k. If necessary, change the variable name and data format.

SPSS makes a guess for the type of each variable. To change a name, select it in the column heading at the bottom of the window, and then type the new name in the Variable Name field at the top. If you need to change the format, use the Data Format drop-down list.

REMEMBER

You can also change the data types later in the Variable View tab of the Data Editor window.

l. Click Continue.

m. In the Would You Like to Save This File Format for Future Use? section, click No.

Saving the file format for future use is something you would do if you were loading more files of this same format into SPSS — it reduces the number of questions to answer and the amount of formatting to do next time.

n. Click the Done button.

o. Look at the data, and correct the data types and formats, if necessary. Then save it all to a file by choosing File ➪ Save As.

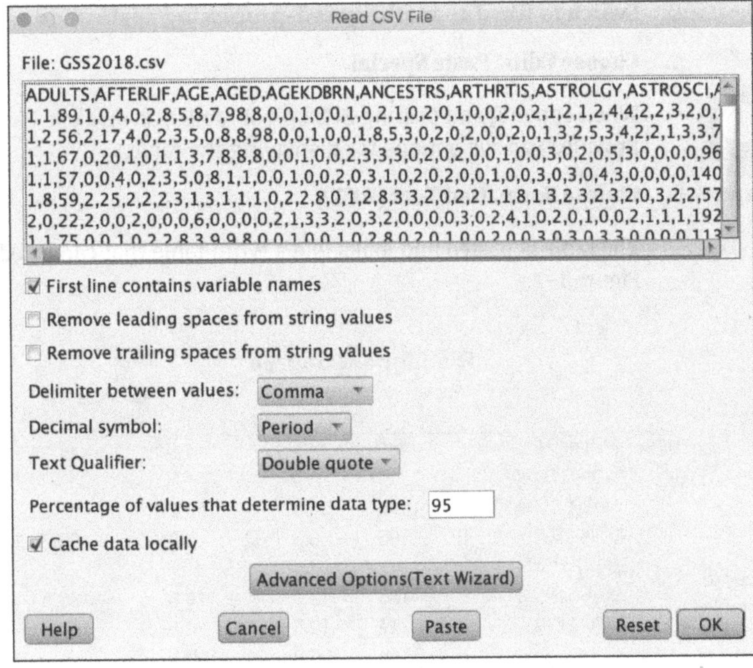

1. Import the GSS2018.xlsx file to SPSS. This example has one worksheet, the variable names are in the first row, and the data begins in row 2.

2. Import the GSS2018 extra title.xlsx file to SPSS. This example has two worksheets in the Excel workbook. The first worksheet is the title page. The second worksheet contains the data, with the variable names in the first row and the data beginning in row 5.

3. Import the GSS2018.dat file to SPSS. This file has a DAT format. The variable names are in the first row and the data begins in row 2.

4. Import the GSS2018 lines.txt file to SPSS. This file has a TXT format. The variable names are in the first row and the data begins in row 3.

Exporting Results

If you have a single table or a small number of tables, you can copy and paste these directly into a file opened in another application. Alternatively, SPSS provides an export facility to export large numbers of tables and charts into a file in a variety of common formats: Excel, Portable Document Format (PDF), HTML, text, Microsoft Word, and PowerPoint files.

1. **Choose File ⇨ Open ⇨ Output and load the Chapter 1 Output.spv file.**

   Download the file at `www.dummies.com/go/spssstatisticsworkbookfd`.

2. **Select the R's highest degree frequency table.**
3. **Choose Edit ⇨ Copy.**
4. **Switch to Word or another word-processing application.**
5. **Choose Edit ⇨ Paste Special.**

TIP

   When copying and pasting SPSS pivot tables, always use Paste Special because it provides various options for displaying the table.

6. **Choose Formatted Text (RFT).**

   The table is pasted and looks like a Word table that can be edited, as shown in Figure 1-2.

R's highest degree

		Frequenc y	Percent	Valid Percent	Cumulative Percent
Valid	LT HIGH SCHOOL	262	11.2	11.2	11.2
	HIGH SCHOOL	1178	50.2	50.2	61.3
	JUNIOR COLLEGE	196	8.3	8.3	69.7
	BACHELOR	465	19.8	19.8	89.5
	GRADUATE	247	10.5	10.5	100.0
	Total	2348	100.0	100.0	

FIGURE 1-2: Pasting a table in Microsoft Word.

See the following for an example of taking SPSS output and bringing into another application.

EXAMPLE

Q. Using the Chapter 1 Output.spv file, copy the graph for R's highest degree and paste it into Word.

A. Select and copy the graph. Then go to Word and choose Edit⇨Paste (or Paste Special).

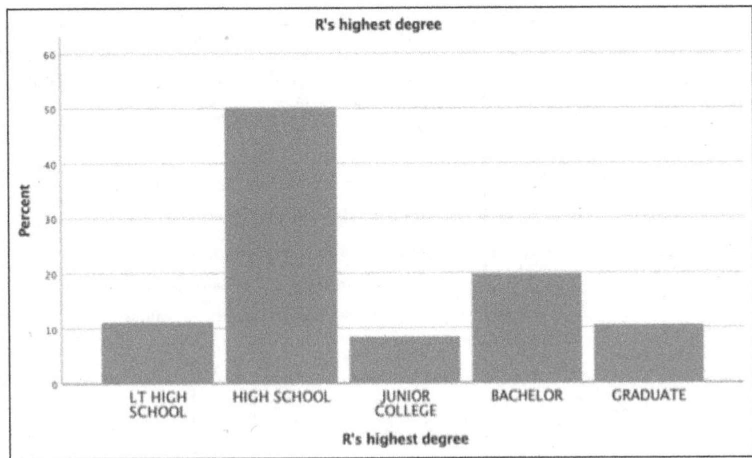

REMEMBER

Charts are always pasted as images that cannot be edited.

5 Using the Chapter 1 Output.spv file, copy the R's highest degree frequency table and paste in Word as a picture.

6 Using the Chapter 1 Output.spv file, copy the R's highest degree frequency table and paste it in Excel as a table using Unicode Text.

7 Using the Chapter 1 Output.spv file, copy the R's highest degree frequency table and paste it in Excel as a picture.

8 Export the Chapter 1 Output.spv file to Word.

Answers to Problems in Working Through Import and Export Challenges

1. Choose File ➪ Import Data ➪ Excel. Click the Read Variable Names from the First Row of Data check box.

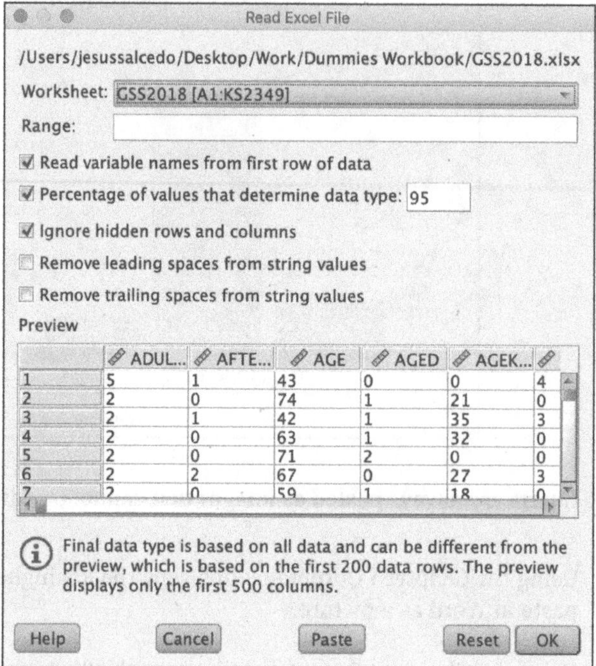

2. Choose File ➪ Import Data ➪ Excel. Select the GSS2018 worksheet. Click the Read Variable Names from the First Row of Data check box. After the data has been imported, delete the extra rows that appear at the beginning of the file.

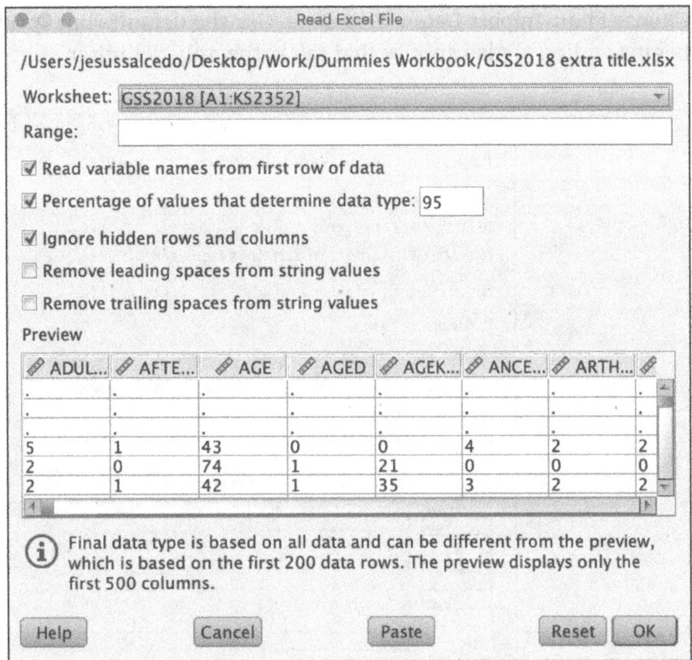

3. Choose File ➪ Import Data ➪ Text Data. Use the defaults but specify that tab is the only delimiter. Deselecting space as a delimiter will remove the extra column.

TIP

Always check the data preview to make sure you do not have any columns that do not have variable names. If this occurs, you did not select the correct delimiter.

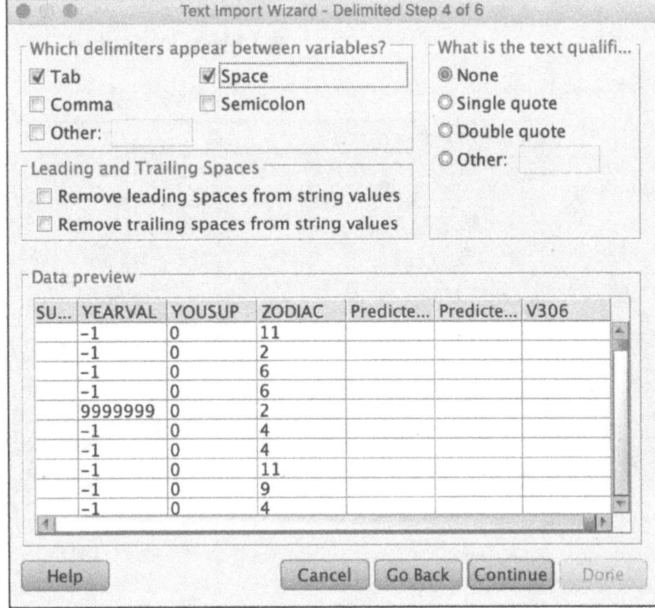

4. Choose File ⇨ Import Data ⇨ Text Data. Use the defaults but specify that the first case of data begins on line 3. Also specify that tab is the only delimiter.

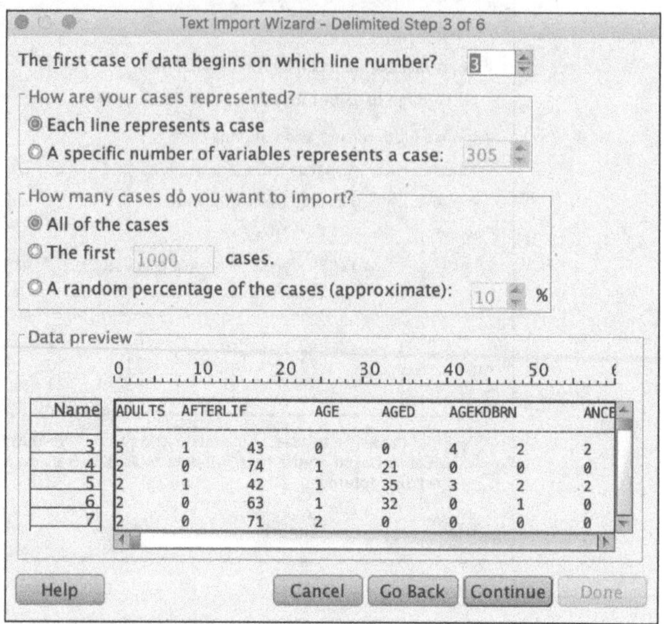

5. Select and copy the R's highest degree frequency table. Then go to Word and choose Edit ⇨ Paste Special and select Picture. You would choose to copy SPSS output as a picture if you did all your editing in SPSS and would like the output to appear exactly as it appeared in SPSS.

R's highest degree

		Frequency	Percent	Valid Percent	Cumulative Percent
Valid	LT HIGH SCHOOL	262	11.2	11.2	11.2
	HIGH SCHOOL	1178	50.2	50.2	61.3
	JUNIOR COLLEGE	196	8.3	8.3	69.7
	BACHELOR	465	19.8	19.8	89.5
	GRADUATE	247	10.5	10.5	100.0
	Total	2348	100.0	100.0	

6. Select and copy the R's highest degree frequency table. Then go to Excel, choose Edit ⇨ Paste Special, and select Unicode Text. You would use the Unicode Text option so that you can further edit your tables in Excel.

	A	B	C	D	E	F	G
1	R's highest degree						
2			Frequency	Percent	Valid Percen	Cumulative Percent	
3	Valid	LT HIGH SCH	262	11.2	11.2	11.2	
4		HIGH SCHOO	1178	50.2	50.2	61.3	
5		JUNIOR COLL	196	8.3	8.3	69.7	
6		BACHELOR	465	19.8	19.8	89.5	
7		GRADUATE	247	10.5	10.5	100	
8		Total	2348	100	100		
9							

7. Select and copy the R's highest degree frequency table. Then go to Excel, choose Edit ⇨ Paste Special, and select Picture. You would choose to copy SPSS output as a picture if you did all your editing in SPSS and would like the output to appear exactly as it appeared in SPSS.

R's highest degree

		Frequency	Percent	Valid Percent	Cumulative Percent
Valid	LT HIGH SCHOOL	262	11.2	11.2	11.2
	HIGH SCHOOL	1178	50.2	50.2	61.3
	JUNIOR COLLEGE	196	8.3	8.3	69.7
	BACHELOR	465	19.8	19.8	89.5
	GRADUATE	247	10.5	10.5	100.0
	Total	2348	100.0	100.0	

8. To export your output from SPSS to Word:

   a. Choose File ⇨ Export.

   b. In the Objects to Export section, select the items to include in the output. For the example, select All Visible.

   c. In the Document section's Type drop-down list, choose Word/RTF.

   d. Click the Browse button, select the directory and name of the exported file (SPSS Output Export to Word), and then click Save.

   e. Click OK.

f. To view the exported file, open Microsoft Word.

The exported Word file contains the tables and graph that make up the Chapter 1 Output.spv file. The tables can be edited with the Microsoft Word table editor because the tables were exported as text, not as a picture.

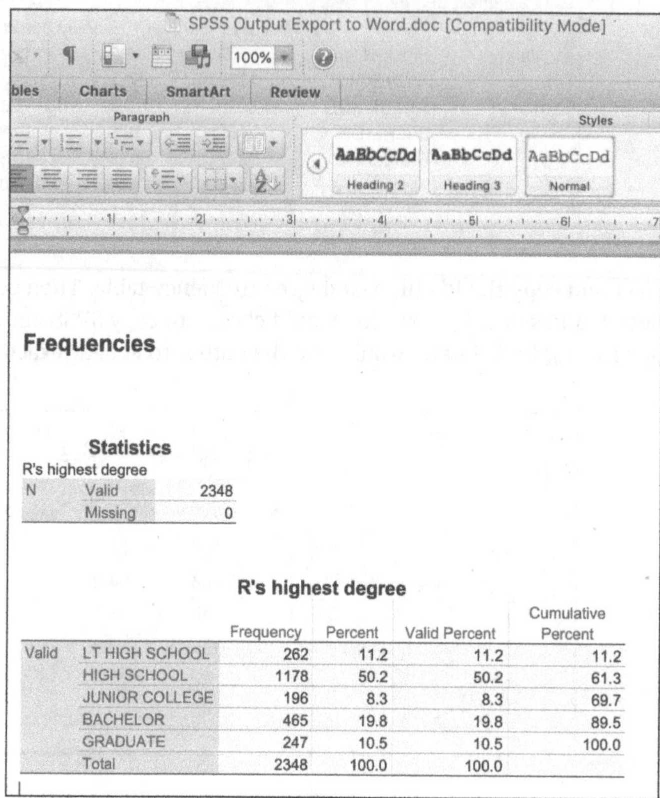

Frequencies

Statistics

R's highest degree

N	Valid	2348
	Missing	0

R's highest degree

		Frequency	Percent	Valid Percent	Cumulative Percent
Valid	LT HIGH SCHOOL	262	11.2	11.2	11.2
	HIGH SCHOOL	1178	50.2	50.2	61.3
	JUNIOR COLLEGE	196	8.3	8.3	69.7
	BACHELOR	465	19.8	19.8	89.5
	GRADUATE	247	10.5	10.5	100.0
	Total	2348	100.0	100.0	

IN THIS CHAPTER

» Defining variables

» Seeing how to work with dates and times

» Finding out how to copy data properties

Chapter 2
Defining Data

SPSS data has three major components: cases (Chapter 3), variables (Chapters 4 and 5), and metadata (Chapter 2). *Metadata* consists of your variable attributes, or definitions. Metadata tells SPSS how a variable is defined and how it can be used. Without a definition, a number serves no purpose. For example, the number 34 could be the number of tickets purchased or the number of sales in the last hour; therefore, a variable's definition is important.

In this chapter, you learn about variable definitions, or attributes. You also work with date and time variables. Finally, because defining data can be time-consuming, we demonstrate a special shortcut menu for copying your data and variable definitions from one dataset to another.

Defining Metadata

Entering data into SPSS is a two-step process. First, you define what sort of data you'll be entering. Then you enter the values. In this section, we briefly describe some of the most important variable attributes in the Variable View tab of Data Editor.

Each variable characteristic has a default, so if you don't specify a characteristic, SPSS uses the default:

» **Name:** Every variable must have a unique name. Following are some handy hints about names:

  • You can use characters in a name, such as @, #, and $, as well as the underscore character (_) and numbers. However, variable names can't start with these special characters.

- Be sure to start every name with a letter, not a number.
- You can't include spaces anywhere in a name, but an underscore is a good substitute.

» **Type:** Most data you enter will be regular numbers. Data such as currency must be displayed in a special format, and data such as dates require special formats so they can be used in calculations. For this type of data, you simply specify what type of data you have, and SPSS takes care of the details for you.

» **Width:** The Width column in the definition of a variable determines the number of characters used to display the value. If the value to be displayed is not large enough to fill the space, the output will be padded with blanks. If it's larger than you specify, it either will be reformatted to fit or asterisks will be displayed.

» **Decimals:** The Decimals column contains the number of digits that appear to the right of the decimal point when the value appears onscreen.

» **Label:** The name and the label serve the same basic purpose: They're descriptors that identify the variable. The difference is that the *name* is the short identifier and the *label* is the long one.

» **Values:** The Values column is where you assign labels to all possible values of a variable. Normally, you make one entry for each possible value that a variable can assume. For example, for a variable named Sex, you could assign the value 1 to the Male label and the value 2 to the Female label.

» **Missing:** You can specify codes for missing data and SPSS will ignore this value when performing calculations.

» **Measure:** Your value for the Measure attribute specifies the level of measurement of your variable. Following are the level of measurement options in SPSS:

- *Nominal:* A value that specifies a category or type of thing. You can have 0 represent No and 1 represent Yes.
- *Ordinal:* A value that specifies the position, or order, of something in a list. For example, first, second, and third are ordinal numbers.
- *Scale:* A number that specifies a magnitude. The scale can be distance, weight, age, or a count of something.

Follow these steps to define a label for a value:

1. **Choose File ⇨ Open ⇨ Data and load the Happy.sav file.**

   You can download the file from the book's companion website at `www.dummies.com/go/spssstatisticsworkbookfd`.

2. **Click the Variable View tab of Data Editor.**
3. **Click in the Value cell for the Happy variable to open Value Labels dialog.**

4. **Click in the Value box and enter the value 0. Click in the Label box and enter the label** IAP. **Click the Add button.**
5. **Click in the Value box and enter the value 1. Click in the Label box and enter the label** VERY HAPPY. **Click the Add button.**
6. **Click in the Value box and enter the value 2. Click in the Label box and enter the label** PRETTY HAPPY. **Click the Add button.**
7. **Click in the Value box and enter the value 3. Click in the Label box and enter the label** NOT TOO HAPPY. **Click the Add button.**
8. **Click in the Value box and enter the value 8. Click in the Label box and enter the label** DK. **Click the Add button.**
9. **Click in the Value box and enter the value 9. Click in the Label box and enter the label** NA, **Click the Add button.**

   The screen should look like Figure 2-1.
10. **To save the value labels and close the dialog, click OK.**

    You've added value labels to the HAPPY variable.

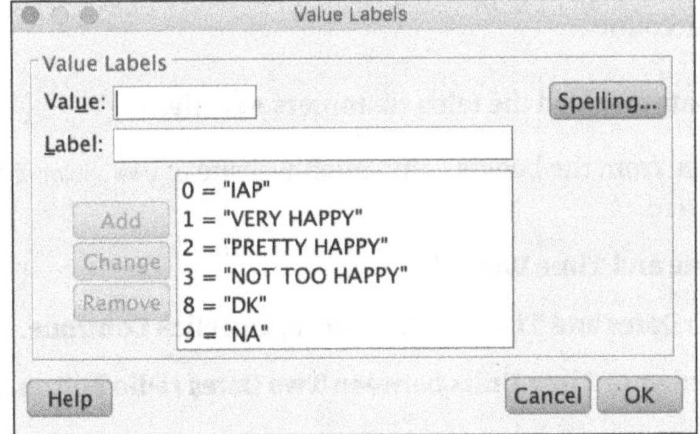

FIGURE 2-1: The completed Value Labels dialog.

See the following for an example of adding metadata.

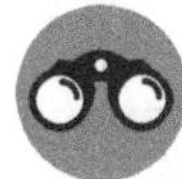

EXAMPLE

Q. Using the Happy.sav file, add the label General Happiness to the HAPPY variable.

A. Click in the Label cell and type **General Happiness.**

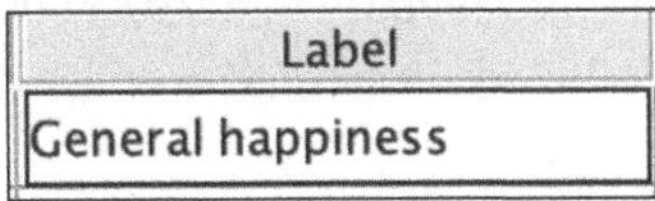

1. Using the Happy.sav file, specify that the HAPPY variable should have no decimals.

2. Using the Happy.sav file, specify that the HAPPY variable should have a width of 1.

3. Using the Happy.sav file, classify the values 0, 8, and 9 for the HAPPY variable as codes for missing values.

4. Using the Happy.sav file, specify that the HAPPY variable should have ordinal for its level of measurement.

Working with Dates and Times

Date and time arithmetic can be tricky, but SPSS can handle it all for you. Just enter date and time variables in whatever formats you specify, and SPSS will convert the values internally to do the calculations. SPSS displays the newly created date and time variables in your specified format, so the variable will be easy to read.

If you have dates that have been properly declared, you can easily do numerous types of calculations. In this example, you calculate the number of months that passed when someone opened and then closed an account:

1. **Choose File ⇨ Open ⇨ Data and load the telco customers.sav file.**

   You can download the file from the book's companion website at `www.dummies.com/go/spssstatisticsworkbookfd`.

2. **Choose Transform ⇨ Date and Time Wizard.**
3. **Select the Calculate with Dates and Times radio button, and click Continue.**
4. **Select the Calculate Number of Time Units between Two Dates radio button, and click Continue.**
5. **Select CLOSURE_DATE and move it to the Date1 field. Then select CONNECT_DATE and move it to the minus Date2 field.**
6. **Specify Months as the Unit.**
7. **Select the Truncate to Integer radio button (see Figure 2-2), and then click Continue.**
8. **Name the Result variable** Duration**, and click Done.**

   You can now see how long (in months) someone has had an account.

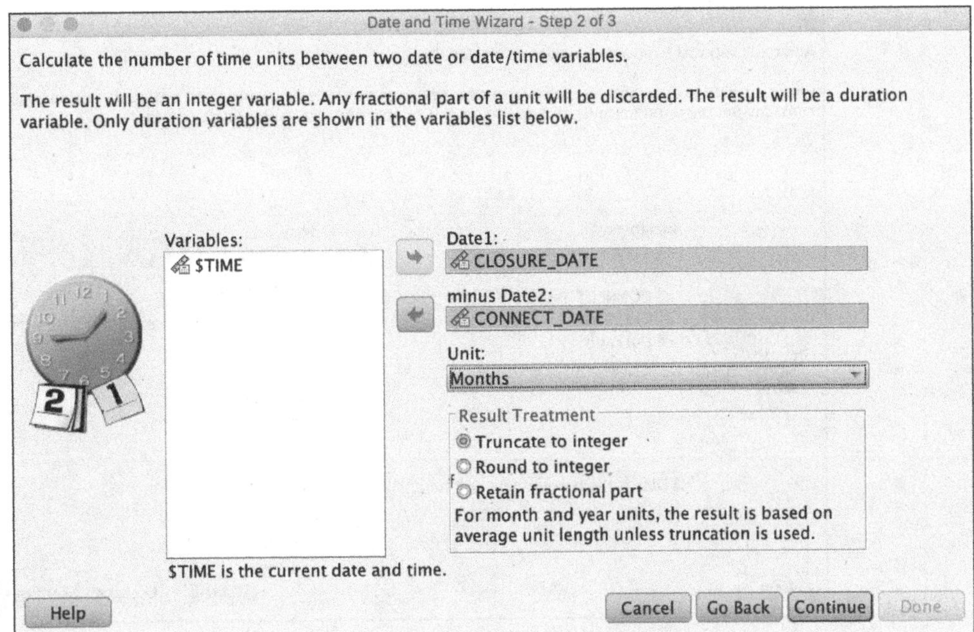

FIGURE 2-2: Date and Time wizard.

See the following for an example of working with date and time data.

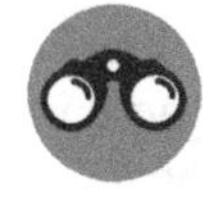

EXAMPLE

Q. Suppose you want to contact customers six months after an account has been closed. Using the telco customers.sav file, add six months to the CLOSURE_DATE variable so that you'll know when to contact customers.

A. Do the following:

a. Use the Date and Time Wizard.

b. Select the Calculate with Dates and Times option, and then select the Add or Subtract a Duration from a Date option.

c. Specify the CLOSURE_DATE variable as the Date variable.

d. Change the Units to months and specify 6 as the Duration Constant (to add six months to every closure date).

e. Name the new variable.

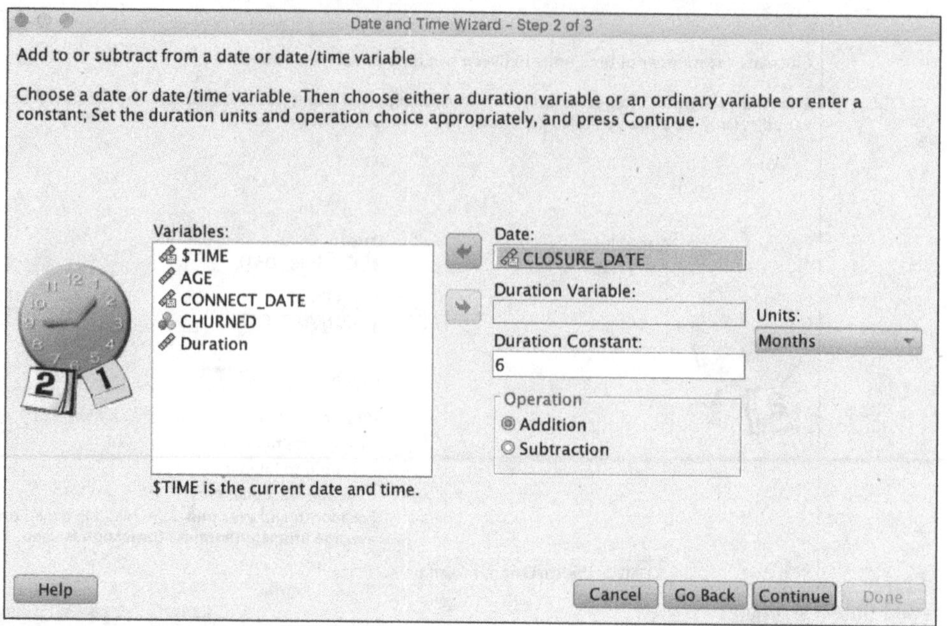

5. Using the telco customers.sav file, determine whether customers close their accounts in a particular month. You need to extract the month from the CLOSURE_DATE variable so you can see the distribution of months.

6. Using the telco customers.sav file, determine whether customers start their accounts on a particular day of the week. You need to extract the day of the week from the CONNECT_DATE variable so you can see the distribution of days of the week.

7. Using the telco customers.sav file, figure out the number of months since a customer closed their account. You need to calculate the difference between today's date and the CLOSURE_DATE variable.

8. Using the rfm_ aggregated.sav dataset, determine whether customers had a purchase in a specific month. Note that the Most_Recent_Date variable is not a numeric value. You need to extract the month from the Most_Recent_Date variable so you can see the distribution of months.

Copying Data Properties

Suppose you have some data definitions in another SPSS file, and you want to copy one or more of those definitions but you don't want the data. All you want is the metadata. SPSS enables you to copy only the variable definitions you want into your current dataset.

TIP

If you have a variable of *the same name* defined in your dataset before you run the Copy Data Properties facility, you can choose to change the existing variable definition by loading new information from another file. The copied definition simply overwrites the previous information. Otherwise, the copying facility creates a new variable.

The following steps show you how to copy data properties:

1. **Choose File ➪ Open ➪ Data and load the GSS2018.sav file.**

   You can download the file from the book's companion website at `www.dummies.com/go/spssstatisticsworkbookfd`.

2. **Choose File ➪ New ➪ Data.**

   You now have a new dataset with no data and no variable information. Make sure this is the active dataset.

3. **Choose Data ➪ Copy Data Properties.**

   The Copy Data Properties facility in the Data menu copies variable properties from one dataset to another. Metadata information can be copied to the active data file from an open dataset or from an external SPSS data file.

4. **In the An Open Dataset box, select GSS2018.sav and click Continue.**

5. **Use the default option (Apply Properties from Selected Source Dataset Variables to Matching Active Dataset Variables).**

   Now you need to specify from which variables you want to copy properties.

6. **Select the variables from which you want to copy information, and then click Continue.**

   To follow along with the example, select all the variables, as shown in Figure 2-3.

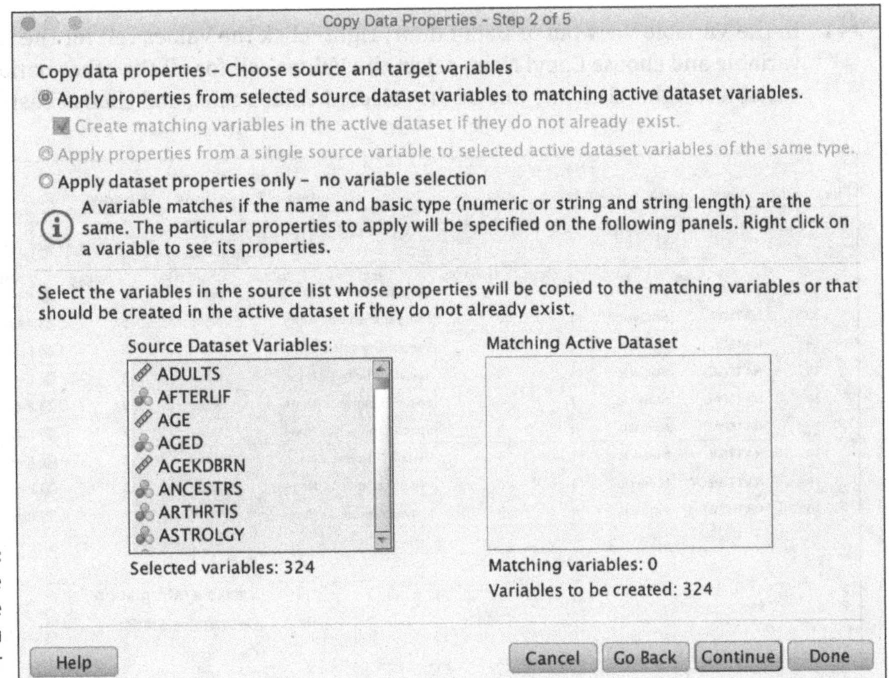
Copy Data Properties - Step 2 of 5
Copy data properties - Choose source and target variables
Apply properties from selected source dataset variables to matching active dataset variables.
Create matching variables in the active dataset if they do not already exist.
Apply properties from a single source variable to selected active dataset variables of the same type.
Apply dataset properties only - no variable selection
A variable matches if the name and basic type (numeric or string and string length) are the same. The particular properties to apply will be specified on the following panels. Right click on a variable to see its properties.
Select the variables in the source list whose properties will be copied to the matching variables or that should be created in the active dataset if they do not already exist.
Source Dataset Variables:
ADULTS
AFTERLIF
AGE
AGED
AGEKDBRN
ANCESTRS
ARTHRTIS
ASTROLGY
Selected variables: 324
Matching Active Dataset
Matching variables: 0
Variables to be created: 324
Help
Cancel
Go Back
Continue
Done

FIGURE 2-3: Select the source variable names you want to use for definitions.

7. **Select the properties of the existing variable definitions you want to copy to the variables you're modifying, and then click Continue.**

   Everything is selected by default, but you can deselect any properties you don't want. These selections apply to all variables you've chosen. If you want to handle each variable separately, you'll have to run through this facility again for each one, selecting different variables each time.

8. **Specify any dataset properties you'd like to copy.**
9. **Click Continue to move to the final dialog.**

   The screen displays the number of existing variable definitions to be changed, the number of new variables to be created, and the number of other properties that will be copied.

10. **To execute the copy facility immediately, click Done.**

    You've copied over all the variable properties from the GSS2018 dataset without copying any of the data.

See the following for an example of copying metadata.

EXAMPLE

Q. Use the National_Interest.sav file. All of the metadata has been defined for the NATACCESS variable, but no metadata has been defined for any of the other national interest variables. Using the Variable View tab of Data Editor, copy the Values information for the NATACCESS variable, and paste this information in the Values cell for all the other national interest variables.

A. In the Variable View tab of Data Editor, right-click the Values cell for the NATACCESS variable and choose Copy. Next, select the Values cell for all the other national interest variables, right-click any one of the selected Values cells, and choose Paste.

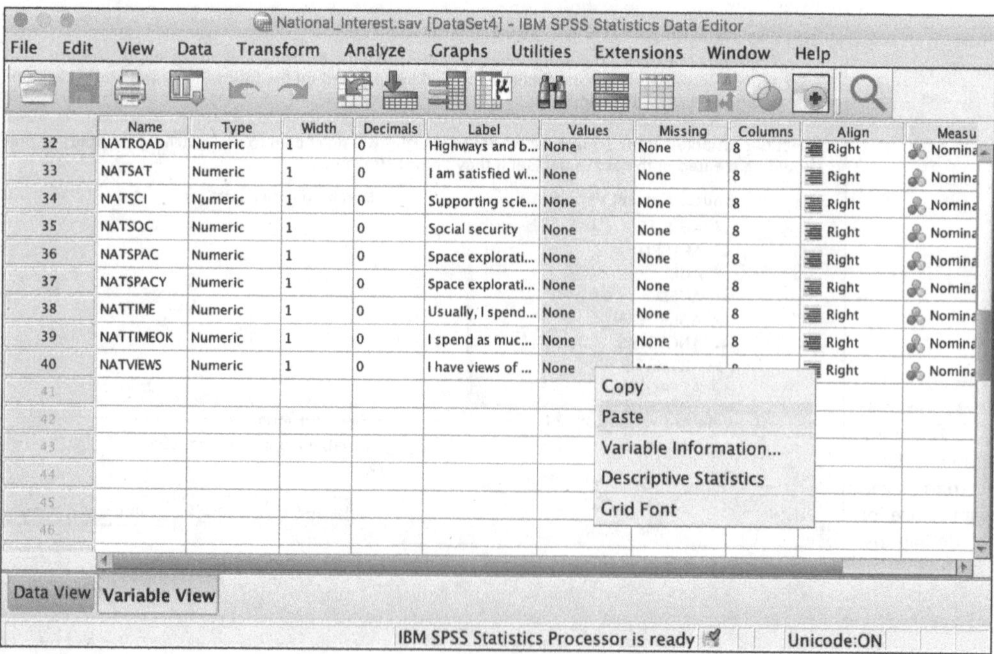

National_Interest.sav [DataSet4] - IBM SPSS Statistics Data Editor
File Edit View Data Transform Analyze Graphs Utilities Extensions Window Help
Name Type Width Decimals Label Values Missing Columns Align Measu
32 NATROAD Numeric 1 0 Highways and b... None None 8 Right Nomina
33 NATSAT Numeric 1 0 I am satisfied wi... None None 8 Right Nomina
34 NATSCI Numeric 1 0 Supporting scie... None None 8 Right Nomina
35 NATSOC Numeric 1 0 Social security None None 8 Right Nomina
36 NATSPAC Numeric 1 0 Space explorati... None None 8 Right Nomina
37 NATSPACY Numeric 1 0 Space explorati... None None 8 Right Nomina
38 NATTIME Numeric 1 0 Usually, I spend... None None 8 Right Nomina
39 NATTIMEOK Numeric 1 0 I spend as muc... None None 8 Right Nomina
40 NATVIEWS Numeric 1 0 I have views of ... None Right Nomina
Copy
Paste
Variable Information...
Descriptive Statistics
Grid Font
Data View Variable View
IBM SPSS Statistics Processor is ready
Unicode:ON

9. Use the National_Interest.sav file. The NATACCESS variable has all of its metadata defined, but no metadata has been defined for any of the other national interest variables. Using the Variable View tab of Data Editor, copy the Missing Values information for the NATACCESS variable, and paste this information in the Missing Values cell for all the other national interest variables.

10. Reopen the National_Interest.sav file. Using the Copy Data Properties transformation, copy all the metadata from the NATACCESS variable to all the other national interest variables.

11. Using the GSS2018.sav dataset, click the Variable View tab of Data Editor. Copy the metadata for the AGE variable, and paste that metadata in a new empty dataset.

12. Using the GSS2018.sav dataset, go to the Variable View tab of Data Editor. Copy the metadata for all the variables, and then paste the metadata in a new empty dataset.

Answers to Problems in Defining Data

1. Click in the Decimals cell and change the value to 0.

	Name	Type	Width	Decimals	Label	Values	Missing	Columns	Align	Measure	Role
1	HAPPY	Numeric	8	0		None	None	8	Right	Ordinal	Input

2. Click in the Width cell and change the value to 1.

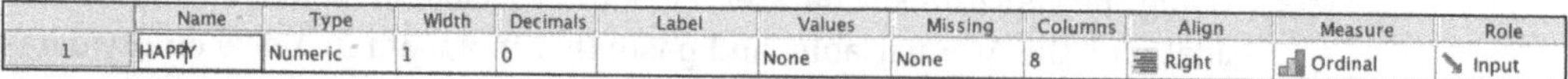

	Name	Type	Width	Decimals	Label	Values	Missing	Columns	Align	Measure	Role
1	HAPPY	Numeric	1	0		None	None	8	Right	Ordinal	Input

3. Click in the Missing values cell, and then select the Discrete Missing Values option. Specify that the values 0, 8, and 9 should be classified as codes for missing values. Click OK.

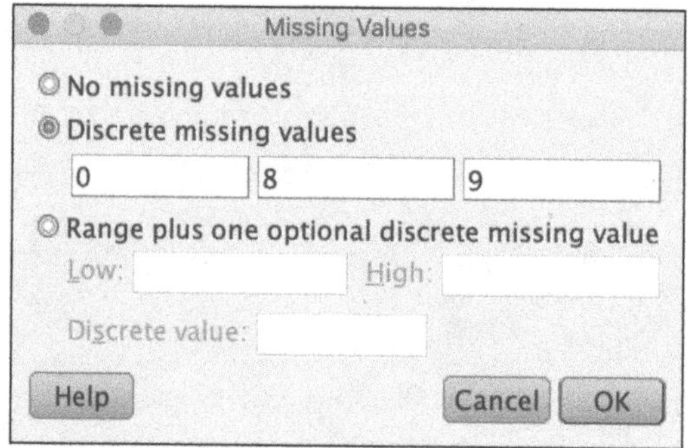

4. Click in the Measure cell and change the value to Ordinal.

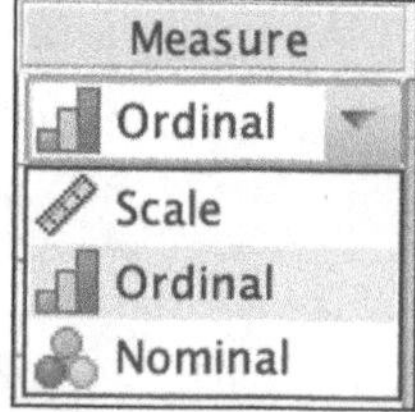

5. Do the following:

   a. Use the Date and Time Wizard.

   b. Select the Extract a Part of a Date or Time Variable option.

   c. Specify the CLOSURE_DATE variable as the Date or Time.

   d. Change the Units to Extract to Months.

   e. Name the new variable.

   f. Use the Frequencies procedure to determine whether customers close their accounts in a particular month.

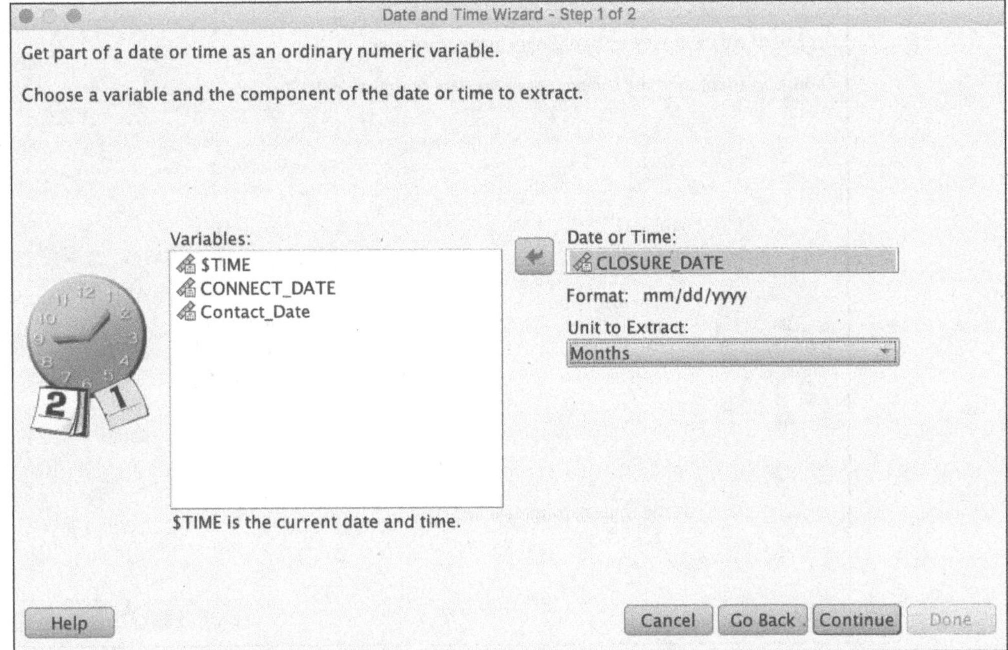

6. Do the following:

   a. Use the Date and Time Wizard.

   b. Select the Extract a Part of a Date or Time Variable option.

   c. Specify the CONNECT_DATE variable as the Date or Time.

   d. Change the Units to Extract to Day of Week.

   e. Name the new variable.

   f. Use the Frequencies procedure to determine whether customers open their accounts on a particular day of the week.

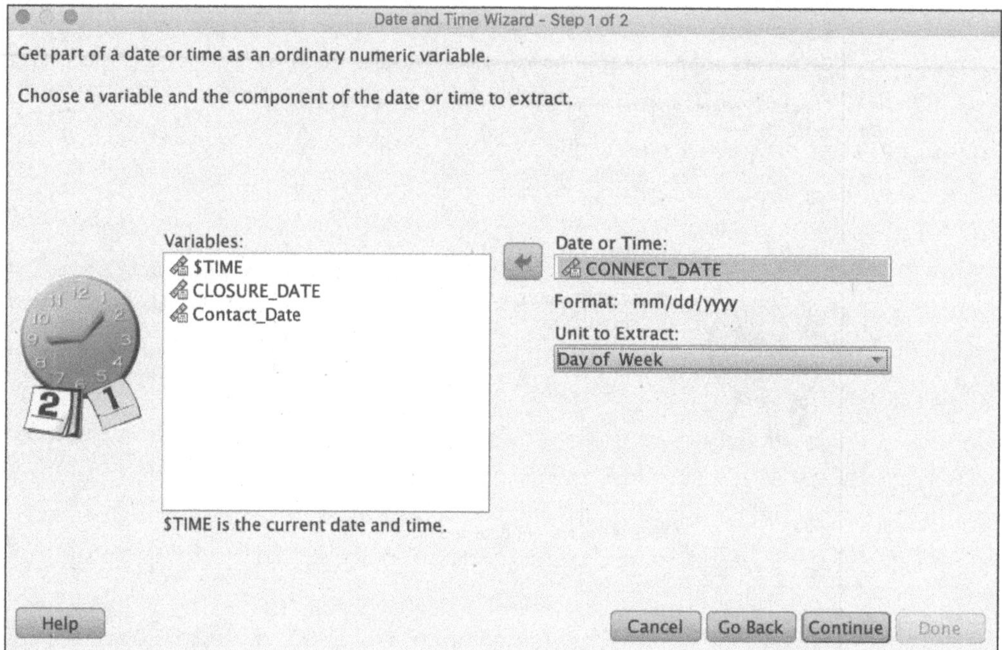

7. Do the following:

   a. Use the Date and Time Wizard.

   b. Select Calculate with Dates and Times option, and then select the Calculate the Number of Time Units between Two Dates option.

   c. Place the $TIME variable in the Date1 box. (The special variable, $TIME, uses the current date and time.) Place the CLOSURE_DATE variable in the minus Date2 box.

   d. Change the Units to Months.

   e. In the Result Treatment box, select Truncate to Integer.

   f. Name the new variable.

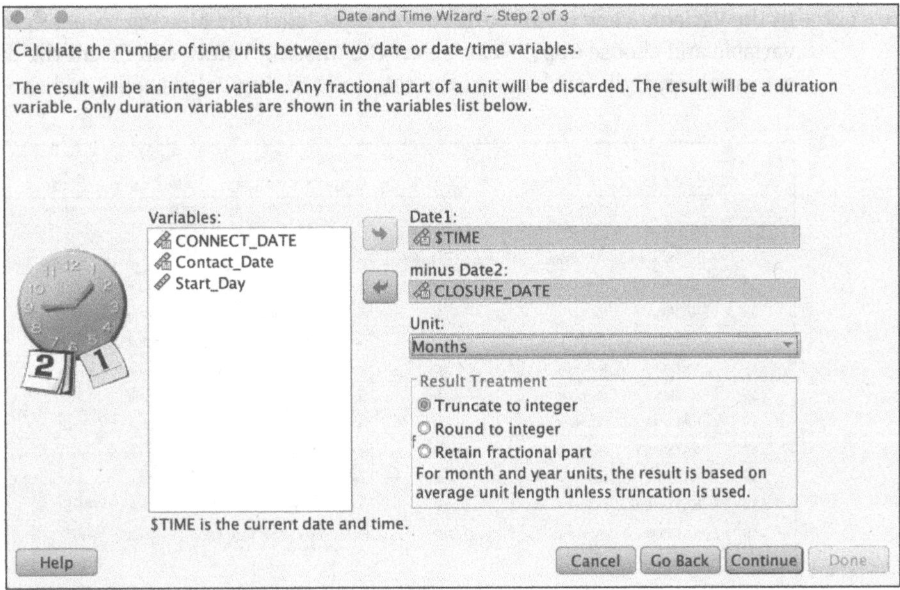

8 Do the following:

a. Use the Date and Time Wizard.
b. Select the Extract a Part of a Date or Time Variable.
c. Specify the Most_Recent_Date variable as the Date or Time.
d. Change the Units to Extract to Months.
e. Name the new variable.
f. Use the Frequencies procedure to determine whether customers had their most recent purchase in a particular month.

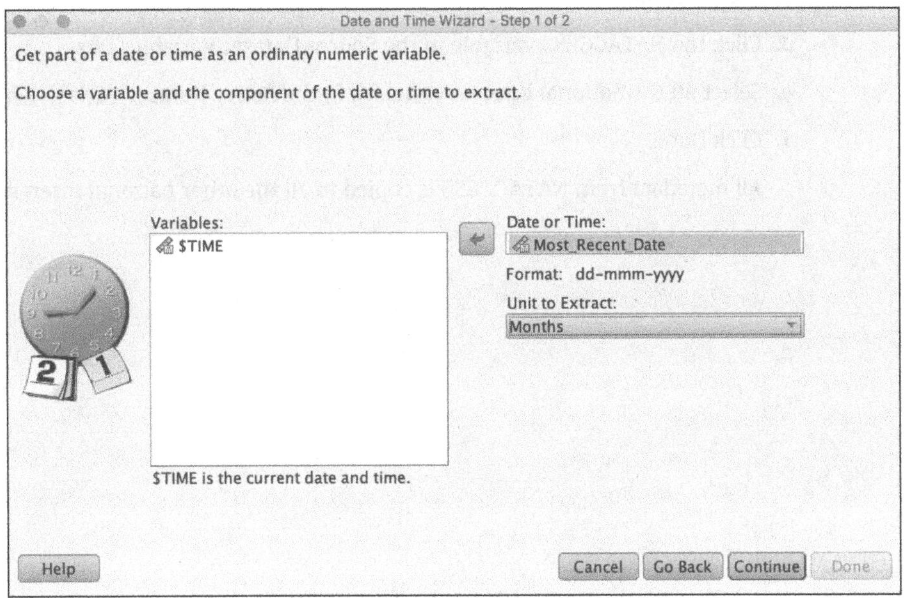

9. In the Variable View tab of Data Editor, right-click the Missing values cell for the NATACCESS variable and choose Copy. Next, select the Missing Values cell for all the other national interest variables, right-click one of the selected Missing Values cells, and choose Paste.

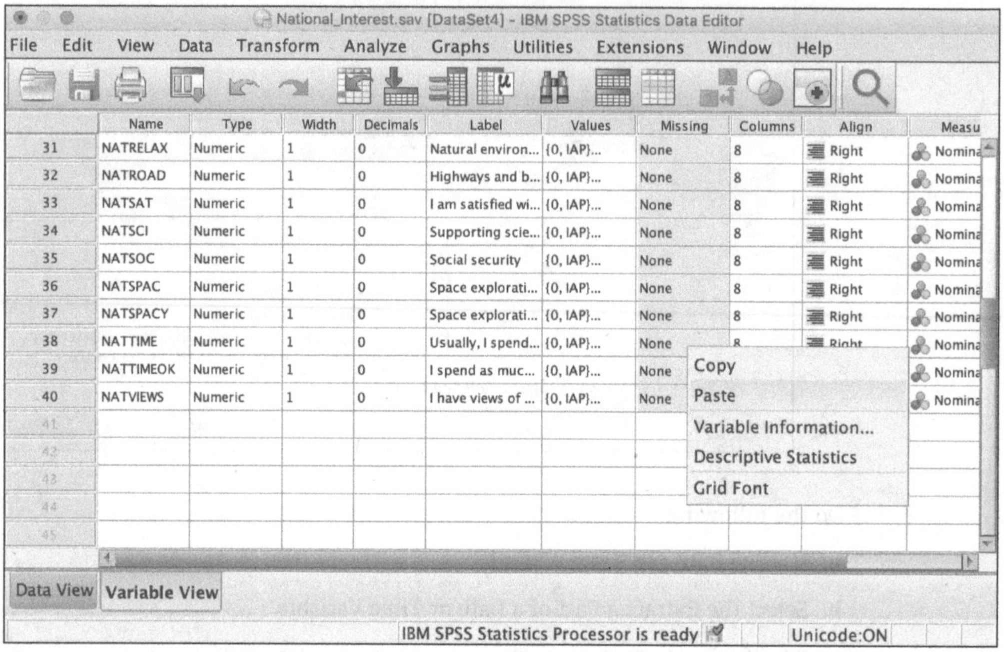

National_Interest.sav [DataSet4] - IBM SPSS Statistics Data Editor

File Edit View Data Transform Analyze Graphs Utilities Extensions Window Help

	Name	Type	Width	Decimals	Label	Values	Missing	Columns	Align	Measu
31	NATRELAX	Numeric	1	0	Natural environ...	{0, IAP}...	None	8	Right	Nomina
32	NATROAD	Numeric	1	0	Highways and b...	{0, IAP}...	None	8	Right	Nomina
33	NATSAT	Numeric	1	0	I am satisfied wi...	{0, IAP}...	None	8	Right	Nomina
34	NATSCI	Numeric	1	0	Supporting scie...	{0, IAP}...	None	8	Right	Nomina
35	NATSOC	Numeric	1	0	Social security	{0, IAP}...	None	8	Right	Nomina
36	NATSPAC	Numeric	1	0	Space explorati...	{0, IAP}...	None	8	Right	Nomina
37	NATSPACY	Numeric	1	0	Space explorati...	{0, IAP}...	None	8	Right	Nomina
38	NATTIME	Numeric	1	0	Usually, I spend...	{0, IAP}...	None	8	Right	Nomina
39	NATTIMEOK	Numeric	1	0	I spend as muc...	{0, IAP}...	None			Nomina
40	NATVIEWS	Numeric	1	0	I have views of ...	{0, IAP}...	None			Nomina
41										
42										
43										
44										
45										

Copy
Paste
Variable Information...
Descriptive Statistics
Grid Font

Data View | Variable View

IBM SPSS Statistics Processor is ready | Unicode:ON

10. Do the following:

    a. Reopen the National_Interest.sav data file.

    b. Use the Copy Data Properties transformation.

    c. Select the Active Dataset option in the Choose the Source of the Properties box and then click Continue.

    d. Click the NATACCESS variable in the Source Dataset Variables box.

    e. Select all the national interest variables in the Active Dataset Variables box.

    f. Click Done.

    All metadata from NATACCESS is copied to all the other national interest variables.

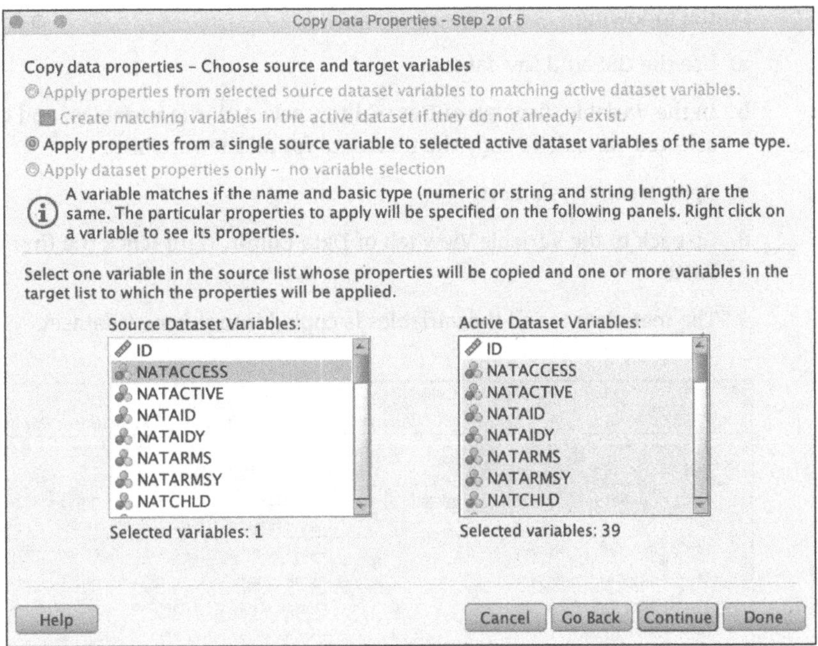
Copy Data Properties - Step 2 of 5
Copy data properties - Choose source and target variables
Apply properties from selected source dataset variables to matching active dataset variables.
Create matching variables in the active dataset if they do not already exist.
Apply properties from a single source variable to selected active dataset variables of the same type.
Apply dataset properties only -- no variable selection
A variable matches if the name and basic type (numeric or string and string length) are the same. The particular properties to apply will be specified on the following panels. Right click on a variable to see its properties.
Select one variable in the source list whose properties will be copied and one or more variables in the target list to which the properties will be applied.
Source Dataset Variables:
ID
NATACCESS
NATACTIVE
NATAID
NATAIDY
NATARMS
NATARMSY
NATCHLD
Selected variables: 1
Active Dataset Variables:
ID
NATACCESS
NATACTIVE
NATAID
NATAIDY
NATARMS
NATARMSY
NATCHLD
Selected variables: 39
Help
Cancel
Go Back
Continue
Done

11 Do the following:

a. Open the GSS2018.sav dataset.

b. In the Variable View tab of Data Editor, right-click the AGE variable and choose Copy.

c. Open an empty dataset.

d. Go back to the Variable View tab of Data Editor, right-click the first row, and choose Paste.

The metadata for the AGE variable is copied over to the new dataset.

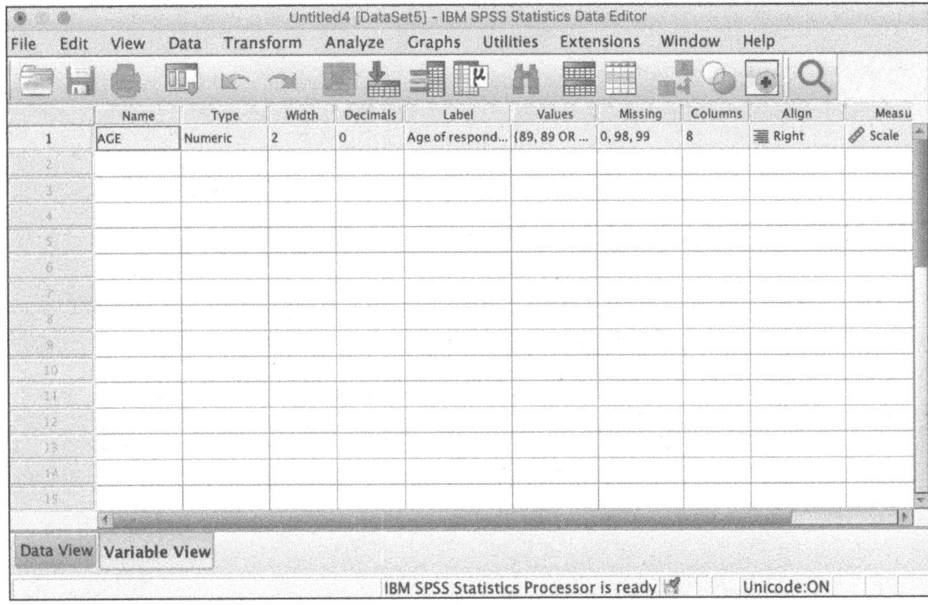
Untitled4 [DataSet5] - IBM SPSS Statistics Data Editor
File Edit View Data Transform Analyze Graphs Utilities Extensions Window Help
Name Type Width Decimals Label Values Missing Columns Align Measu
1 AGE Numeric 2 0 Age of respond... {89, 89 OR ... 0, 98, 99 8 Right Scale
Data View
Variable View
IBM SPSS Statistics Processor is ready
Unicode:ON

12. Do the following:

    a. Use the GSS2018.sav dataset.

    b. In the Variable View tab of Data Editor, select all the variables, and then right-click any selected variable to copy the metadata for all the variables.

    c. Open an empty dataset.

    d. Go back to the Variable View tab of Data Editor, right-click the first row, and choose Paste.

    The metadata for all the variables is copied into this new dataset.

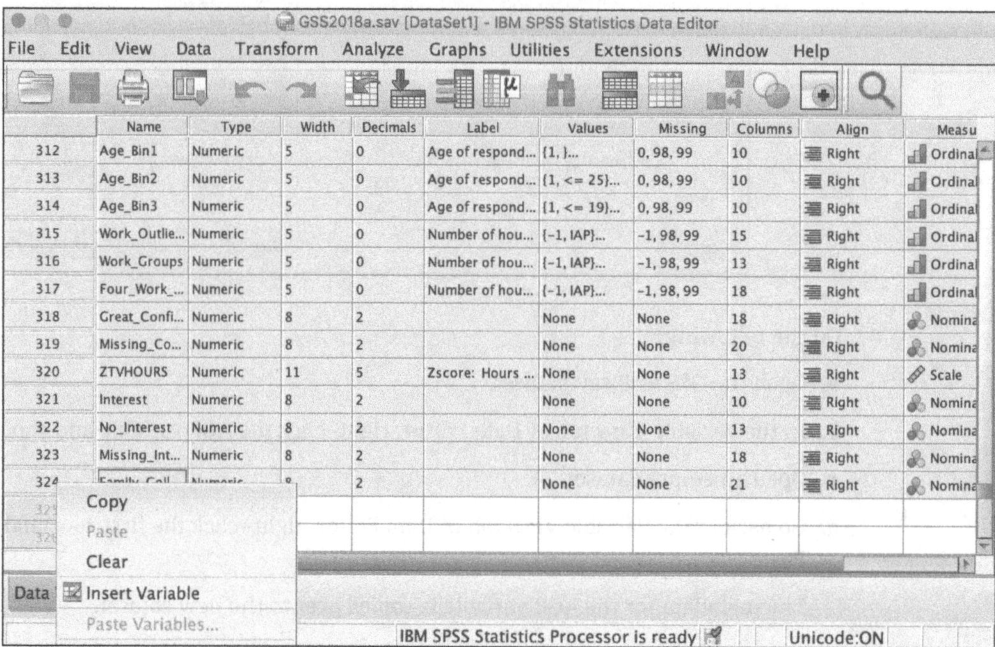

2 Messing with Data in SPSS

IN THIS PART . . .

Specify the data you want to work with.

Transform variables so that they answer your questions.

Create variables.

IN THIS CHAPTER

» Selecting cases

» Splitting files

» Adding cases

» Adding variables

Chapter 3

Using the Data Menu

Once your data is in SPSS, you may find that you want to analyze only some of your data or perform the same analysis on various subsets. For example, you may want to do separate analyses for new customers and returning customers.

In addition, sometimes data is kept in different files, and these files can be similar (for example, the same customer information just separated by location) or very different (for example, customer satisfaction information in one file and purchasing behavior in another). In this chapter, we discuss various data management techniques.

Selecting Data

The Select Cases transformation allows you to specify a group of interest and then run your analyses on only that group.

Follow these steps to select a subset of cases:

1. **Choose File ⇨ Open ⇨ Data and load the GSS2018.sav file.**

   You can download the file from the book's companion website at `www.dummies.com/go/spssstatisticsworkbookfd`. This file contains data from the General Social Survey (GSS), a nationally representative survey of adults in the United States that collects data on contemporary opinions, attitudes, and behaviors.

2. **Choose Data ➪ Select Cases.**
3. **Select the If Condition Is Satisfied radio button, and then click the If button.**

   Now you can specify the selection criteria.

4. **Move the MARITAL variable from the list on the left to the expression box (in the top left).**

   You can move the variable by either dragging it or by selecting it and then clicking the arrow button.

5. **Use your keyboard or the on-screen keypad to enter =5 in the expression box, as shown in Figure 3-1.**

   You have just told SPSS that you want to select only cases that have a value of 5 for the MARITAL variable. You type 5, and not the word *NEVER MARRIED*, because the stored value is 5 even though the label for a value of 5 is NEVER MARRIED.

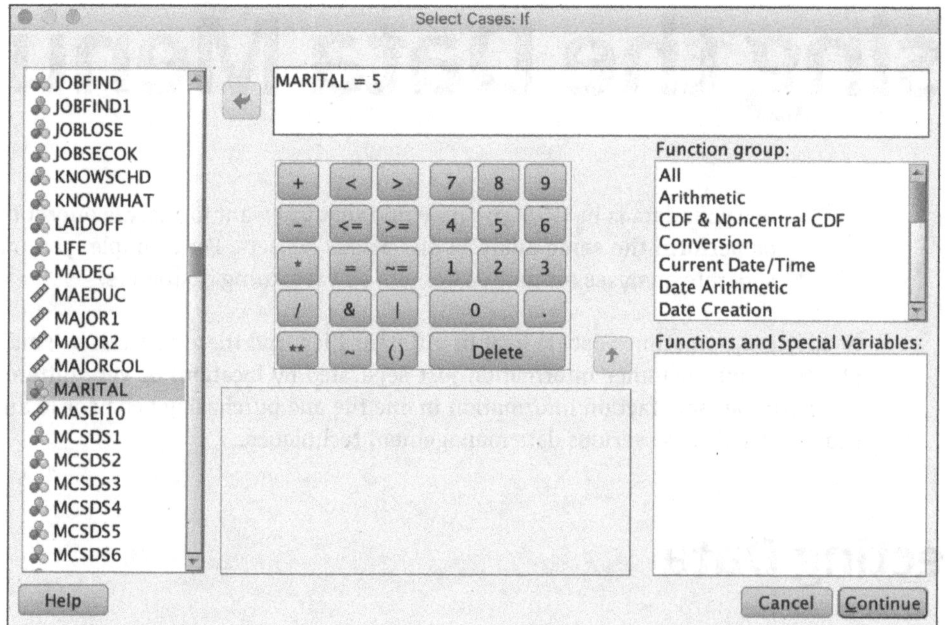

FIGURE 3-1: The If dialog.

6. **Click Continue, and then click OK.**
7. **Make sure you're looking at the Data View tab of Data Editor.**

   The slashes over some row IDs (in the first column) indicate that non-NEVER MARRIED people are being ignored (for the time being) and only NEVER MARRIED are being analyzed. The filter_$ variable is also created and is comprised of 0 and 1 for Not Selected cases and Selected cases, respectively.

WARNING

From this point forward, every piece of output that you generate will use only NEVER MARRIED.

8. **To return to using all cases, click the All Cases radio button in the main Select Cases dialog.**

See the following for an example of selecting cases.

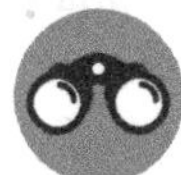

EXAMPLE

Q. Using the GSS2018.sav file, select anyone who has a value of NEVER MARRIED for the MARITAL variable and a value of 0 for the CHILDS variable.

A. Do the following:

a. Use the Select Cases transformation.

b. Select the If Condition Is Satisfied option and then click the If button.

c. Write the expression **(MARITAL = 5) & (CHILDS =0)** in the Numeric Variables box.

You'll now see data for only those people who have never been married and don't have children.

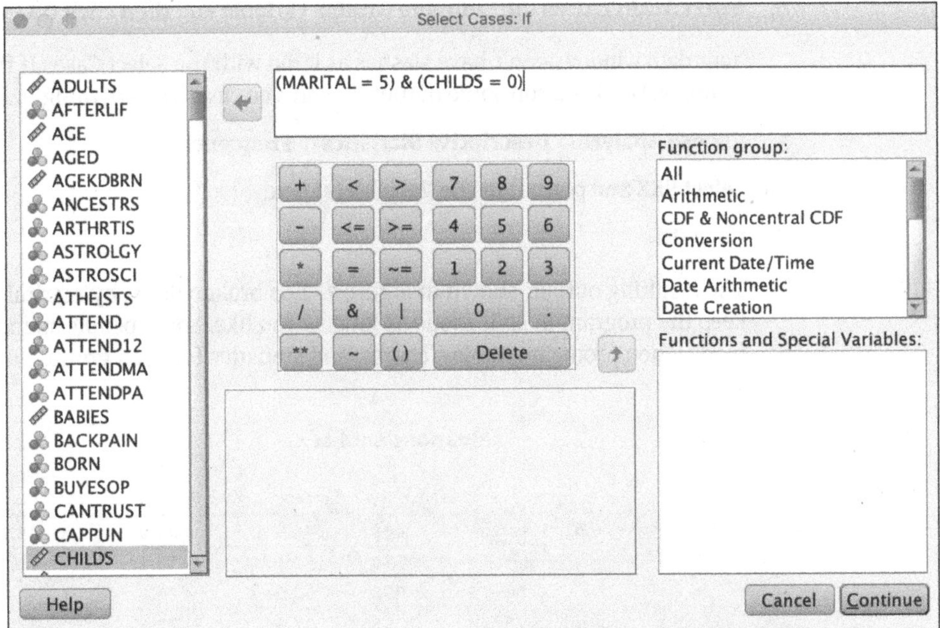

1 Using the GSS2018.sav file, select anyone who has a value of NEVER MARRIED for the MARITAL variable and a value of 0 for the CHILDS variable, but this time copy the selected cases to a new dataset.

2 Using the GSS2018.sav file, use the filter_$ variable (which you created in the previous exercise) to select anyone who has a value of NEVER MARRIED for the MARITAL variable and a value of 0 for the CHILDS variable.

3 Using the GSS2018.sav file, select a random sample of 10% of the cases in the dataset.

4 Using the GSS2018.sav file, select a range of cases from the 10th to the 100th case.

Splitting Data

Under some conditions, you may want to run a series of analyses on one group of cases, and then select another group of cases and rerun the same analyses on them. The Split File transformation allows you to select each group, one at a time, and run all your analyses on each separate group.

1. **Choose File ⇨ Open ⇨ Data and load the GSS2018.sav file.**

   You can download the file from the book's companion website at `www.dummies.com/go/spssstatisticsworkbookfd`.

2. **Choose Data ⇨ Split File.**

   The Split File dialog appears.

3. **Select the Compare Groups radio button.**
4. **Select MARITAL as the Compare Groups variable and then click OK.**

   Your data window won't have slashes as it did with the Select Cases If filter in the preceding example. Until you run some output, it won't be clear that anything has changed.

5. **Choose Analyze ⇨ Descriptive Statistics ⇨ Frequencies.**
6. **Select SEX and place it in the Variable(s) box.**
7. **Click OK.**

   The resulting output, shown in Figure 3-2, is broken down by marital status. You can keep the program in split mode as long as you like. Some people spend hours in split mode when producing tables, charts, and statistics for each group.

Respondents sex

Marital status			Frequency	Percent	Valid Percent	Cumulative Percent
MARRIED	Valid	MALE	468	46.9	46.9	46.9
		FEMALE	530	53.1	53.1	100.0
		Total	998	100.0	100.0	
WIDOWED	Valid	MALE	55	27.5	27.5	27.5
		FEMALE	145	72.5	72.5	100.0
		Total	200	100.0	100.0	
DIVORCED	Valid	MALE	161	40.0	40.0	40.0
		FEMALE	242	60.0	60.0	100.0
		Total	403	100.0	100.0	
SEPARATED	Valid	MALE	32	42.7	42.7	42.7
		FEMALE	43	57.3	57.3	100.0
		Total	75	100.0	100.0	
NEVER MARRIED	Valid	MALE	335	50.0	50.0	50.0
		FEMALE	335	50.0	50.0	100.0
		Total	670	100.0	100.0	
NA	Valid	MALE	1	50.0	50.0	50.0
		FEMALE	1	50.0	50.0	100.0
		Total	2	100.0	100.0	

FIGURE 3-2: The frequency of genders while in split mode.

REMEMBER

When you're finished with the Split File (or Select Cases) transformation, turn it off by choosing the Analyze All Cases, Do Not Produce Groups radio button in the Split File dialog.

See the following for an example of splitting data.

EXAMPLE

Q. Using the GSS2018.sav file, split the data using the MARITAL variable; however, this time select the Organize Output by Groups option. Then run the Frequencies procedure on the SEX and BORN variables.

A. Do the following:

a. Use the Split File transformation.

b. Place the MARITAL variable in the Groups Based On box.

c. Select the Organize Output by Groups option.

d. Run Frequencies on the SEX and BORN variables.

Note how the output is organized so that the results for the MARRIED group appear first, followed by the results of the WIDOWED group, and so on.

Frequency Table

Respondents sex[a]

		Frequency	Percent	Valid Percent	Cumulative Percent
Valid	MALE	468	46.9	46.9	46.9
	FEMALE	530	53.1	53.1	100.0
	Total	998	100.0	100.0	

a. Marital status = MARRIED

Was R born in this country[a]

		Frequency	Percent	Valid Percent	Cumulative Percent
Valid	YES	840	84.2	84.2	84.2
	NO	158	15.8	15.8	100.0
	Total	998	100.0	100.0	

a. Marital status = MARRIED

5 Using the GSS2018.sav file, split the data using the RACE and SEX variables, and select the Compare groups option. Then run the Frequencies procedure on the DEGREE variable.

6 Using the GSS2018.sav file, split the data using the RACE and SEX variables, and select the Organize Output by Groups option. Then run the Frequencies procedure on the DEGREE and CLASS variables.

7. Using the GSS2018.sav file, split the data using the SEX and AGE variables, and select the Compare Groups option. Then run the Frequencies procedure on the BORN variable.

8. Using the GSS2018.sav file, split the data using the SEX and AGE variables, and select the Organize Output by Groups option. Then run the Frequencies procedure on the BORN and HAPPY variables.

Merging Files by Adding Cases

The Add Cases transformation appends two data files with the same or similar variables, thus creating a combined file with the total number of cases from both files. Oftentimes both files have the same variables and the variables must have the same name, coded values, and type (for example, string versus numeric).

The following example shows you how to combine data files:

1. **Make sure that no other data files are open before you begin the merge operation.**
2. **Choose File ➪ Open ➪ Data and load the stroke_invalid file.**

   Download the stroke_invalid file and the stroke_valid file from the book's companion website at `www.dummies.com/go/spssstatisticsworkbookfd`. The stroke_invalid data file has 39 variables and 1,183 cases. This file doesn't contain information on whether the patient had a stroke.
3. **Choose File ➪ Open ➪ Data and load the stroke_valid file.**

   As mentioned, download the file from the book's companion website. This data file has 42 variables and 1,048 cases. This file contains information on whether the patient had a stroke, which is why it has three additional variables.

   Now you want to combine these two files.
4. **Using the Window menu, ensure that the stroke_valid file is the active dataset (make sure it's selected).**

   Technically it doesn't matter if you add the second file to the first file or vice versa, but your choice will determine which data is at the top of the combined file (data from the active dataset appears first in the combined file).
5. **Choose Data ➪ Merge Files ➪ Add Cases.**

   At this point, you can combine the active dataset (stroke_valid) with any files open in SPSS Statistics or files saved as an SPSS Statistics data file. If you want to combine files in other formats, you must first read the files into SPSS.
6. **Select the stroke_invalid file and then click Continue.**

   The dialog shown in Figure 3-3 appears.

   Variables that have the same names in both files are listed in the Variables in New Active Dataset box. Variables that do not match are listed in the Unpaired Variables box.

The variables are combined by variable name, and the variable formats should be the same. For example, a variable such as gender should not be coded as 1 and 2 in one file and M and F in the second file.

When matching variables that don't have the same name across both files, you have three options:

- Change the name of one of the variables before combining the files.
- Use the rename option in the Add Cases transformation.
- Pair any unpaired variables by clicking the Pair button; the new variable's name is taken from the variable in the active data file.

The file legend in the lower-left corner lists the symbol corresponding to each file, which is used to designate the source for unpaired variables.

Variables that are unpaired and don't measure the same thing can be moved to the Variables in New Active Dataset list; they'll be retained in the combined file.

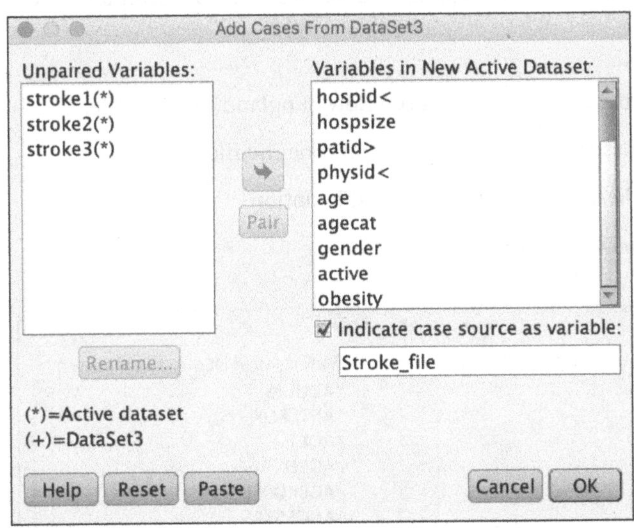

FIGURE 3-3: Unpaired variables.

7. **Select the stroke1, stroke2, and stroke3 variables, and then click the arrow to move them to the box on the right.**
8. **Select the Indicate Case Source as Variable option and rename the new variable Stroke_file.**

   The Indicate Case Source as Variable option allows you to create a new variable, named source01 by default, which will be coded 0 if the case comes from the active dataset or 1 if the case comes from the other data file. This case source indicator variable can be useful if you don't have a variable in the files that uniquely identifies that file.

9. **Click OK.**

   The combined file, with the new Stroke_file variable, is generated. The result is a new file with 43 variables and 2,231 cases. Note that the stroke1, stroke2, and stroke3 variables have missing information for the cases that came from the stroke_invalid file, which makes sense because these are the unpaired variables. Note also that the name of the file is still stroke_valid.

10. **Choose File ⇨ Save As. Name the file combined_stroke and click Save.**

WARNING

When merging, don't be casual about saving the file after the merge. If you just click Save, the new combined file will overwrite the original active file and you will have lost the original file! Instead, make sure you use Save As (not Save). Remember to do this as soon as you confirm that the new combined file looks correct.

Now you can perform analyses on the combined data file or compare the people in the first file with the people in the second file using the new variable you just created.

Only two data files can be combined simultaneously when using dialogs. However, you can merge an unlimited number of files by using Syntax.

EXAMPLE

Q. Use the Add Cases transformation to combine the Pacific.sav and New_England.sav data files. Create a new source variable called location to distinguish the two files.

A. Do the following:

a. Open the Pacific.sav data file and the New_England.sav data file.

b. Use the Add Cases transformation to combine the files.

c. Select the Indicate Case Source as Variable option.

d. Rename the new variable **Location.**

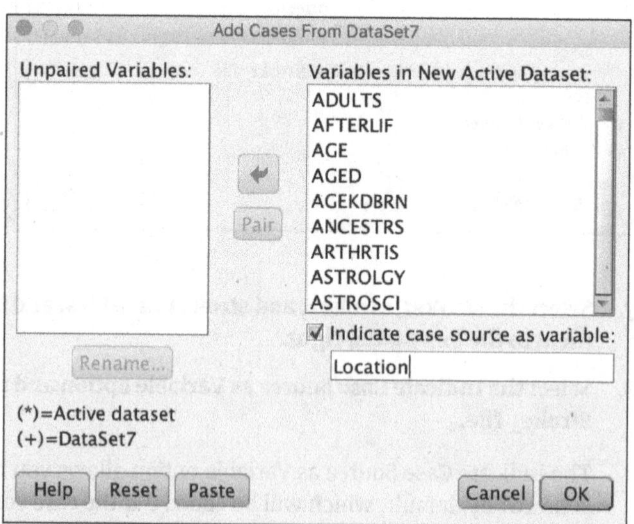

9 Use the Add Cases transformation to combine the Mountain.sav and Middle_Atlantic.sav data files. Create a new source variable called location to distinguish the two files.

10 Use the Add Cases transformation to combine the newly combined Mountain and Middle_Atlantic data files with the combined file of Pacific and New_England (these combined files were created in prior examples).

11. Use the Add Cases transformation to combine the world 1995 other countries.sav and world 1995 americas.xlsx data files.

12. Use the Chapter3.sps file to combine the New_England.sav, Middle_Atlantic.sav, Mountain.sav, and Pacific.sav datasets.

Merging Files Adding Variables

The Add Variables transformation joins two data files so that information held for an individual in different locations can be analyzed together. You can perform two types of Add Variables merges: one-to-one and one-to-many. Both types add variables to cases matched on key variables. *Key variables* are case identifiers that exist in both files (for example, a variable such as ID for the customer ID number).

WARNING

Both input files must be sorted in ascending order on the key variables to get a one-to-one match to work properly.

In a *one-to-one merge*, one and only one case from the first file will be combined with one case from the second file. In a *one-to-many merge*, one file is designated as the table file and cases from that file can match multiple cases in the case file. The case file defines the cases in the merged file. The values of the key variable(s) must define unique cases in the table file but not in the case file.

The following example of a one-to-one merge uses the electronics_company_info and electronics_complete data files. Before starting any merge example, close all other data files. Merges can be confusing if you have a lot of unrelated files open.

Follow these steps to perform a one-to-one merge:

1. **Choose File ➪ Open ➪ Data and load the electronics_company_info file.**

   Download this file and the electronics_complete file from the book's companion website at `www.dummies.com/go/spssstatisticsworkbookfd`. The data file has 5 variables and 5,003 cases, and contains information on each customer's company.

   WARNING

   Input files must be sorted on key variables. In this example, the data has already been sorted on the ID key variable.

2. **Choose File ➪ Open ➪ Data and load the electronics_complete file.**

   As mentioned, download the file from the book's companion website. The data file has 12 variables and 5,003 cases, and contains each customer's purchase history.

   Now you'll combine these two files.

3. **Choose Data ➪ Merge Files ➪ Add Variables.**

   The Add Variables dialog appears, as shown in Figure 3-4. At this point, you can combine the active dataset (electronics_complete) with any files open in SPSS Statistics or saved as an SPSS Statistics data file.

REMEMBER

If you want to combine files in a format other than a .sav file, you must first read the files into SPSS Statistics.

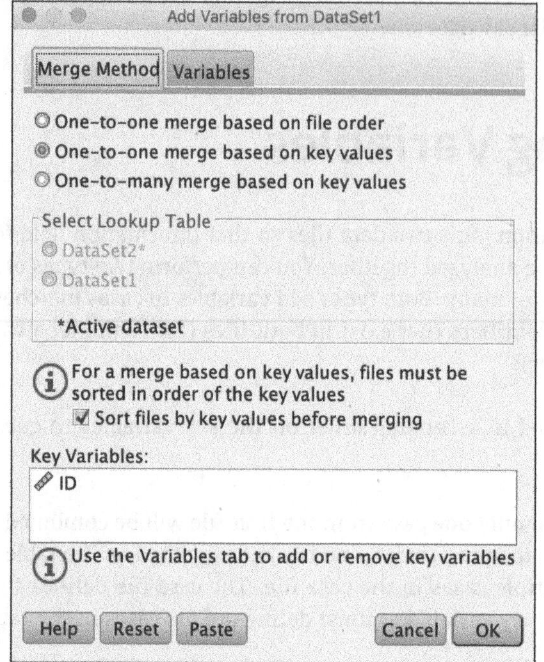

FIGURE 3-4: The Merge method dialog.

4. **Select the electronics_company_info file, and then click Continue.**
5. **Make sure that the One-to-One Merge Based on Key Values option is selected.**

   Note that the Sort Files by on Key Values before Merging option has also been selected, so technically SPSS will sort both files again. If you have previously sorted your files (as we have done), this option does not need to be selected. Also note that SPSS has identified the ID variable as the key variable.

WARNING

The One-to-One Merge Based on File Order option combines files based on order. The first record in the first dataset is joined with the first record in the second dataset, and so on. When a dataset runs out of records, no further output records are produced. This order method can provide inaccurate results if cases are missing from a file or if files have been sorted differently.

6. **Click the Variables tab of the dialog.**

   Before running the merge, make sure there are no problems with the variables.

   Variables that have unique names are listed in the Included Variables box. If the same variable name is used in both files, only one set of data values can be retained — these variables will appear in the Excluded Variables box. Although a renaming facility is available in the Add Variables dialog, it's safer to use unique names from the beginning.

If you're merging two files from two time periods, some of the variables may have the same name because they measure the same concept. In this case, each variable should be given a unique name — perhaps numbered or based on the date of the survey — to differentiate the different time periods.

You can rename variables in two ways:

- Change the name of one of the variables before adding the files.
- Use the rename option in the Add Cases facility.

7. **Click OK.**

   The new combined file is generated. The result is a new file with 16 variables and 5,003 cases.

8. **Choose File ⇨ Save As. Name the file** combined_electronics **and click Save.**

Now you can perform analyses on the combined data file and investigate relationships that wouldn't have been possible without first performing the merge.

REMEMBER

Only two data files can be combined simultaneously when using dialogs. However, you can merge an unlimited number of files by using Syntax.

EXAMPLE

Q. Use a one-to-many match to combine the rfm_transactional1.sav and rfm_aggregated.sav files.

A. Use the following steps to complete a one-to-many match:

a. Choose File ⇨ Open ⇨ Data and load the rfm_aggregated file.

   You can download this file and the rfm_transactions1 file from the book's companion website at `www.dummies.com/go/spssstatisticsworkbookfd`. The rfm_aggregated data file has 4 variables and 995 cases. It contains the customers' purchase history, with each row representing a customer.

WARNING

   Input files must be sorted on key variables. In this example, the data has already been sorted on the ID key variable.

b. Choose File ⇨ Open ⇨ Data and load the rfm_transactions1 file.

   As mentioned, download the file from the book's companion website. The data file has 5 variables and 4,906 cases. This file contains customer transactional data, with each row representing a transaction.

c. Choose Data ⇨ Merge Files ⇨ Add Variables.

d. Select the rfm_aggregated file, and then click Continue.

e. Make sure that the One-to-Many Merge Based on Key Values option is selected.

   When doing a one-to-many merge, you must identify which file is the aggregated file, or *lookup table* (the file that will have a case merge with several cases in the other file). In our case, the non-active dataset (rfm_aggregated) is the aggregated file.

f. Make sure that the non-active dataset (the dataset without the asterisk) is chosen as the Select Lookup table.

As in the preceding example, note that the Sort Files on Key Values before Merging option has also been selected, and that SPSS has identified the ID variable as the key variable.

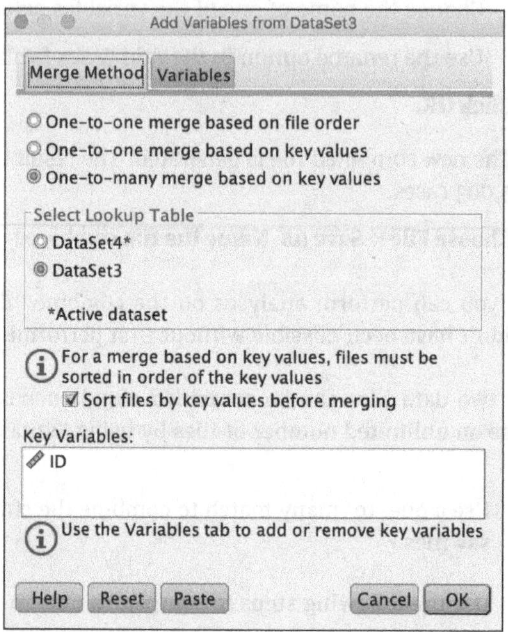

g. Click the Variables tab of the dialog.

Before running the merge, make sure there are no problems with the variables. Everything looks good in this example.

h. Click OK.

The new combined file is generated, with 8 variables and 4,906 cases.

i. Choose File ⇨ Save As. Name the file **combined_rfm** and then click Save.

Now you can perform analyses on the combined data file and investigate relationships that wouldn't have been possible without first performing the merge.

13 Combine the telco customers.sav and telco customer satisfaction.sav files.

14 Combine the telco customers_products_revenue.sav file and the combined file (telco customers and telco customer satisfaction) you created in preceding example.

15 Combine the customer_revenue.sav and cust_survey.sav files.

16 Use the file Chapter3.sps to combine the following datasets: telco customers.sav, telco customers_products_revenue.sav, and telco customer satisfaction.sav. Use CUSTOMER_ID as the key variable.

Answers to Problems in Using the Data Menu

1. Do the following:

   a. Use the Select Cases transformation.

   b. Select the If Condition Is Satisfied option and then click the If button.

   c. Write the expression **(MARITAL = 5) & (CHILDS =0)** in the Numeric Variables box.

   d. Click in the Output section and select Copy Selected Cases to a New Dataset.

   e. Name the dataset.

   This new file will have data for only those people who have never been married and do not have children.

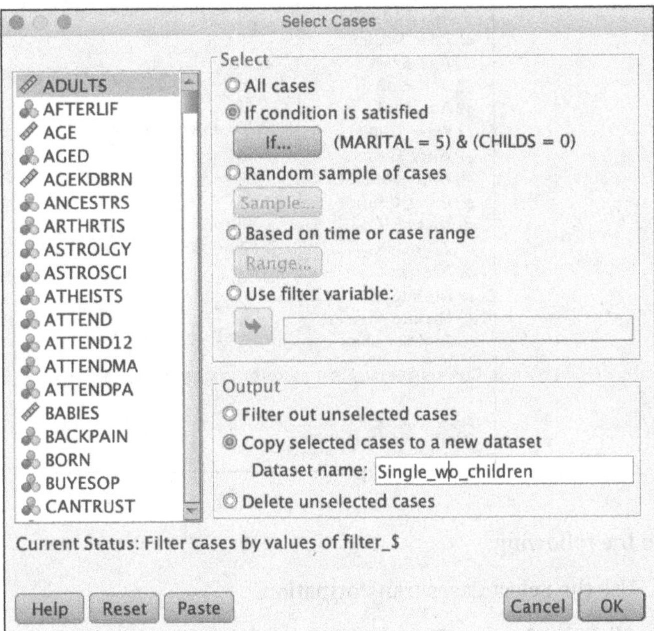

2. Do the following:

   a. Use the Select Cases transformation.

   b. Click the Use Filter Variable option.

   c. Select the filter_$ variable.

   This variable, which was automatically created in the preceding exercise, selects anyone who has a value of NEVER MARRIED for the MARITAL variable and a value of 0 for the CHILDS variable.

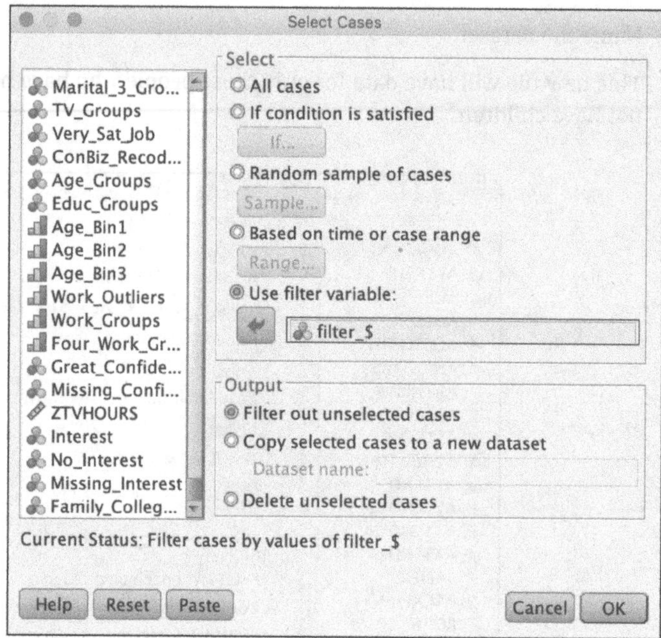

3. Do the following:

   a. Use the Select Cases transformation.

   b. Click the Random Sample of Cases option and then click the Sample button.

   c. Specify that 10% of the cases will be selected.

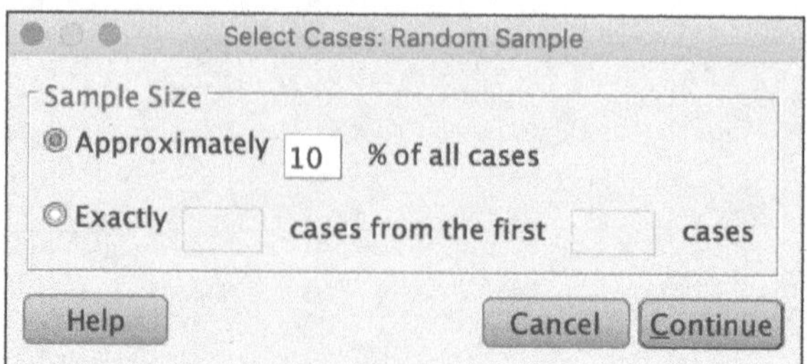

4 Do the following:

a. Use the Select Cases transformation.

b. Click the Based on Time or Case Range option and then click the Range button.

c. Specify that you will select from the 10th to the 100th case.

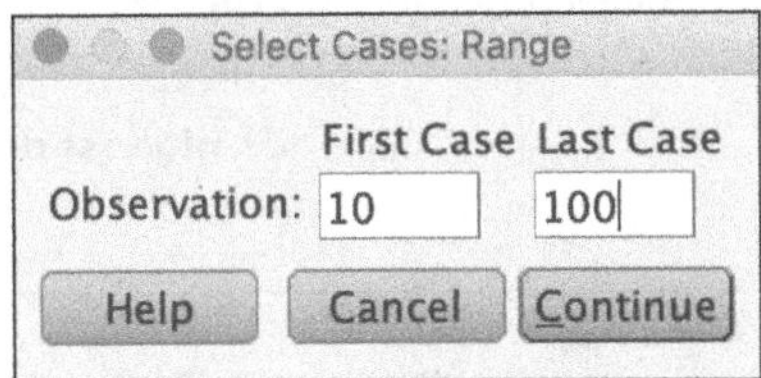

5 Do the following:

a. Use the Split File transformation.

b. Place the RACE and SEX variables in the Groups Based On box.

c. Select the Compare Groups option.

d. Run Frequencies on the DEGREE variable.

R's highest degree

Race of respondent	Respondents sex			Frequency	Percent	Valid Percent	Cumulative Percent
WHITE	MALE	Valid	LT HIGH SCHOOL	70	9.1	9.1	9.1
			HIGH SCHOOL	397	51.6	51.6	60.7
			JUNIOR COLLEGE	57	7.4	7.4	68.1
			BACHELOR	158	20.5	20.5	88.7
			GRADUATE	87	11.3	11.3	100.0
			Total	769	100.0	100.0	
	FEMALE	Valid	LT HIGH SCHOOL	72	7.8	7.8	7.8
			HIGH SCHOOL	449	48.6	48.6	56.4
			JUNIOR COLLEGE	96	10.4	10.4	66.8
			BACHELOR	200	21.6	21.6	88.4
			GRADUATE	107	11.6	11.6	100.0
			Total	924	100.0	100.0	
BLACK	MALE	Valid	LT HIGH SCHOOL	20	12.7	12.7	12.7
			HIGH SCHOOL	92	58.6	58.6	71.3
			JUNIOR COLLEGE	8	5.1	5.1	76.4
			BACHELOR	31	19.7	19.7	96.2
			GRADUATE	6	3.8	3.8	100.0
			Total	157	100.0	100.0	
	FEMALE	Valid	LT HIGH SCHOOL	35	15.4	15.4	15.4
			HIGH SCHOOL	115	50.4	50.4	65.8
			JUNIOR COLLEGE	23	10.1	10.1	75.9
			BACHELOR	36	15.8	15.8	91.7
			GRADUATE	19	8.3	8.3	100.0
			Total	228	100.0	100.0	

6 Do the following:

a. Use the Split File transformation.

b. Place the RACE and SEX variables in the Groups Based On box.

c. Select the Organize Output by Groups option.

d. Run the Frequencies procedure on the DEGREE and CLASS variables.

R's highest degree[a]

		Frequency	Percent	Valid Percent	Cumulative Percent
Valid	LT HIGH SCHOOL	70	9.1	9.1	9.1
	HIGH SCHOOL	397	51.6	51.6	60.7
	JUNIOR COLLEGE	57	7.4	7.4	68.1
	BACHELOR	158	20.5	20.5	88.7
	GRADUATE	87	11.3	11.3	100.0
	Total	769	100.0	100.0	

a. Race of respondent = WHITE, Respondents sex = MALE

Subjective class identification[a]

		Frequency	Percent	Valid Percent	Cumulative Percent
Valid	LOWER CLASS	56	7.3	7.3	7.3
	WORKING CLASS	293	38.1	38.3	45.6
	MIDDLE CLASS	383	49.8	50.0	95.6
	UPPER CLASS	34	4.4	4.4	100.0
	Total	766	99.6	100.0	
Missing	DK	3	.4		
Total		769	100.0		

a. Race of respondent = WHITE, Respondents sex = MALE

7 Do the following:

a. Use the Split File transformation.

b. Place the SEX and AGE variables in the Groups Based On box.

c. Select the Compare Groups option.

d. Run Frequencies on the BORN variable.

Was R born in this country

Respondents sex	Age of respondent			Frequency	Percent	Valid Percent	Cumulative Percent
MALE	18	Valid	YES	6	100.0	100.0	100.0
	19	Valid	YES	12	85.7	85.7	85.7
			NO	2	14.3	14.3	100.0
			Total	14	100.0	100.0	
	20	Valid	YES	7	100.0	100.0	100.0
	21	Valid	YES	13	92.9	92.9	92.9
			NO	1	7.1	7.1	100.0
			Total	14	100.0	100.0	
	22	Valid	YES	25	96.2	96.2	96.2
			NO	1	3.8	3.8	100.0
			Total	26	100.0	100.0	

8. Do the following:

   a. Use the Split File transformation.

   b. Place the SEX and AGE variables in the Groups Based On box.

   c. Select the Organize output by groups option.

   d. Run the Frequencies procedure on the BORN and HAPPY variables.

Was R born in this country[a]

		Frequency	Percent	Valid Percent	Cumulative Percent
Valid	YES	6	100.0	100.0	100.0

a. Respondents sex = MALE, Age of respondent = 18

General happiness[a]

		Frequency	Percent	Valid Percent	Cumulative Percent
Valid	VERY HAPPY	1	16.7	16.7	16.7
	PRETTY HAPPY	4	66.7	66.7	83.3
	NOT TOO HAPPY	1	16.7	16.7	100.0
	Total	6	100.0	100.0	

a. Respondents sex = MALE, Age of respondent = 18

9. Do the following:

   a. Open the Mountain and Middle_Atlantic data files.

   b. Use the Add Cases transformation to combine the files.

   c. Select the Indicate Case Source as Variable option.

   d. Rename the new variable Location.

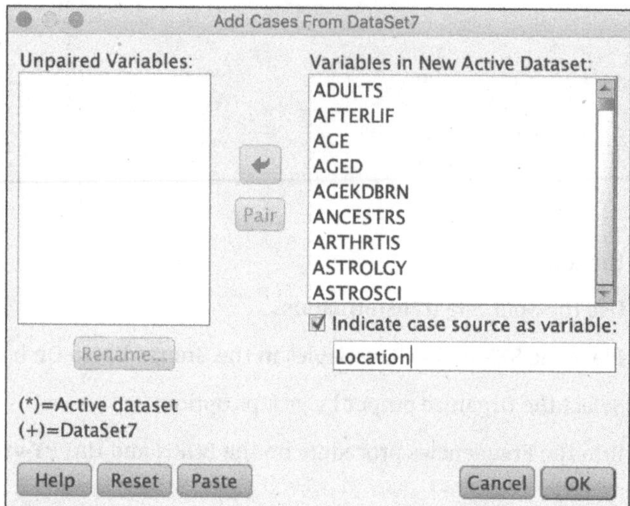

10. Use the Add Cases transformation to combine both combined files created in prior examples (Pacific and New_England, and Mountain and Middle_Atlantic). Note that answer 12 will show you how to combine all four files by using Syntax.

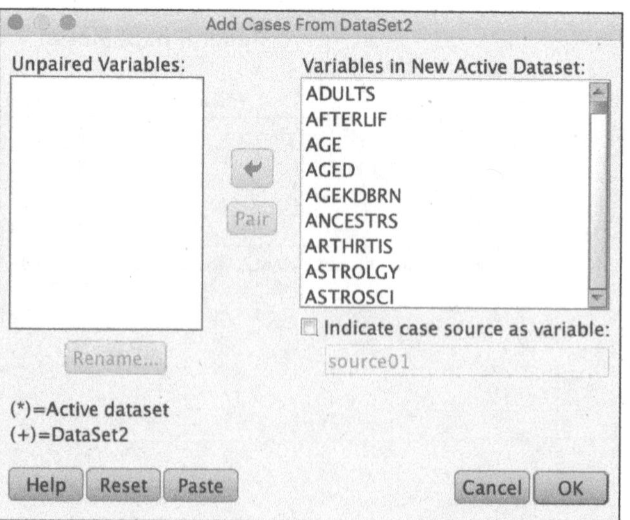

11 Do the following:

a. Open the world 1995 americas.xlsx file and then the world 1995 other countries.sav data file.

b. Use the Add Cases transformation to try to combine these files.

You'll see that you can't combine these files because string variables must have the same variable width, but the COUNTRY and REGION variables have different variable widths.

c. To fix the problem, change the width of the COUNTRY and REGION variables in both files to a value of 30.

d. Now use the Add Cases transformation to combine the files.

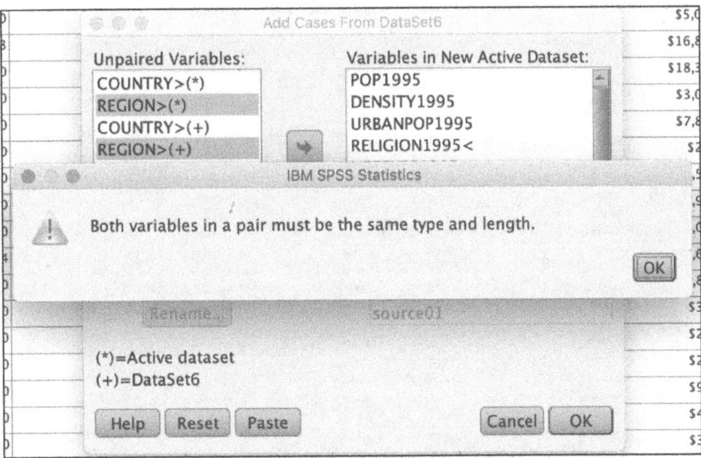

12 Open the New_England.sav data file and then the Chapter3.sps Syntax file. Run the Syntax code for the Add Files command. Note that you need to specify where the data files are located on your computer.

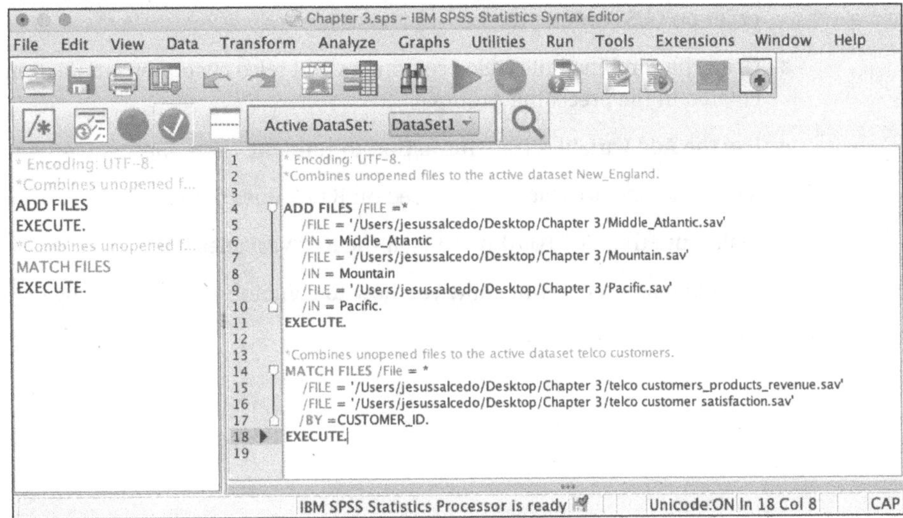

13. Do the following:

    a. Open the telco customer satisfaction.sav data file, and sort the file in ascending order on CUSTOMER_ID.

    b. Open the telco customers.sav data file, and sort the file in ascending order on CUSTOMER_ID.

    c. Use the Add Variables transformation to combine these files.

    d. Select the One-to-One Merge Based on Key Values option.

    e. Make sure that CUSTOMER_ID is in the Key Variables box.

telco customers.sav [DataSet6] - IBM SPSS Statistics Data Editor

File Edit View Data Transform Analyze Graphs Utilities Extensions Window Help

386 : CONNECT_D... 03/09/1998

Visible: 22 of 22 Variables

	CUSTOMER_ID	GENDER	AGE	TARIFF	HANDSET	CONNECT_DATE	CLOSURE_DATE	CHURNED	sat1	sat2	sat3	sat4
382	K103770	F	32	CAT 10	ANF481	02/17/1998	.	0	.	.	.	.
383	K103780	F	33	CAT 10	ANF481	12/27/1998	.	0	.	.	.	.
384	K103790	M	42	CAT 10	ASAD1	05/19/1996	.	0	.	.	.	.
385	K103800	F	43	CAT 10	ANF481	01/25/1998	.	0	.	.	.	.
386	K103810	M	32	CAT 10	ASAD9	03/09/1998	03/09/2001	1	.	.	.	.
387	K103820	F	43	CAT 10	ASAD9	06/20/1998	02/04/2001	1	5	5	5	5
388	K103830	F	54	CAT 10	ASAD1	11/20/1999	.	0	.	.	.	.
389	K103840	F	39	CAT 10	ASAD1	10/21/1996	.	0	.	.	.	.
390	K103850	M	39	CAT 10	ANF100	10/23/1997	.	0	.	.	.	.
391	K103860	M	40	CAT 10	ANF481	12/23/1998	.	0	.	.	.	.
392	K103870	F	29	CAT 10	ASAD9	09/15/1996	11/03/1998	1	.	.	.	.
393	K103880	M	42	CAT 10	ANF481	04/12/1997	07/04/1999	1	.	.	.	.
394	K103890	F	27	CAT 10	ANF100	04/27/1996	.	0	.	.	.	.
395	K103900	M	41	CAT 10	ANF100	02/12/1996	.	0	.	.	.	.
396	K103910	F	28	CAT 10	ASAD1	05/20/1998	.	0	.	.	.	.
397	K103920	M	31	CAT 10	ANF481	10/18/1996	.	0	.	.	.	.
398	K103930	M	57	CAT 10	ANF481	01/05/1998	.	0	5	5	4	5
399	K103940	F	42	CAT 10	ASAD1	04/04/1996	.	0	.	.	.	.
400	K103950	F	31	CAT 10	ASAD9	02/15/1996	12/14/1998	1	.	.	.	.
401	K103960	F	70	CAT 10	ANF100	06/26/1999	.	0	.	.	.	.
402	K103970	F	48	CAT 10	ANF100	08/17/1996	.	0	.	.	.	.

Data View Variable View

IBM SPSS Statistics Processor is ready Unicode:ON

14. Do the following:

    a. Open the telco customers_products_revenue.sav data file, and sort the file in ascending order on CUSTOMER_ID.

    b. Open the combined file (telco customers and telco customer satisfaction), which you created in the preceding example.

    c. Use the Add Variables transformation to combine these files.

    d. Select the One-to-One Merge Based on Key Values option.

    e. Make sure that CUSTOMER_ID is in the Key Variables box.

    Note that answer 16 will show you how to combine all three files by using Syntax.

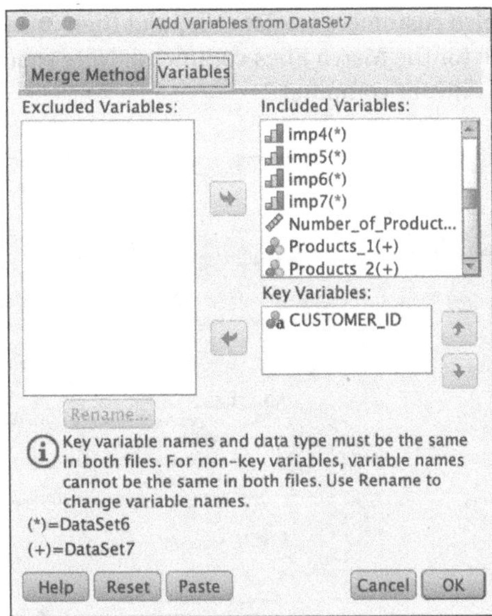

15 Do the following:

a. Open the customer_revenue.sav data file, and sort the file in ascending order on the Orgntype and HowlongasCustomer variables.

b. Open the cust_survey.sav data file, and sort the file in ascending order on the Orgntype and HowlongasCustomer variables.

c. Use the Add Variables transformation to combine these files.

d. Select the One-to-One Merge Based on Key Values option.

e. Make sure that Orgntype and HowlongasCustomer are in the Key Variables box.

f. Make sure that the customer_revenue.sav data file has been specified as the Lookup Table.

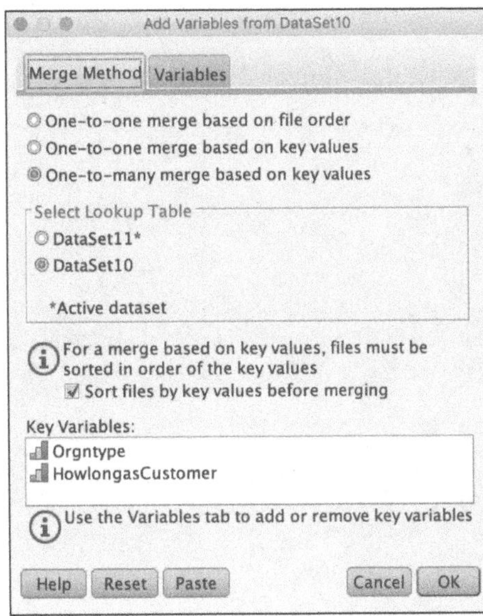

16 Open the telco customers.savdata file, and then open the Chapter3.sps Syntax file. Run the Syntax code for the Match Files command. Note that you need to specify where the data files are located on your computer.

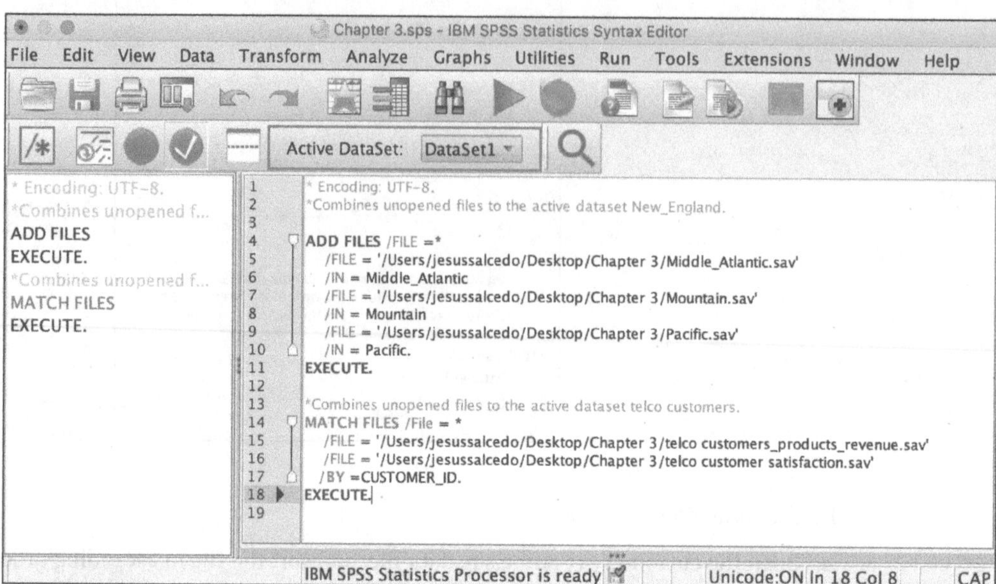

IN THIS CHAPTER

» Counting values within cases

» Recoding variables

» Using visual binning

Chapter 4

Putting the Transform Menu to Work

After your data is in SPSS, you may find that it contains errors or isn't organized appropriately. A way to fix these problems is to make modifications to the data so that the values are easier to work with and read. This chapter contains some methods you can use to modify your data so that you can better answer your research questions.

Counting Case Occurrences

If your data keeps track of similar multiple occurrences, such as who has visited various countries or how several satisfaction questions were answered, you can generate a count of the occurrences for each case. You specify what value(s) cause a variable to qualify, and SPSS creates a variable and counts the number of qualifying variables from among those you choose. For example, if you have a number of expenses for each case, you could have SPSS count the number of expenses that exceed a certain threshold.

In the following example, we count when people have a lot of confidence in various institutions:

1. **Choose File ⇨ Open ⇨ Data and load the GSS2018.sav file.**

   You can download the file from the book's companion website at `www.dummies.com/go/spssstatisticsworkbookfd`. This file contains data from the General Social Survey

(GSS), a nationally representative survey of adults in the United States that collects data on contemporary opinions, attitudes, and behaviors.

2. **Choose Transform ⇨ Count Values within Cases.**
3. **Select the variables to use in the count: CONARMY, CONBUS, CONCLERG, CONEDUC, CONFED, CONFINAN, CONJUDGE, CONLABOR, CONLEGIS, and CONMEDIC.**
4. **In the Target Variable box, name your new variable** Great_Confidence.

   When you've finished, the screen will look like Figure 4-1. This operation works with only numeric variables because it must perform numeric matches on the values.

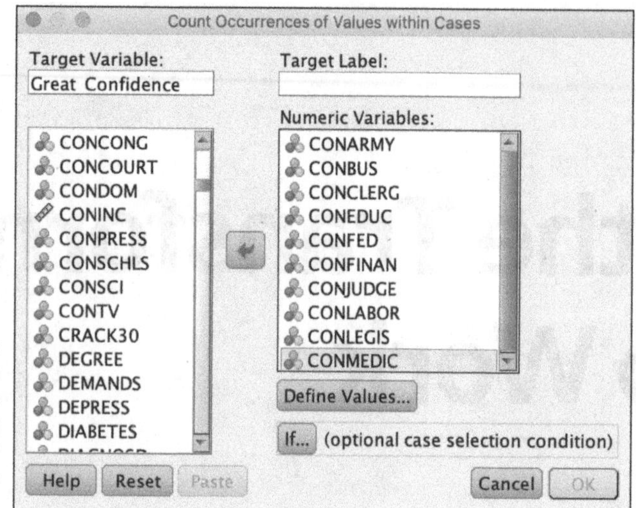

FIGURE 4-1: The chosen variables to be counted and the name of the new variable.

5. **Click the Define Values button.**
6. **In the Value area, type** 1 **and then click the Add button to move the** 1 **to the Values to Count box on the right.**

   After adding the value, you'll have the result shown in Figure 4-2. The new variable will contain a count of the variables you named that have a value that matches at least one of the criteria you specified. Each case is counted separately.

   As you can see in figure, the total can also be based on missing values and ranges of values. In the ranges, you can specify both the high and low values, or you can specify one end of the range and have the other end be either the largest or smallest value in the set. If you select a number of criteria, SPSS will check each variable against them all.

7. **Click Continue, and then click OK.**

   Values appear for the new variable.

See the following for an example of counting case occurrences.

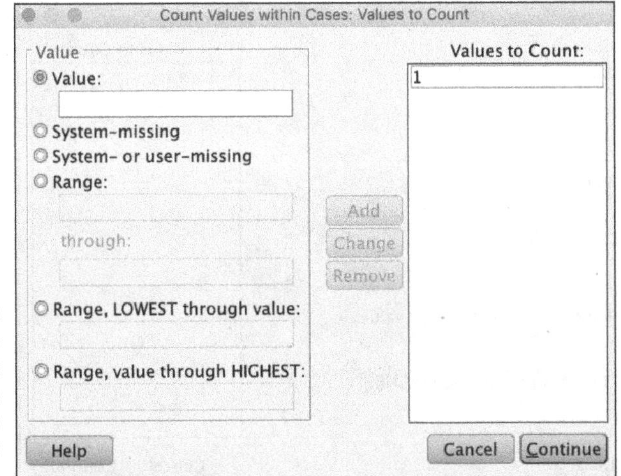

FIGURE 4-2: Define the criteria that determine which values are included in the count.

EXAMPLE

Q. Using the GSS2018.sav file, count when a missing value is present in the following variables: CONARMY, CONBUS, CONCLERG, CONEDUC, CONFED, CONFINAN, CONJUDGE, CONLABOR, CONLEGIS, and CONMEDIC.

A. Do the following:

a. Use the Count Values within Cases transformation.

Place the variables (CONARMY, CONBUS, CONCLERG, CONEDUC, CONFED, CONFINAN, CONJUDGE, CONLABOR, CONLEGIS, and CONMEDIC) in the Numeric Variables box.

b. Type **Missing_Confidence** in the Numeric Variables box.

c. Click the Define Values button, select the System- or User-Missing value, and then click Add.

This new variable counts the number of times someone had a missing value for the ten confidence variables. After you create a variable like this, you can decide to exclude cases that have too much missing data.

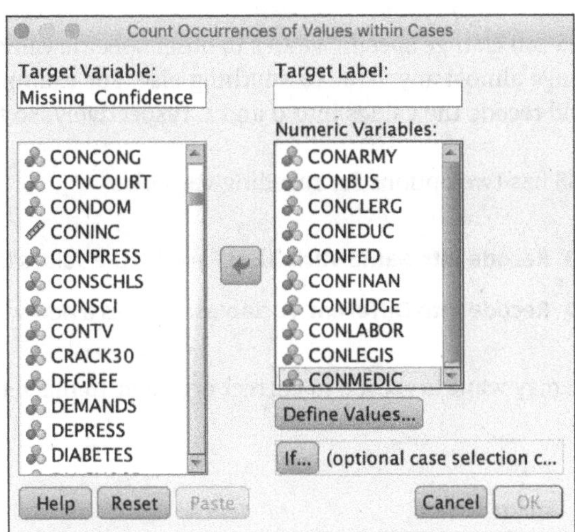

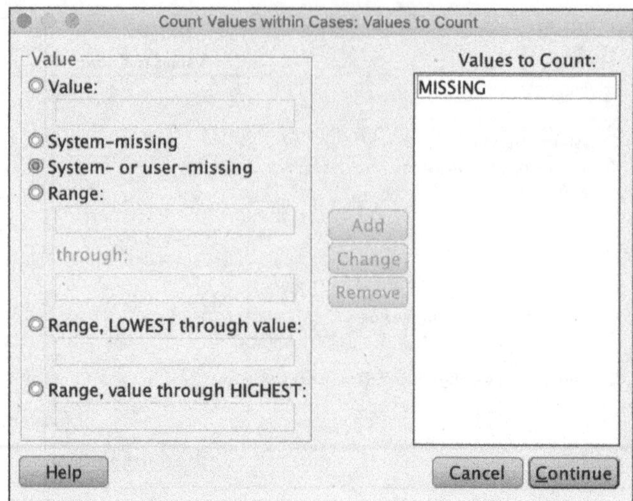

1. Using the GSS2018.sav file, count when someone is very interested in the following professions: INTECON, INTEDUC, INTENVIR, INTFARM, INTINTL, INTMED, INTMIL, INTSCI, INTSPACE, and INTTECH.

2. Using the GSS2018.sav file, count when someone has no interest in the following professions: INTECON, INTEDUC, INTENVIR, INTFARM, INTINTL, INTMED, INTMIL, INTSCI, INTSPACE, and INTTECH.

3. Using the GSS2018.sav file, count when a missing value is present in the following variables: INTECON, INTEDUC, INTENVIR, INTFARM, INTINTL, INTMED, INTMIL, INTSCI, INTSPACE, and INTTECH.

4. Using the GSS2018.sav file, count when someone has at least a college education in the following variables: EDUC, MAEDUC, PAEDUC, and SPEDUC.

Recoding Variables

SPSS can change specific values to other specific values according to rules you provide. You can change almost any value to anything else. For example, if you have Yes and No responses, you could recode the values into 0 and 1, respectively, so that you can better analyze the data.

SPSS has two options for recoding variables:

- **Recode into Same Variables:** Recodes the values in place without creating a new variable.
- **Recode into Different Variables:** Creates a new variable and keeps the original variable.

You may want to recode to correct errors or to make the data easier to use.

WARNING

Changes to your data can't be automatically reversed, so you could destroy information. For this reason, don't use the Recode into Same Variables option unless you're sure that you no longer want to use the original variable. The main reason to use this option is when you want to change a bunch of variables all at once. It's safer to stick with Recode into Different Variables because you're keeping the original variable and the revised version.

Recoding into different variables

In the next example, we recode the MARITAL variable, which has five categories (Married, Widowed, Divorced, Separated, and Never Married) into a new variable that has only three categories (Married, Never Married, and Other). The Divorced, Widowed, and Separated groups have only a few cases, so we are combining those groups before using them in an analysis. The following steps create the recoded values and store them in a new variable:

1. **Choose File ⇨ Open ⇨ Data and load the GSS2018.sav file.**

   You can download the file from the book's companion website at `www.dummies.com/go/spssstatisticsworkbookfd`.

2. **Choose Transform ⇨ Recode into Different Variables.**
3. **In the left panel, move the MARITAL variable, which holds the values you want to change, to the center panel.**
4. **In the Output Variable area, enter the name** Marital_3_Groups **for the new variable.**

   WARNING

   For the output variable, if you select a new variable name, a new variable is created. If you select an existing variable name, its values will be overwritten. Here, you chose a new name to protect the existing data.

5. **Click the Change button.**
6. **Click the Old and New Values button.**
7. **In the Old Value text box, enter 1, which is the existing value. In the New Value text box, enter 1, which is the new recoded value. Click the Add button.**
8. **In the Old Value text box, enter 5, which is the existing value. In the New Value text box, enter 2, which is the new recoded value. Click the Add button.**
9. **In the Old Value section, click Range and enter 2 in the first text box and 4 in the second. In the New Value text box, enter 3, which is the new recoded value. Click the Add button.**
10. **In the Old Value section, click System- or User-Missing. In the New Value text box, enter 9, which is the new recoded value. Click the Add button.**

    Be sure to map all values (as shown in Figure 4-3) — even the ones that don't change — because you're creating a new variable and it has no preset values.

11. **Click Continue, and then click OK.**

    You now have a new variable and the values have been coded in a more useful manner.

See the following for an example of recoding into different variables.

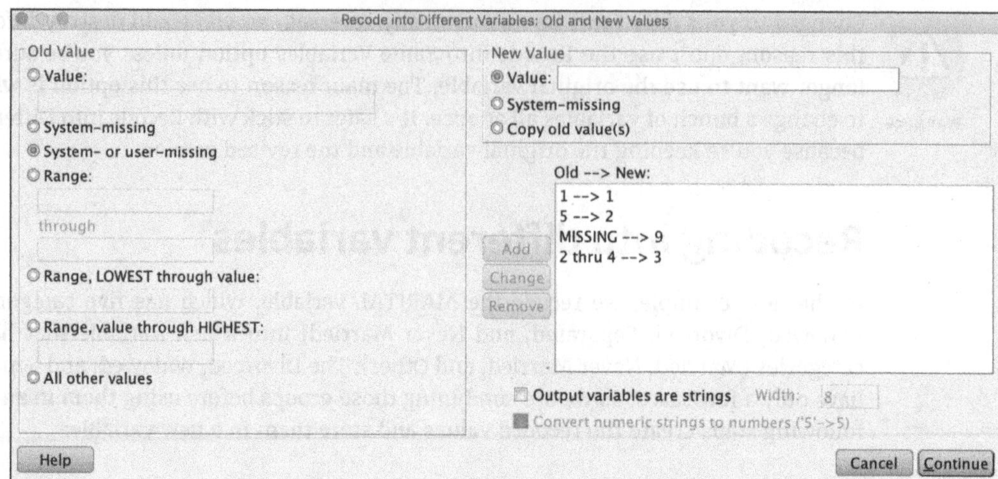

FIGURE 4-3: All possible values recoded for a new variable.

EXAMPLE

Q. The TVHOURS variable records the number of hours people watch TV per day. The values range from 0 to 24. We want to create groups so that we can easily identify when someone watches no TV (0), some TV (1-3), a lot of TV (4-8), and an excessive amount of TV (9 or more). Using the GSS2018.sav file, create this variable.

A. Do the following:

a. Use the Recode into Different Variables transformation.

b. Place the TVHOURS variable in the Input Variable box.

c. Type **TV_Groups** in the Output Variable box.

d. Click the Change button, and then click the Old and New Values button.

e. Specify the old value, 0, and new value, 0, and click Add.

f. Click the Change button and then click the Old and New Values button.

g. Specify the old value, which is Range 1 through 3, and the new value, which is 1, and then click Add.

h. Specify the old value, which is Range 4 through 8, and the new value, which is 2, and then click Add.

i. Specify the old value, which is Range, Value through Highest 9, and the new value, which is 4, and then click Add.

This new variable will group respondents based on how much TV they watch per day.

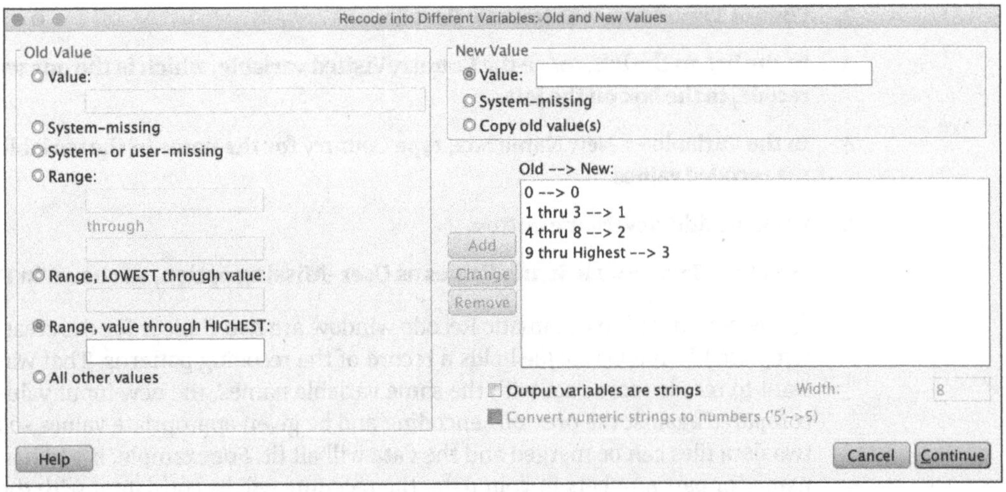

5 The SATJOB variable records the level of job satisfaction. We want to identify those people who are Very Satisfied (1) so that we can compare them to everyone else (0). Using the GSS2018.sav file, create this variable.

6 The CONBIZ variable was coded so that higher values represent lower confidence while lower values represent higher confidence. Intuitively, this doesn't make much sense, so we might want to recode the values for this variable so that higher values represent higher confidence and lower values represent lower confidence. Using the GSS2018.sav file, create this variable.

7 The AGE variable has values that range from 18 to 90. We want to create the following age groups: 19 and younger, 20-29, 30-39, 40-49, 50-59, 60-69, 70-79, 80-89, and 90 and above. Using the GSS2018.sav file, create this variable.

8 The EDUC variable has values that range from 0 to 20. We want to create the following education groups to represent the level of education: 0-5, 6-8, 9-12, 13-16, and 17 and above. Using the GSS2018.sav file, create this variable.

Automatic recoding

Automatic recoding converts string values into numeric values with labels. String variables can sometimes create confusing behaviors in SPSS because some dialogs don't recognize string variables, even though these dialogs accept categorical variables. The purpose of the Automatic Recode transformation is to transform a string variable into a numeric variable that gives you more options for analysis.

1. **Choose File ⇨ Open ⇨ Data and load the Census.sav file.**

   You can download the file from the book's companion website at `www.dummies.com/go/spssstatisticsworkbookfd`.

2. **Choose Transform ⇨ Automatic Recode.**
3. **In the list on the left, move the CountryVisited variable, which is the one we want to recode, to the box on the left.**
4. **In the Variable -> New Name box, type** Country **for the name of the variable to receive the recoded values.**
5. **Click the Add New Name button.**
6. **Select the Treat Blank String Values as User-Missing option, as shown in Figure 4-4.**

   At the bottom of the Automatic Recode window are two choices for the creation of a template file. A *template file* holds a record of the recoding patterns. That way, if you want to recode more data with the same variable names, the new input values will be compared against the previous encoding and be given appropriate values so that the two data files can be merged and the data will all fit. For example, if you have brand names or part numbers in your data, the recoding will be consistent with the original values because it will be assigned the same *pattern* of recoded values.

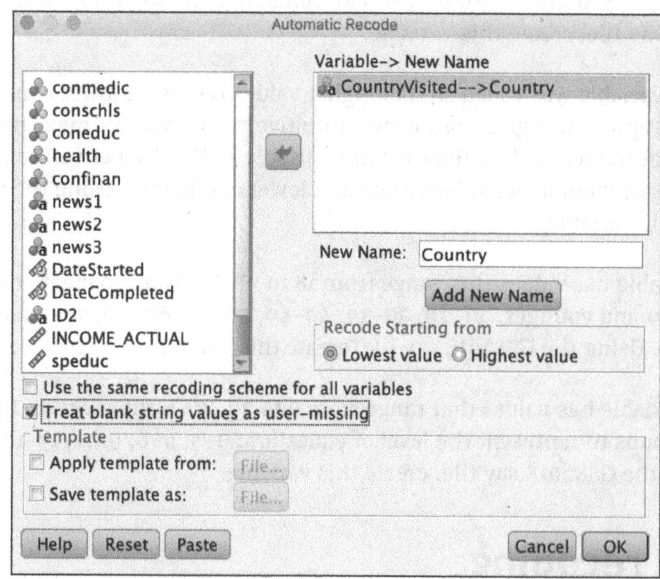

FIGURE 4-4: The dialog for automatic recoding.

7. **Click OK.**

   Recoding takes place, as shown in the output window in Figure 4-5. The values in the new variable come about from sorting the values of the original variable and then assigning numbers to them in the new sorted order. If the input values are a string of characters instead of digits, the strings are sorted alphabetically (well, almost — uppercase letters come before lowercase).

```
CountryVisited into Country
Old Value                    New Value  Value Label

Afghanistan                          1  Afghanistan
Albania                              2  Albania
Algeria                              3  Algeria
Angola                               4  Angola
Argentina                            5  Argentina
Australia                            6  Australia
Austria                              7  Austria
Bahrain                              8  Bahrain
Bangladesh                           9  Bangladesh
Barbados                            10  Barbados
Belgium                             11  Belgium
Benin                               12  Benin
Bhutan                              13  Bhutan
Bolivia                             14  Bolivia
Botswana                            15  Botswana
Brazil                              16  Brazil
Bulgaria                            17  Bulgaria
Burkina Faso                        18  Burkina Faso
Burundi                             19  Burundi
Cameroon                            20  Cameroon
Canada                              21  Canada
Cape Verde                          22  Cape Verde
Central African Rep                 23  Central African Rep
Chad                                24  Chad
Chile                               25  Chile
```

FIGURE 4-5: Autorecoded values.

See the following for an example of using automatic recode.

EXAMPLE

Q. Using the Census.sav file, use the Automatic Recode transformation on the news1 variable and treat blank values as missing.

A. Do the following:

a. Place the news1 variable in the Variable box.

b. Type **News_num1** in the New Name box.

c. Click the Add New Name button.

d. Select the Treat Blank String Values as User-Missing option.

e. The new numeric variable is created.

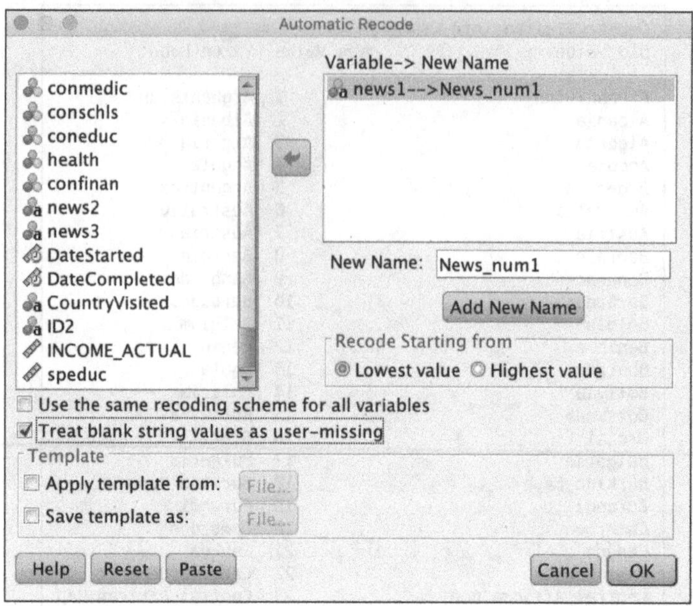


```
Old Value  New Value  Value Label

Internet           1  Internet
Newspaper          2  Newspaper
Other              3  Other
Radio              4  Radio
TV                 5  TV
```

9 Using the Census.sav file, use the Automatic Recode transformation on the news1, news2, and news3 variables. Make sure you use the same recoding scheme for all the variables and treat blank values as missing.

10 Using the Census.sav file, use the Automatic Recode transformation on the news1, news2, and news3 variables. Make sure you use the same recoding scheme for all the variables and treat blank values as missing, but this time save the template.

11 Using the Census.sav file, use the Automatic Recode transformation on the news1, news2, and news3 variables. Make sure you use the same recoding scheme for all the variables and treat blank values as missing, but this time apply the template you created in Exercise 10.

12 Using the Census.sav file, use the Automatic Recode transformation on the ID2 variable.

Binning

If you're using a scale variable, which contains a range of values, you can create groups of those values and organize them into bins. For example, you could use the income variable to create income groups. You can specify the size and content of bins in several ways. The binning process is automatic, which is different than the options that you have with the Recode transformation:

1. **Choose File ⇨ Open ⇨ Data and load the GSS2018.sav file.**

   You can download the file from the book's companion website at `www.dummies.com/go/spssstatisticsworkbookfd`.

2. **Choose Transform ⇨ Visual Binning.**
3. **Move AGE from the Variables box to the Variables to Bin box.**
4. **Click Continue.**
5. **Click the Make Cutpoints button.**
6. **Select the points at which you want to cut the data into parts to create the bins.**

   In this example, we divide the data into even percentiles of numbers of cases — that is, each bin will contain approximately the same number of cases, as shown in Figure 4-6. Note that nine cutpoints divide the data into 10 bins, each holding 10% of the cases.

   We could have divided the data into equal-width intervals — that is, each bin would contain a range of the same magnitude, which would put different numbers of cases in each bin. Or the cutpoints could have been based on standard deviations.

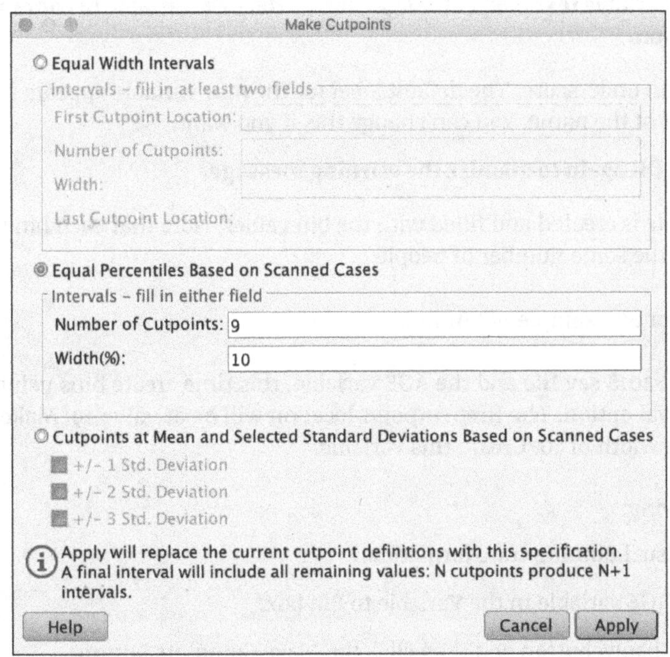

FIGURE 4-6: Specify how you want the data divided into bins.

7. **Click the Apply button.**

   The cutpoints appear as vertical lines on the bar graph, as shown in Figure 4-7. You can click the Make Cutpoints button repeatedly and cut the data different ways until you get the cutpoints the way you like. Any new cutpoints you define replace previous ones.

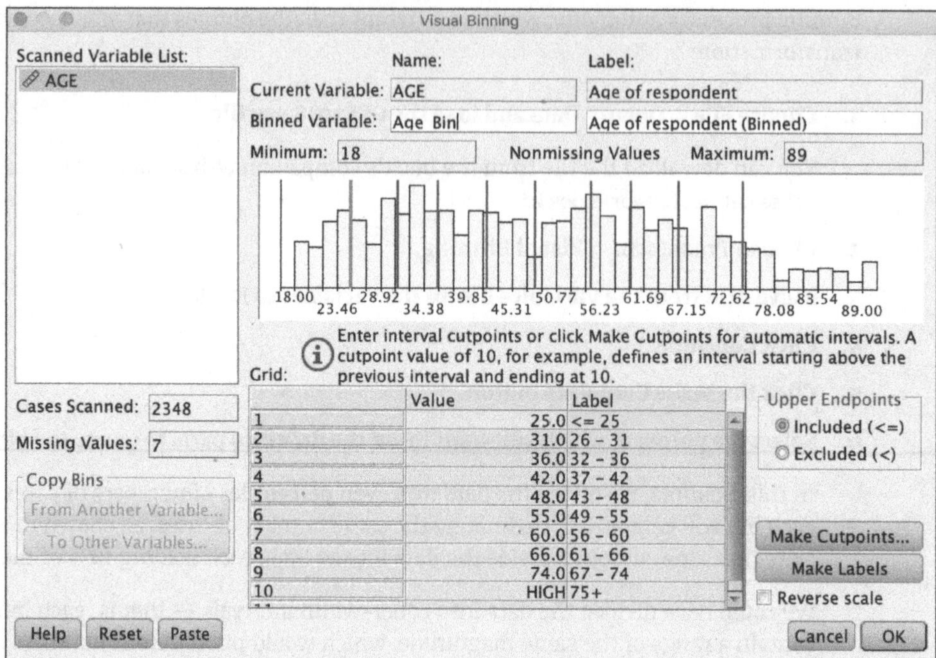

FIGURE 4-7: A bar graph of the data with cutpoints for binning.

8. **In the Binned Variable text box, enter Age_Bin as the new variable to contain the binning information.**

   Be sure to use an underscore. The default label for the new variable appears in the text box to the right of the name. You can change this if you want.

9. **Click OK. Click OK again to dismiss the warning message.**

   The new variable is created and filled with the bin values. Note that each bin has approximately the same number of people.

See the following for an example of binning.

EXAMPLE

Q. Using the GSS2018.sav file and the AGE variable, this time create bins using the Equal Widths Interval option. The first cutpoint location will be at value 19. Make seven cutpoints with a width of 10. Create this variable.

A. Do the following:

   a. Use the Visual Binning transformation.

   b. Place the AGE variable in the Variable to Bin box.

   c. Click the Change button and then click the Make Cutpoints button.

   d. Select the Equal Width Intervals option.

   e. Make the first cutpoint location at value 19, and then make seven cutpoints with a width of 10.

   f. Name the variable whatever you want.

This new variable will have the same width for each group, but the percentage of people in each group will be different.

Make Cutpoints

Equal Width Intervals

Intervals – fill in at least two fields

First Cutpoint Location: 19

Number of Cutpoints: 7

Width: 10

Last Cutpoint Location: 79

Equal Percentiles Based on Scanned Cases

Intervals – fill in either field

Number of Cutpoints:

Width(%):

Cutpoints at Mean and Selected Standard Deviations Based on Scanned Cases

+/- 1 Std. Deviation

+/- 2 Std. Deviation

+/- 3 Std. Deviation

Apply will replace the current cutpoint definitions with this specification. A final interval will include all remaining values: N cutpoints produce N+1 intervals.

Help Cancel Apply

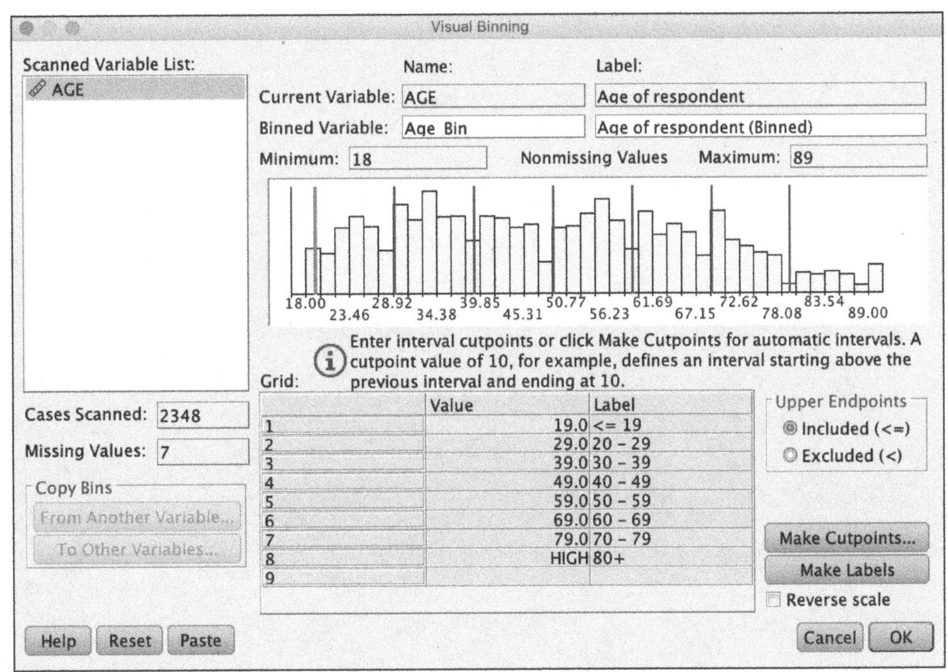

13. Using the GSS2018.sav file and the AGE variable, this time create bins by selecting the Mean and Standard Deviation option. Select all three standard deviation boxes in this section. Create this variable.

14. Using the GSS2018.sav file and the HRS1 variable, create bins by selecting the Mean and Standard Deviation option then selecting only the +/- 3 Std. Deviation option. You use this combination so you can identify outliers. Create this variable.

15. Using the GSS2018.sav file and the HRS1 variable, create work groups by selecting the Equal Widths Interval option. The first cutpoint location should be at value 20, and you should make three cutpoints with a width of 10. Create this variable.

16. Using the GSS2018.sav file and the HRS1 variable, create work groups by selecting the Equal Percentiles option. You should make three cutpoints with a width of 25. Create this variable.

Answers to Problems in Putting the Transform Menu to Work

1. Do the following:

   a. Use the Count Values within Cases transformation.

   b. Place the variables (INTECON, INTEDUC, INTENVIR, INTFARM, INTINTL, INTMED, INTMIL, INTSCI, INTSPACE, and INTTECH) in the Numeric Variables box.

   c. Type **Interest** in the Target Variable box.

   d. Click the Define Values button, type 1, and then click Add.

   This new variable counts the number of times someone had interest in these ten professions. Now you can take a closer look at the people with a lot of interest and figure out why that might be.

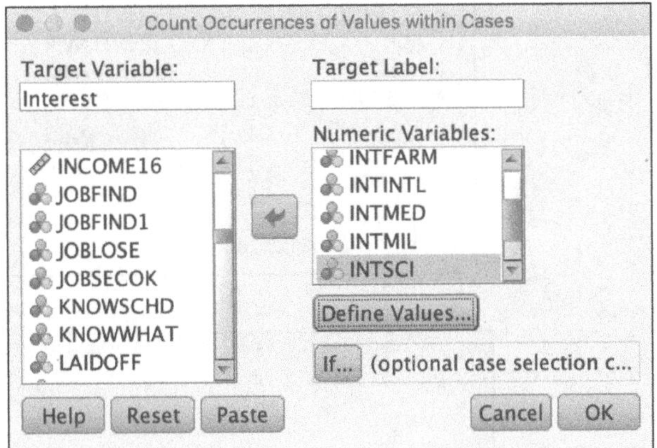

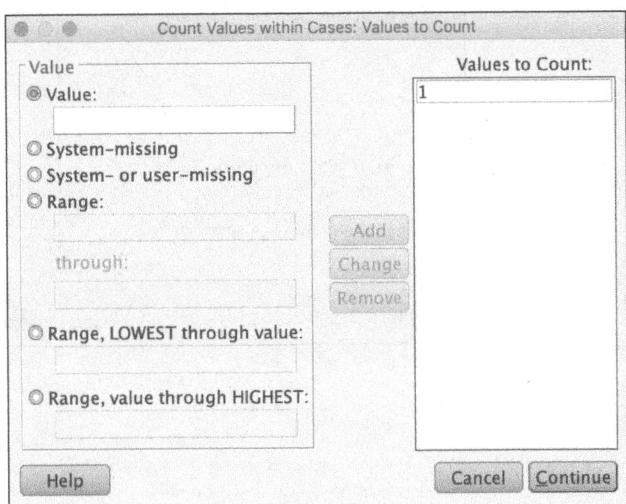

2 Do the following:

a. Use the Count Values within Cases transformation.

b. Place the variables (INTECON, INTEDUC, INTENVIR, INTFARM, INTINTL, INTMED, INTMIL, INTSCI, INTSPACE, and INTTECH) in the Numeric Variables box.

c. Type **No_Interest** in the Target Variable box.

d. Click the Define Values button, type 3, and then click Add.

This new variable counts the number of times someone had no interest in these ten professions. Take a closer look at the people with no interest and figure out why that might be.

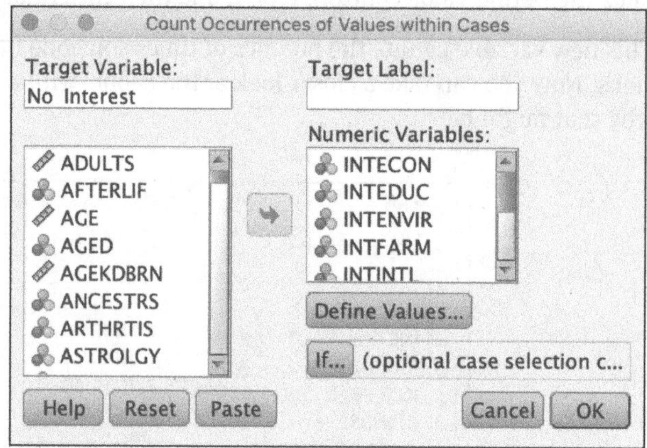

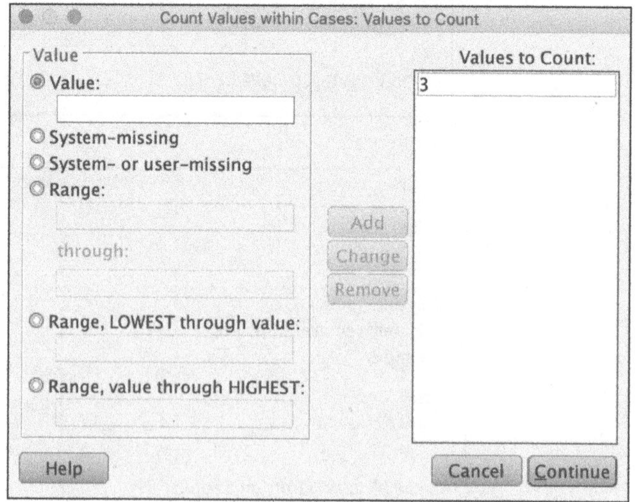

3. Do the following:

   a. Use the Count Values within Cases transformation.

   b. Place the variables (INTECON, INTEDUC, INTENVIR, INTFARM, INTINTL, INTMED, INTMIL, INTSCI, INTSPACE, and INTTECH) in the Numeric Variables box.

   c. Type **Missing_Interest** in the Target Variable box.

   d. Click the Define Values button, select the System- or User-Missing value, and then click Add.

   This new variable counts the number of times someone had missing information regarding these ten professions. Now you can exclude the people with a lot of missing information from further analyses.

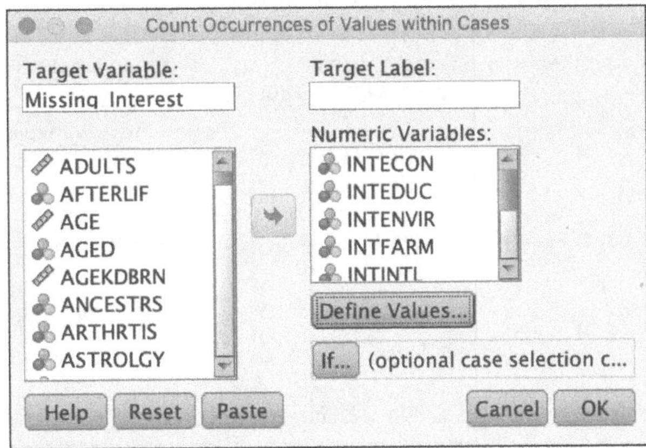

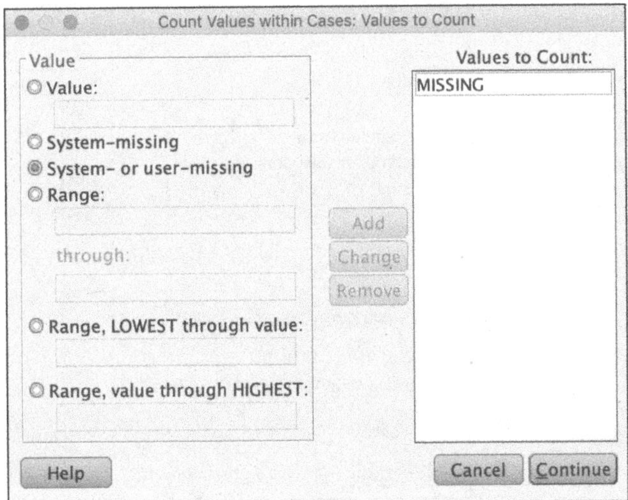

4 Do the following:

a. Use the Count Values within Cases transformation.

b. Place the variables (EDUC, MAEDUC, PAEDUC, and SPEDUC) in the Numeric Variables box.

c. Type **Family_College_Educ** in the Target Variable box.

d. Click the Define Values button, select the Range, Value through Highest option, and type 16. Click Add.

This new variable counts the number of times someone in a family had at least a college education. Now you can focus on the families with either a lot of college graduates or very few college graduates to better understand their characteristics.

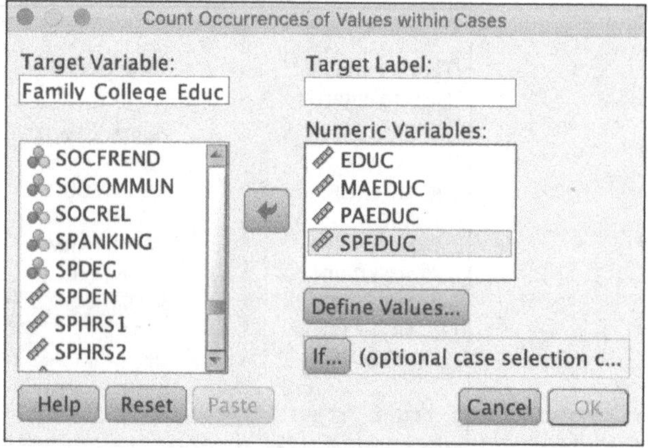

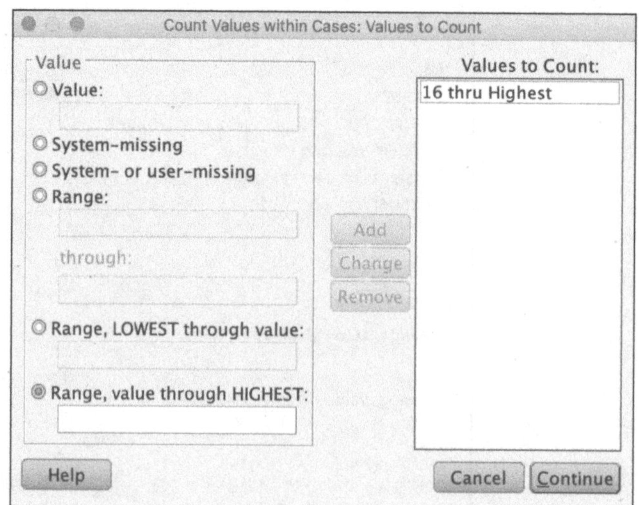

5 Do the following:

a. Use the Recode into Different Variables transformation.

b. Place the SATJOB variable in the Input Variable box, and type **Very_Sat_Job** in the Output Variable box.

c. Click the Change button and then click the Old and New Values button.

d. Specify 1 as the old value, specify 1 as the new value, and then click Add.

e. Specify All Other Values as the old value, specify 0 as the new value, and then click Add.

This new variable will compare those very satisfied to everyone else.

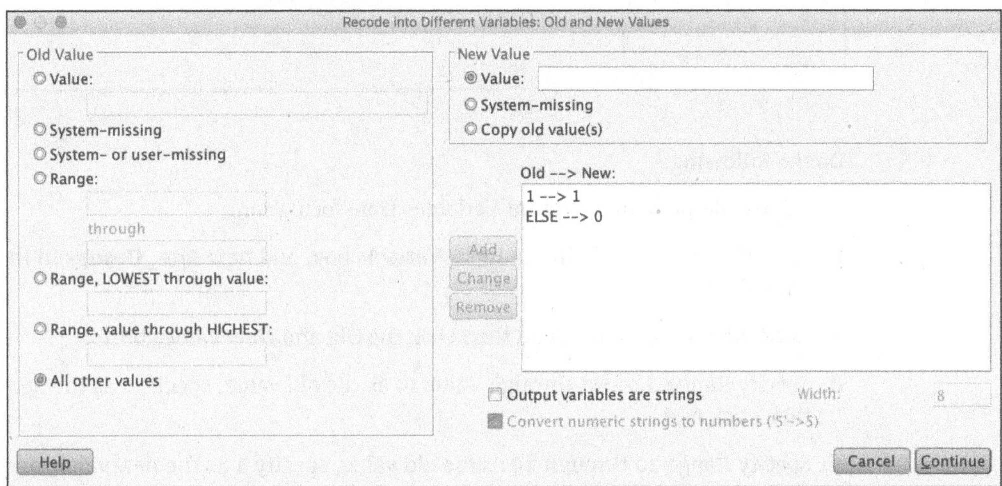

6 Do the following:

a. Use the Recode into Different Variables transformation.

b. Place the CONBIZ variable in the Input Variable box, and type **Conbiz_Recoded** in the Output Variable box.

c. Click the Change button and then click the Old and New Values button.

d. Specify 1 as the old value, specify 5 as the new value, and then click Add.

e. Specify 2 as the old value, specify 4 as the new value, and then click Add.

f. Specify the 3 as old value, specify 3 as the new value, and then click Add.

g. Specify 4 as the old value, specify 2 as the new value, 2, and then click Add.

h. Specify 5 as the old value, 5, specify 1 as the new value, and then click Add.

i. Specify All Other Values as the old value, specify Copy Old Value(s) as the new value, and then click Add.

This new variable has been recoded so that higher values represent greater confidence, making it easier to understand.

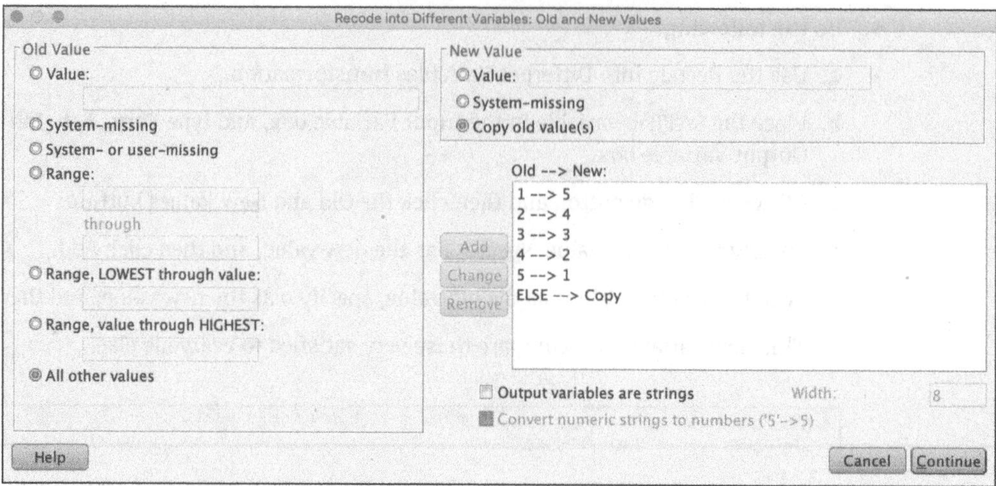

7 Do the following:

a. Use the Recode into Different Variables transformation.
b. Place the AGE variable in the Input Variable box, and type **Age_Groups** in the Output Variable box.
c. Click the Change button and then click the Old and New Values button.
d. Specify Range, Lowest through Value 19 as the old value, specify 1 as the new value, and then click Add.
e. Specify Range 20 through 29 as the old value, specify 2 as the new value, and then click Add.
f. Specify Range 30 through 39 as the old value, specify 3 as the and new value, and then click Add.
g. Specify Range 40 through 49 as the old value, specify 4 as the and new value, and then click Add.
h. Specify Range 50 through 59 as the old value, specify 5 as the new value, and then click Add.
i. Specify Range 60 through 69 as the old value, specify 6 as the new value, and then click Add.
j. Specify Range 70 through 79 as the old value, specify 7 as the new value, and then click Add.
k. Specify Range 80 through 89 as the old value, specify 8 as the new value, and then click Add.
l. Specify Range, Value through Highest 90 as the old value, specify 9 as the new value, and then click Add.

This new variable contains age groups with equal size intervals.

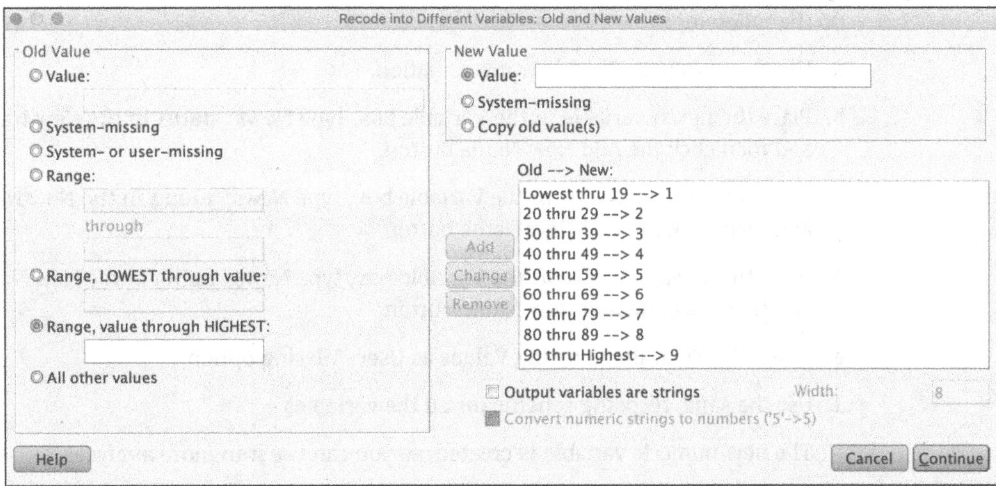

8. Do the following:

   a. Use the Recode into Different Variables transformation.

   b. Place the EDUC variable in the Input Variable box, and type **Educ_Groups** in the Output Variable box.

   c. Click the Change button and then click the Old and New Values button.

   d. Specify Range 0 through 5 as the old value, specify 1 as the new value, and then click Add.

   e. Specify Range 6 through 8 as the old value, specify 2 as the new value, and then click Add.

   f. Specify Range 9 through 12 as the old value, specify 3 as the new value, and then click Add.

   g. Specify Range 13 through 16 as the old value, specify 4 as the new value, and then click Add.

   h. Specify Range, Value through Highest 17 as the old value, specify 5 as the new value, and then click Add.

   This new variable contains education groups that represent the level of education.

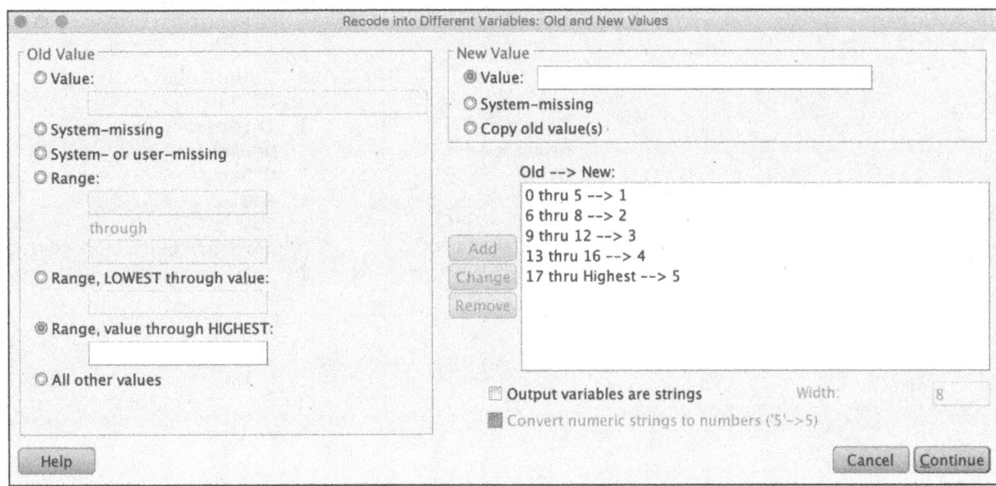

9. Do the following:

   a. Use the Automatic Recode transformation.

   b. Place the news1 variable in the Variable box, type **News_num1** in the New Name box, and then click the Add New Name button.

   c. Place the news2 variable in the Variable box, type **News_num2** in the New Name box, and then click the Add New Name button.

   d. Place the news3 variable in the Variable box, type News_num3 in the New Name box, and then click the Add New Name button.

   e. Select the Treat Blank String Values as User-Missing option.

   f. Use the same recoding scheme for all the variables.

   The new numeric variable is created, so you can use it in more analyses.

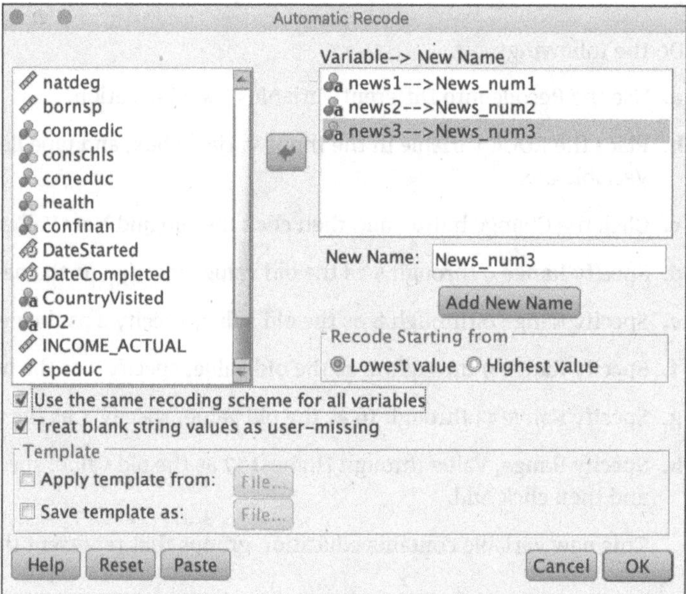


```
Old Value       New Value  Value Label

Internet                1  Internet
Newspaper               2  Newspaper
Other                   3  Other
Radio                   4  Radio
TV                      5  TV
Word of Mouth           6  Word of Mouth
                M       7M
```

10 Do the same as in question 9, except click the Save Template As option.

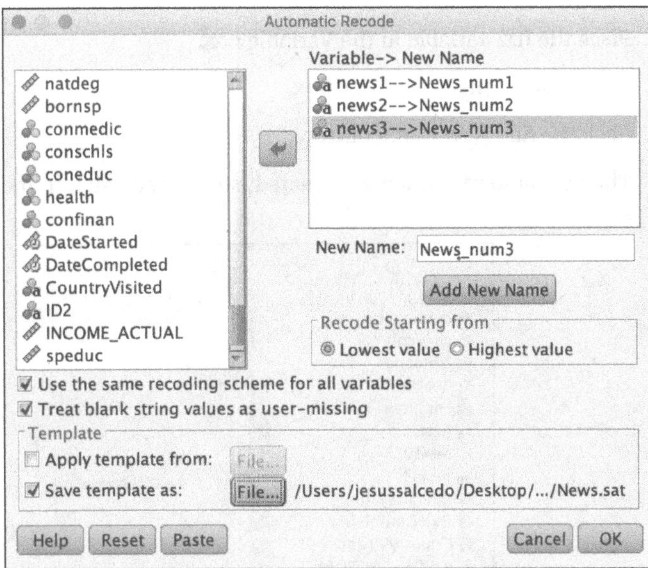

11 Do the same as in question 9, but click the Apply Template As option and select the template you created in question 10.

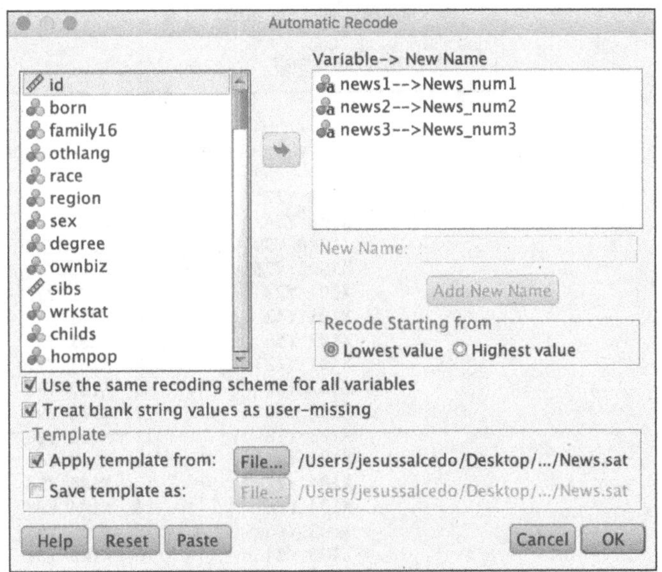


```
Old Value       New Value  Value Label

Internet                1  Internet
Newspaper               2  Newspaper
Other                   3  Other
Radio                   4  Radio
TV                      5  TV
Word of Mouth           6  Word of Mouth
                M       7M
```

12. Do the following:

    a. Use the Automatic Recode transformation.

    b. Place the ID2 variable in the Variable box.

    c. Type **Code** in the New Name box.

    d. Click the Add New Name button.

    The new numeric variable is created, so you can use it in more analyses.

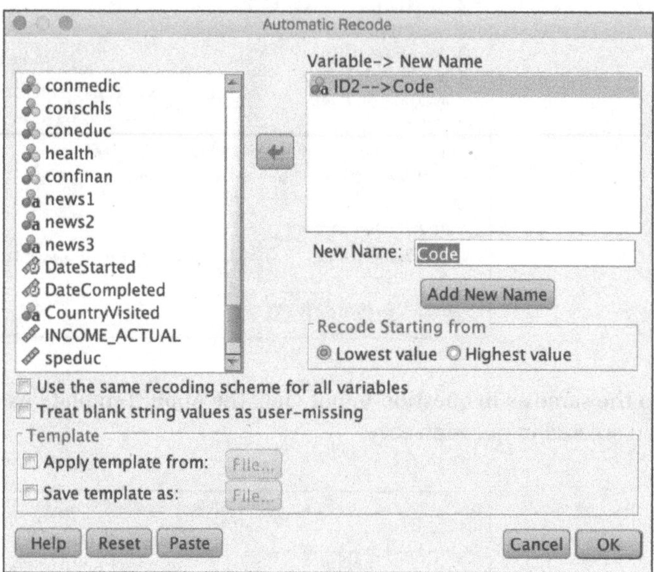

Old Value	New Value	Value Label
X100 Y27	1	X100 Y27
X100 Y44	2	X100 Y44
X1000 Y39	3	X1000 Y39
X1000 Y76	4	X1000 Y76
X101 Y23	5	X101 Y23
X101 Y42	6	X101 Y42
X101 Y56	7	X101 Y56
X101 Y72	8	X101 Y72
X102 Y86	9	X102 Y86
X102 Y99	10	X102 Y99
X103 Y28	11	X103 Y28
X103 Y4	12	X103 Y4
X104 Y22	13	X104 Y22
X104 Y88	14	X104 Y88
X105 Y86	15	X105 Y86
X106 Y34	16	X106 Y34
X106 Y53	17	X106 Y53
X107 Y34	18	X107 Y34
X107 Y40	19	X107 Y40
X107 Y64	20	X107 Y64
X107 Y66	21	X107 Y66
X107 Y76	22	X107 Y76
X109 Y12	23	X109 Y12
X109 Y46	24	X109 Y46
X109 Y77	25	X109 Y77

13 Do the following:

a. Use the Visual Binning transformation.

b. Place the AGE variable in the Variable to Bin box.

c. Click the Change button and then click the Make Cutpoints button.

d. Select the Cutpoints at Mean Selected Standard Deviations option, and then select all three boxes in this section.

e. Name the variable whatever you want.

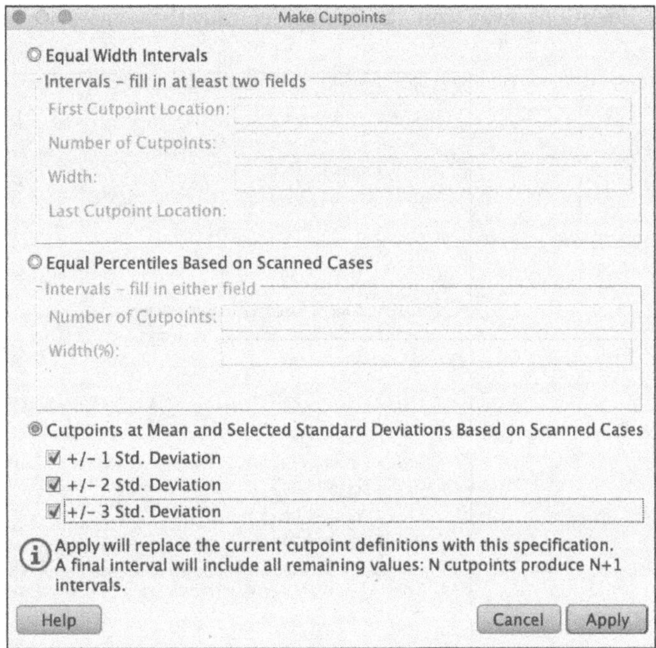

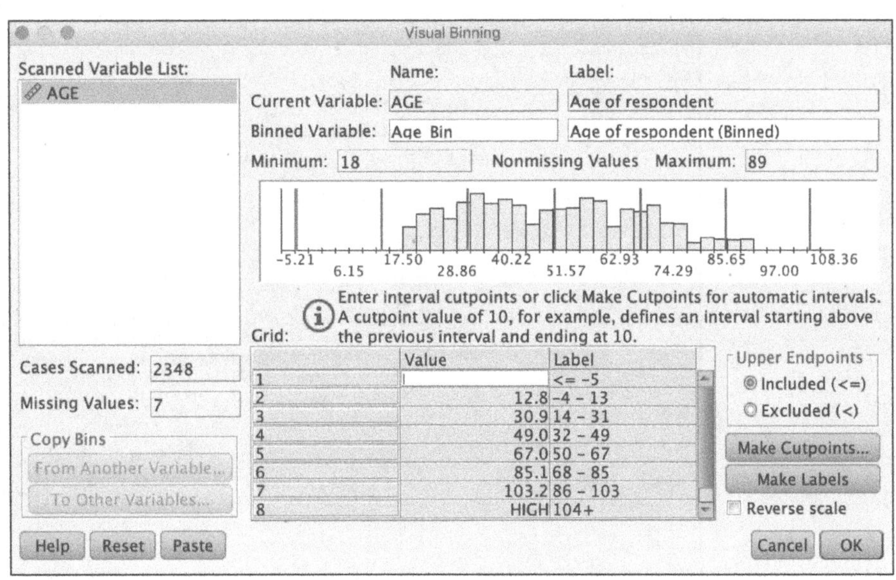

14 Do the following:

a. Use the Visual Binning transformation.

b. Place the HRS1 variable in the Variable to Bin box.

c. Click the Change button and then click the Make Cutpoints button.

d. Select the Cutpoints at Mean Selected Standard Deviations option and then select 3 Std. Deviations.

e. Name the variable whatever you want.

This new variable allows you to identify cases that are more than three standard deviations away from the means; typically these cases are referred to as outliers.

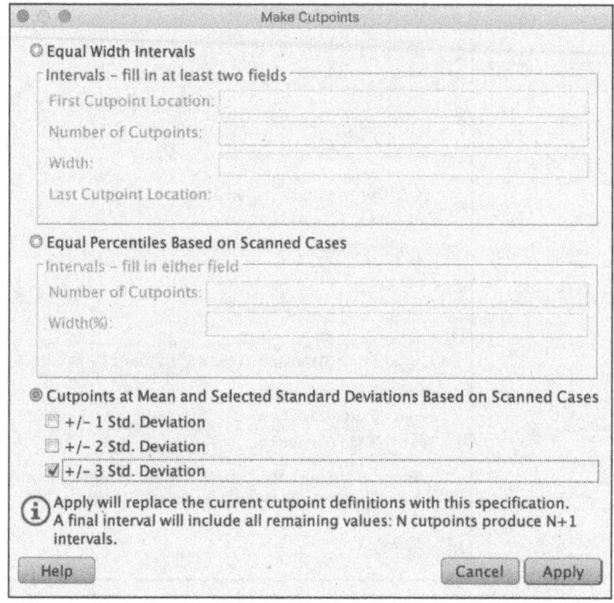

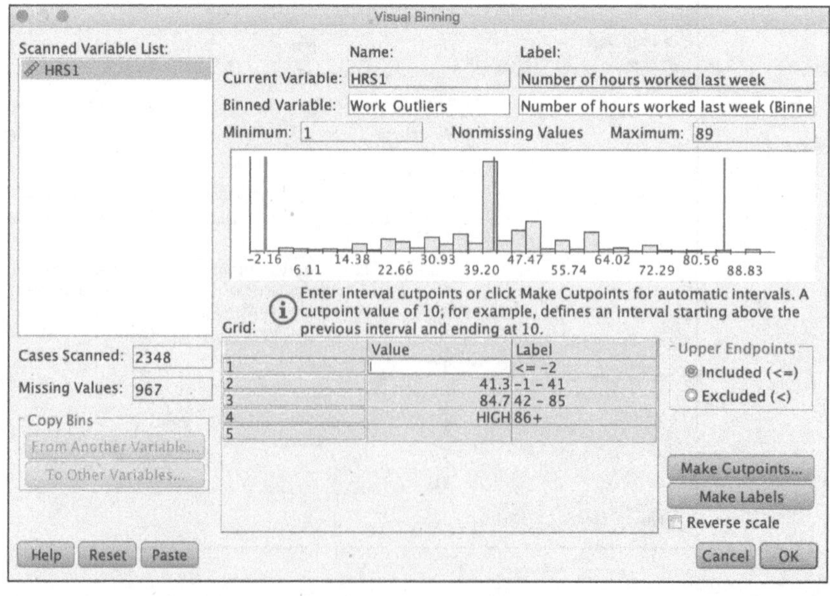

15 Do the following:

a. Use the Visual Binning transformation.

b. Place the HRS1 variable in the Variable to Bin box.

c. Click the Change button and then click the Make Cutpoints button.

d. Select the Equal Widths Interval option.

   The first cutpoint location (of three) will be at value 20.

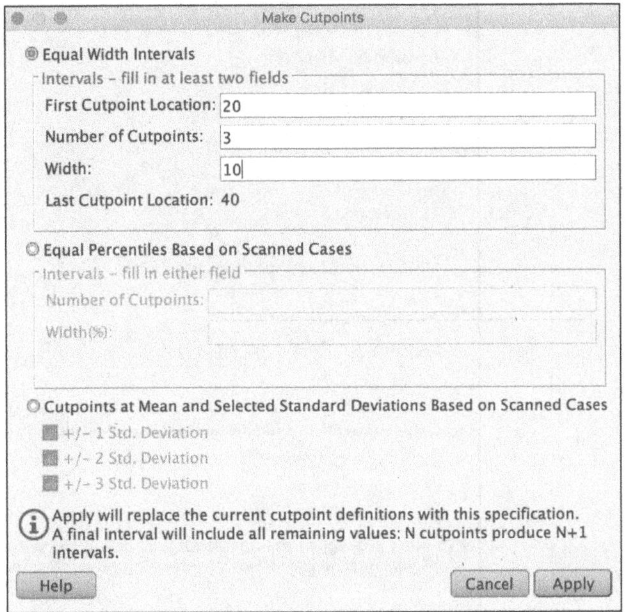

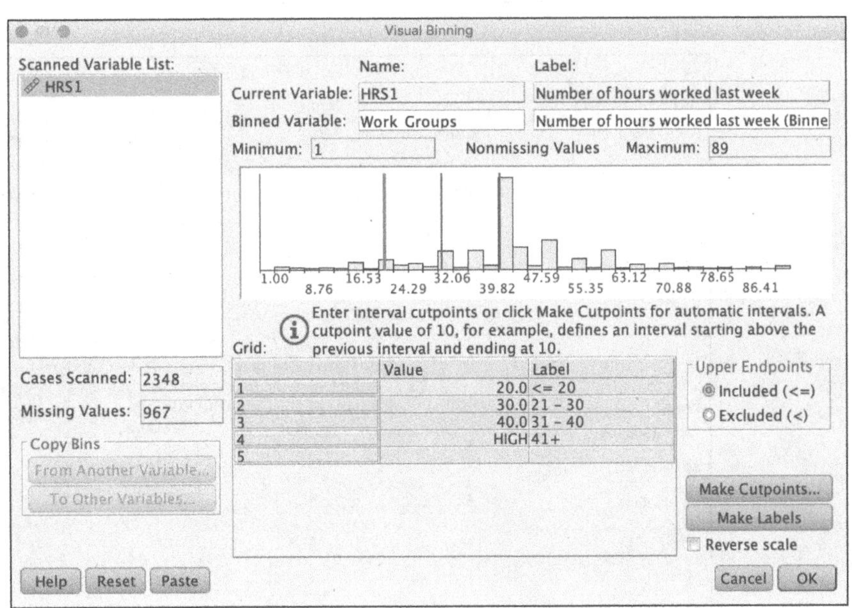

16 Do the following:

a. Use the Visual Binning transformation.

b. Place the HRS1 variable in the Variable to Bin box.

c. Click the Change button and then click the Make Cutpoints button.

d. Select the Equal Percentiles option.

The result is three cutpoints with a width of 25.

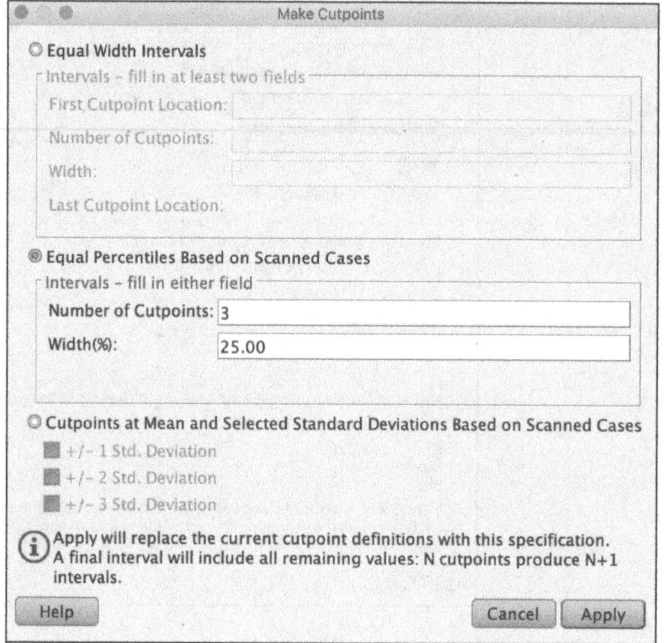

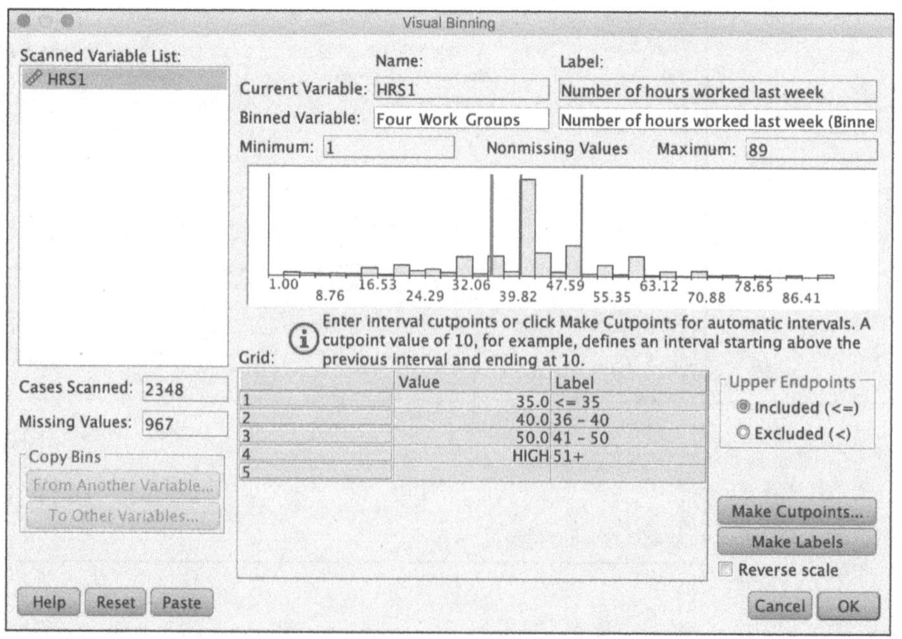

IN THIS CHAPTER

» **Calculating new variables**

» **Computing with more complex variables**

» **Working with system variables**

» **Working with string variables**

» **Using multiple functions**

Chapter 5
Computing Variables

The Compute Variable dialog is one of the most frequently used dialogs in SPSS because it contains dozens of functions as well as numerous arithmetic and logical operators that perform all kinds of calculations. This dialog is where you go to make new variables out of existing ones.

In this chapter, we focus on the variety of functions available in the Compute Variable dialog. We won't discuss them all, but we do provide examples of several of the most useful ones.

SPSS divides functions into function groups. Here's just a sampling of the kinds of functions you'll find:

- Arithmetic functions
- Statistical functions
- String functions
- String and numeric conversion functions
- Date and time functions
- Random variable and distribution functions
- Missing value functions
- Logical functions

When you click a function group name such as Statistical, you see a list of the functions that belong to that function group. In this chapter, you investigate some of these functions to give you an idea of the different kinds of things you can do.

Calculating a New Variable

In this first section, we focus on creating difference and ratio variables, along with using some simple functions. In the following example, you calculate the difference in years of education between the respondent and their spouse:

1. **Choose File ⇨ Open ⇨ Data and load the GSS2018.sav file.**

   You can download the file from the book's companion website at `www.dummies.com/go/spssstatisticsworkbookfd`. This file contains data from the General Social Survey (GSS), a nationally representative survey of adults in the United States that collects data on contemporary opinions, attitudes, and behaviors.

2. **Choose Transform ⇨ Compute Variable.**
3. **Drag EDUC to the Numeric Expression box.**
4. **Add a minus symbol to the expression using either the dialog's keypad or your keyboard.**
5. **Drag SPEDUC to the Numeric Expression box.**
6. **In the Target Variable box, type** ED_DIFF, **which is the name of your new variable.**

   The dialog now looks like Figure 5-1.

7. **Click OK.**

   You've now calculated the difference in years of education between the respondent and their spouse.

See the following for an example of using the compute variable transformation.

EXAMPLE

Q. Use the GSS2018.sav file and the Compute Variable transformation to calculate the ratio of EMAILHR (hours per week using email) to WWWHR (hours per week being online).

A. Do the following:

a. Use the Compute Variable transformation.

b. Type **EMAILHR / WWWHR** in the Numeric Expression box.

c. Name the new variable.

You can now see how much respondents use email relative to being online.

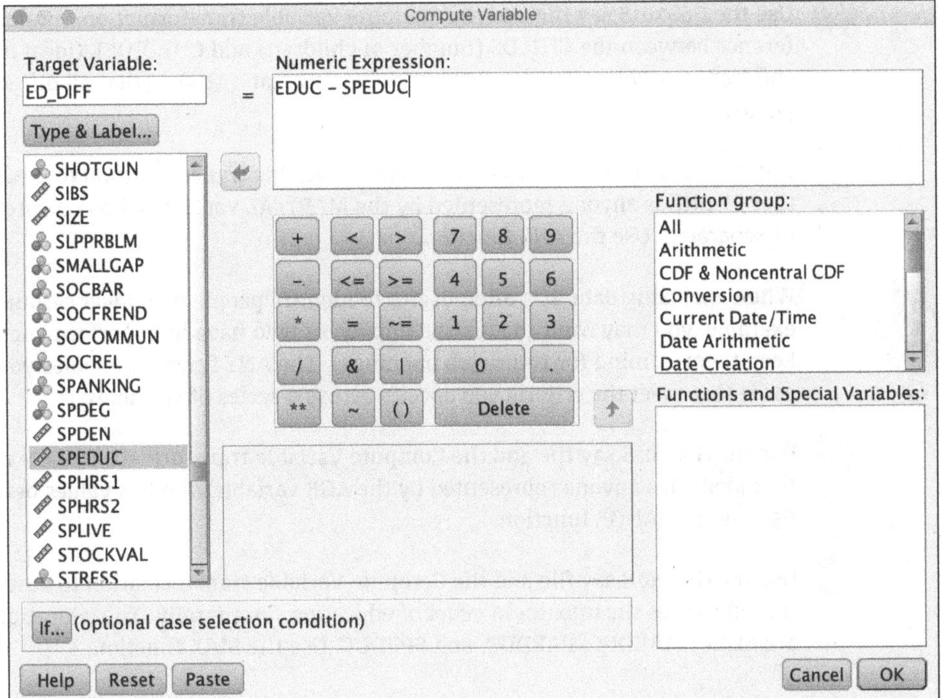

FIGURE 5-1: A simple subtraction of two variables.

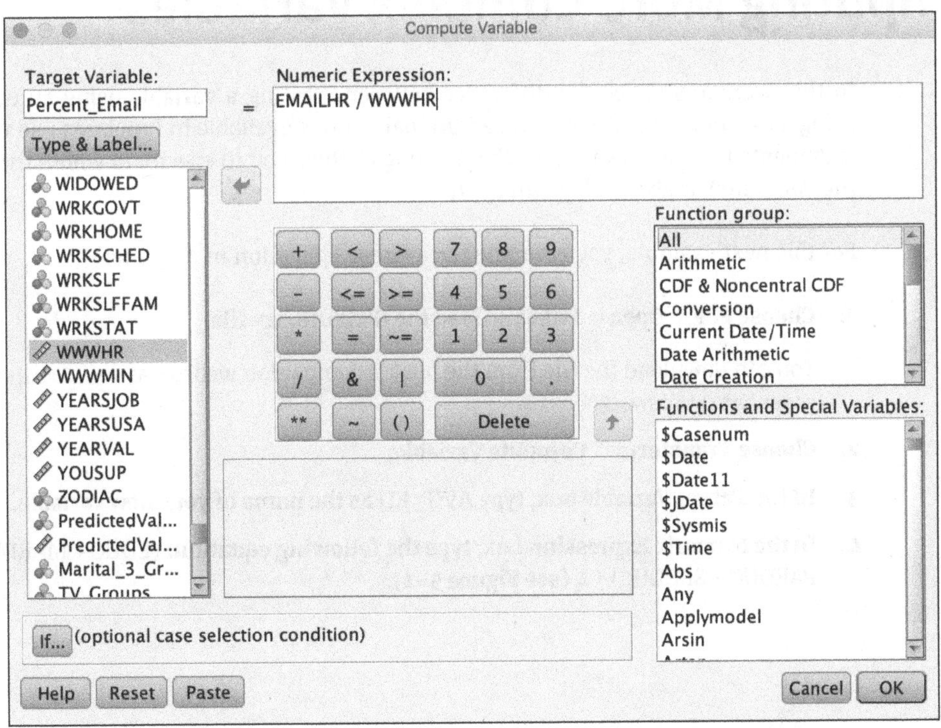

1. Use the GSS2018.sav file and the Compute Variable transformation to calculate the difference between the CHILDS (number of children) and CHLDIDEL (ideal number of children) variables. Use the absolute value function (ABS) so that all differences are positive.

2. Use the GSS2018.sav file and the Compute Variable transformation to create a variable that identifies anyone represented by the MARITAL variable who is divorced, widowed, or separated. Use the ANY function.

TIP

When analyzing data, it's often useful to identify people who meet certain criteria. For example, you may want to identify employees who have high performance ratings (to keep them in mind for future job openings). The ANY function enables you to identify cases that meet the criteria you specify across a series of variables.

3. Use the GSS2018.sav file and the Compute Variable transformation to create a variable that identifies anyone represented by the AGE variable who has values between 18 and 65. Use the RANGE function.

4. Use the GSS2018.sav file and the Compute Variable transformation to create a variable that identifies the maximum years of education for a family. The education variables are EDUC, MAEDUC, PAEDUC, and SPEDUC. Use the MAX function.

Computing More Complex Variables

In this section, you see the difference between creating a variable using an equation versus using a function. You also explore additional options available in functions and use conditional statements to create a variable. The If dialog enables you to specify to whom the expression in the Compute Variable dialog will apply.

For this next example, you compute the average education in a family:

1. **Choose File ⇨ Open ⇨ Data and load the GSS2018.sav file.**

   You can download the file from the book's companion website at `www.dummies.com/go/spssstatisticsworkbookfd`.

2. **Choose Transform ⇨ Compute Variable.**
3. **In the Target Variable box, type** AVG_ED **as the name of your new variable.**
4. **In the Numeric Expression box, type the following equation:** (EDUC + MAEDUC + PAEDUC + SPEDUC) / 4 **(see Figure 5-2).**

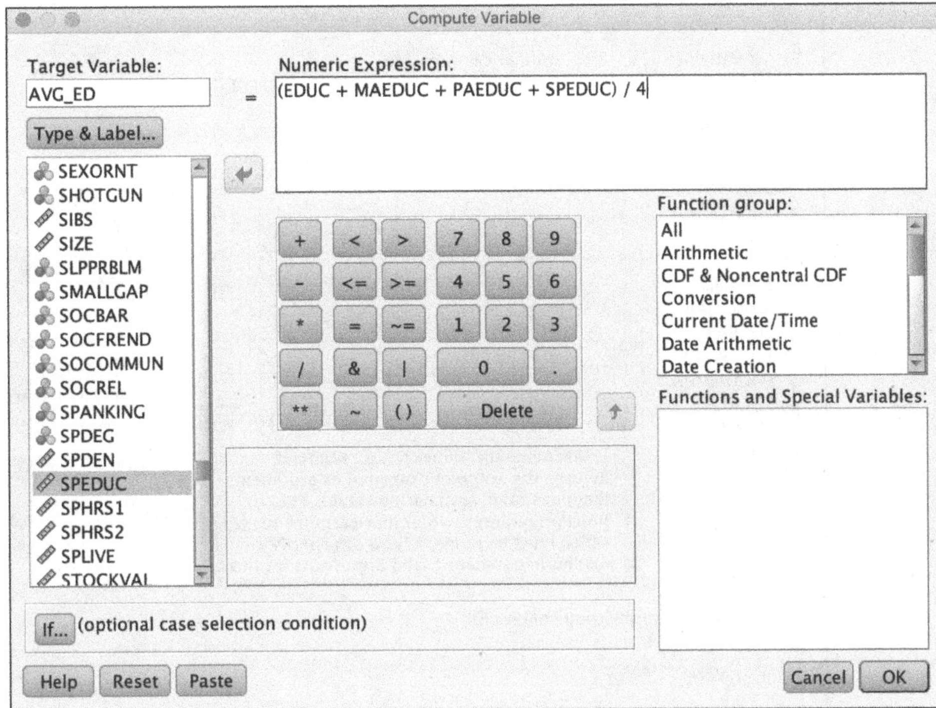

FIGURE 5-2: Calculating an average using an equation.

5. **Click OK.**

   You've calculated the average education per family. This calculation works as long as there is no missing data. If you use an equation, however, and an element in the equation is missing, you'll have missing data for the calculation.

See the following for an example of using the compute variable transformation.

EXAMPLE

Q. Using the GSS2018.sav file, calculate the mean education (EDUC, MAEDUC, PAEDUC, and SPEDUC) per family by using the MEAN function.

A. Do the following:

a. Use the Compute Variable transformation.

b. Type **MEAN(EDUC, MAEDUC, PAEDUC, SPEDUC)** in the Numeric Expression box.

c. Name the new variable.

   You've calculated the mean education per family. When using a function, even if data is missing, the result is not missing because SPSS knows how to calculate the function, in this case a mean.

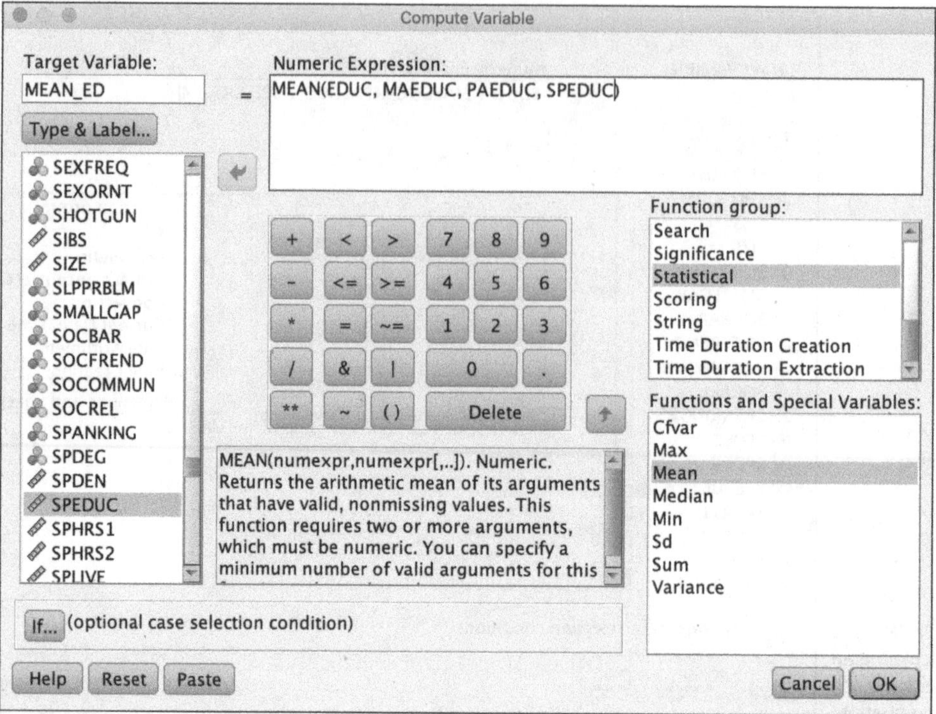

5. Using the GSS2018.sav file, calculate the mean education (EDUC, MAEDUC, PAEDUC, and SPEDUC) per family by using the MEAN function — but only if there are at least three valid values.

TIP

Specifying the minimum number of valid arguments for a function to be evaluated (the .n suffix) works with all statistical functions.

6. Using the GSS2018.sav file, calculate the total number of hours the respondent and their spouse have worked in the last week (HRS1 and SPHRS1) by using the SUM function.

7. Using the GSS2018.sav file and the Compute Variable transformation, count the number of times respondents provided a valid value on the MCSDS variables (MCSDS1, MCSDS2, MCSDS3, MCSDS4, MCSDS5, MCSDS6, and MCSDS7). Use the NVALID function.

8. Using the GSS2018.sav file, calculate the difference in education between the respondent and their same-sex parent. Note that you need to use the IF condition to do this calculation.

Using System Variables

Most values used in SPSS come from the variables in the dataset. You simply specify the variable names, and SPSS knows where to go and get the values.

SPSS has another type of variable that is already defined and can be used anywhere in a program. Predefined variables are called *system variables.* They begin with a dollar sign ($) and already contain values. These next examples use system variables.

In the following, you compute today's date:

1. **Choose File ➪ Open ➪ Data and load the GSS2018.sav file.**

   Download the file from the book's companion website at `www.dummies.com/go/spssstatisticsworkbookfd`.

2. **Choose Transform ➪ Compute Variable.**
3. **In the Target Variable box, type** Date **as the name of your new variable.**
4. **In the Numeric Expression box, use the $DATE system variable (see Figure 5-3).**

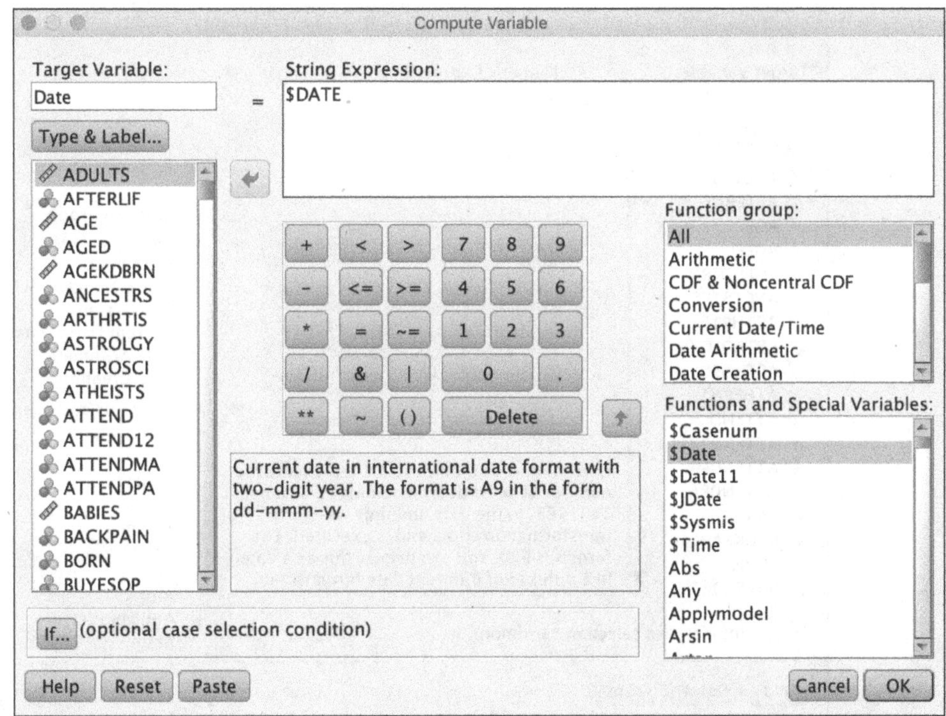

FIGURE 5-3: Calculating today's date.

5. **Click OK.**

   You've calculated today's date. Note that you can use several system date variables, each with its own date format.

See the following for an example of using the compute variable transformation.

EXAMPLE

Q. Using the GSS2018.sav file and the $TIME system variable, calculate the current date and time.

A. Do the following:

a. Use the Compute Variable transformation.

b. Type **$TIME** in the Numeric Expression box.

c. Name the new variable.

You've calculated the current date and time as the number of seconds from the start of the Gregorian calendar! (You can change the format if you want by using either the Compute variable transformation or the Date and Time Wizard or the Variable View tab of Data Editor.)

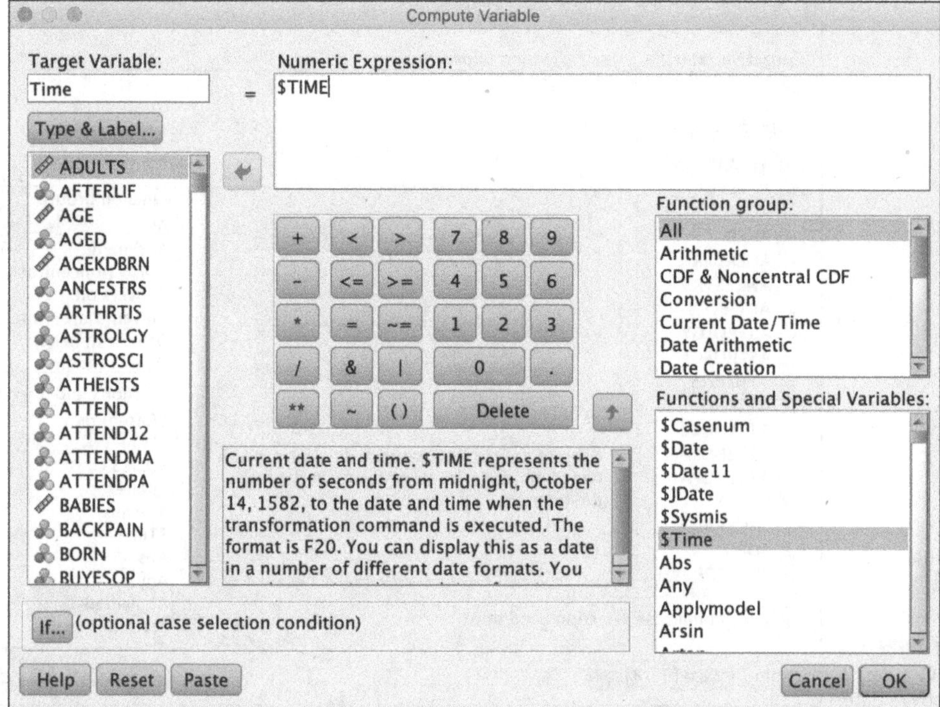

9 Using the GSS2018.sav file, create an ID variable for the dataset by using the $CASENUM system variable.

10 Using the GSS2018.sav file, create an empty variable by using the $SYSMIS system variable.

11 Using the GSS2018.sav file and the Compute Variable transformation, determine where you have a system or user-missing value for the HAPMAR variable. Use the MISSING function.

12 Using the GSS2018.sav file and the Compute Variable transformation, create a variable that has its value labels as the actual values. Use the VALUELABEL function on the DEGREE variable.

Using String Functions

Most of the data you use in SPSS is numeric, but SPSS also has powerful functions you can use on alphanumeric data. In this section, you begin to get familiar with several string functions.

In the following example, you change the birthmonth variable so that all the letters are uppercase. Then you extract the first three characters so that you have an abbreviation for each month. You perform these two modifications in two separate steps, and then you perform them in one step:

1. **Choose File ⇨ Open ⇨ Data and load the String.sav file.**

   You can download the file from the book's companion website at `www.dummies.com/go/spssstatisticsworkbookfd`. The String dataset has various string variables to show some of the functionality of the alphanumeric functions in SPSS.

2. **Choose Transform ⇨ Compute Variable.**
3. **In the Target Variable box, type** Month_Caps **as the name of your new variable.**
4. **Click the Type & Label button.**
5. **The new variable will be a string, so click the String radio button and specify a width of 15.**

   Make sure that the width has a value large enough to capture all the characters of a new variable; otherwise, the new variable will be truncated.

6. **Click Continue.**
7. **In the String Expression box, type the following equation:** UPCASE(birthmonth) **(see Figure 5-4).**
8. **Click OK.**

   You've created a new variable that has a person's birth month in all caps.

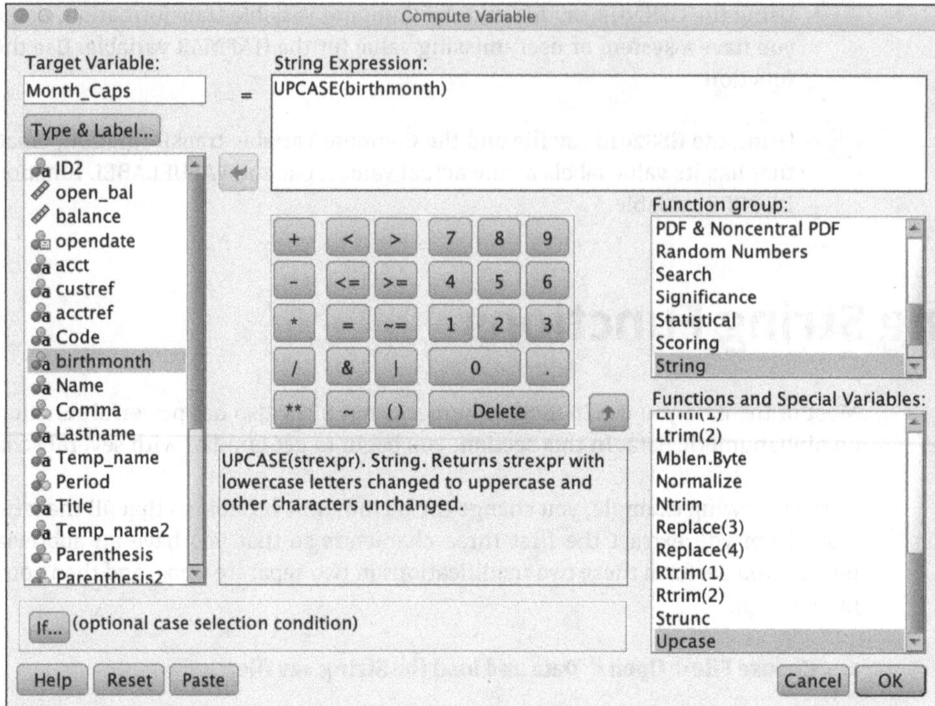

FIGURE 5-4: Creating a variable with all caps.

See the following for an example of using the compute variable transformation.

EXAMPLE

Q. Using the String.sav file, extract the first three characters from the newly created Month_Caps variable. Use the CHAR.SUBSTR function.

A. Use the Compute Variable transformation to do the following:

a. Name the new variable.

b. Click the Type & Label button and change the variable type to String with a width of 3.

c. Type **CHAR.SUBSTR(Month_Caps, 1, 3)** in the String Expression box.

The CHAR.SUBSTR function tells SPSS: Using the Month_Caps variable and a starting position of 1, extract the first three characters. You've created a variable that abbreviates the name of each month.

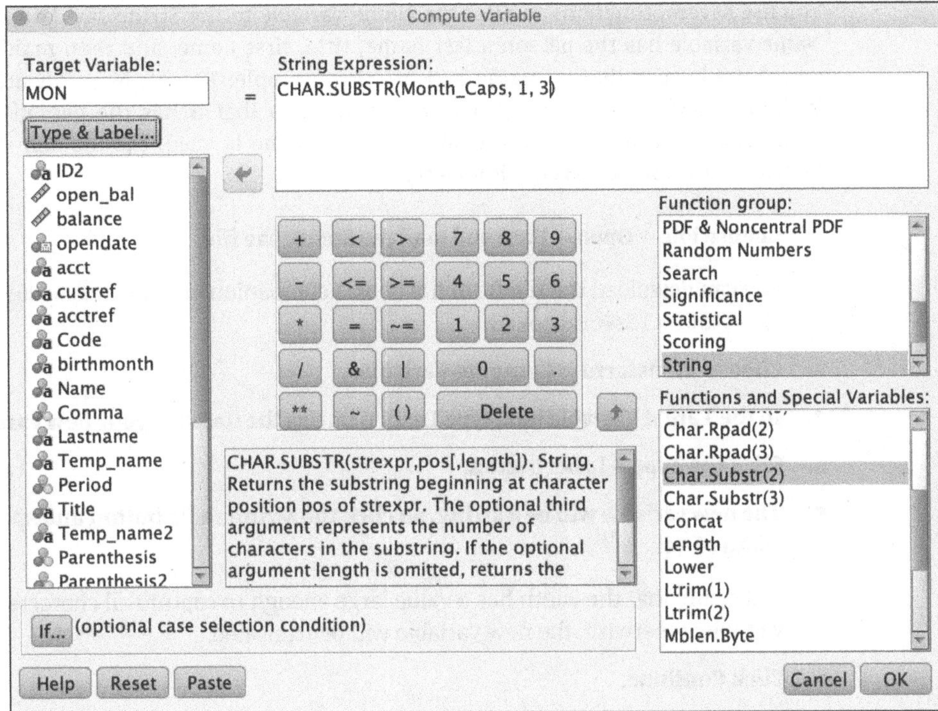

13 Using the String.sav file, extract the first three letters of the birthmonth variable and convert these to uppercase. Use the CHAR.SUBSTR and UPCASE functions in one expression.

14 Using the String.sav file, create a variable called Account_number by concatenating the acctref, acct, and custref variables. Use the CONCAT function.

15 The ID2 variable is comprised of two parts. You want to extract the first portion, but you can't just use the CHAR.SUBSTR function because the first part of ID2 is not a uniform length. However, a space separates the two parts, so you can first count the number of characters until the space. Use the String.sav file and the CHAR.INDEX function.

16 Now that you have counted the number of characters until the space, you can use the String.sav file and the CHAR.SUBSTR function to extract the first part of ID2.

Using Multiple String Functions

In the next example, we introduce functions related to parsing strings and nesting functions. Nesting functions can seem tricky at first because the expression looks complicated, but you're simply putting one function inside another function. The inside function acts as an argument to the outside function.

You use several functions to alter the Name variable in the String.sav data file. Currently, the Name variable has the person's last name, title, first name, and then maiden name in cases where the husband's first name appears (for example: Robert, Mrs. Edward Scott (Elisabeth Walton McMillan). You reorganize this variable so that it has the person's first name, last name, and in cases where the husband's first name is used, the maiden name will be used instead (for example: Elisabeth Robert).

1. **Choose File ⇨ Open ⇨ Data and load the String.sav file.**

   You can download the file from the book's companion website at `www.dummies.com/go/spssstatisticsworkbookfd`.

2. **Choose Transform ⇨ Compute Variable.**
3. **In the Target Variable box, type** Last_name **as the name of your new variable.**
4. **Click the Type & Label button.**
5. **The new variable will be a string, so click the String radio button and specify a width of 50.**

   Make sure that the width has a value large enough to capture all characters of a new variable; otherwise, the new variable will be truncated.

6. **Click Continue.**
7. **In the String Expression box, type the following equation:** (CHAR.SUBSTR(Name, 1, CHAR.INDEX(Name, ",") – 1)) **(see Figure 5-5).**

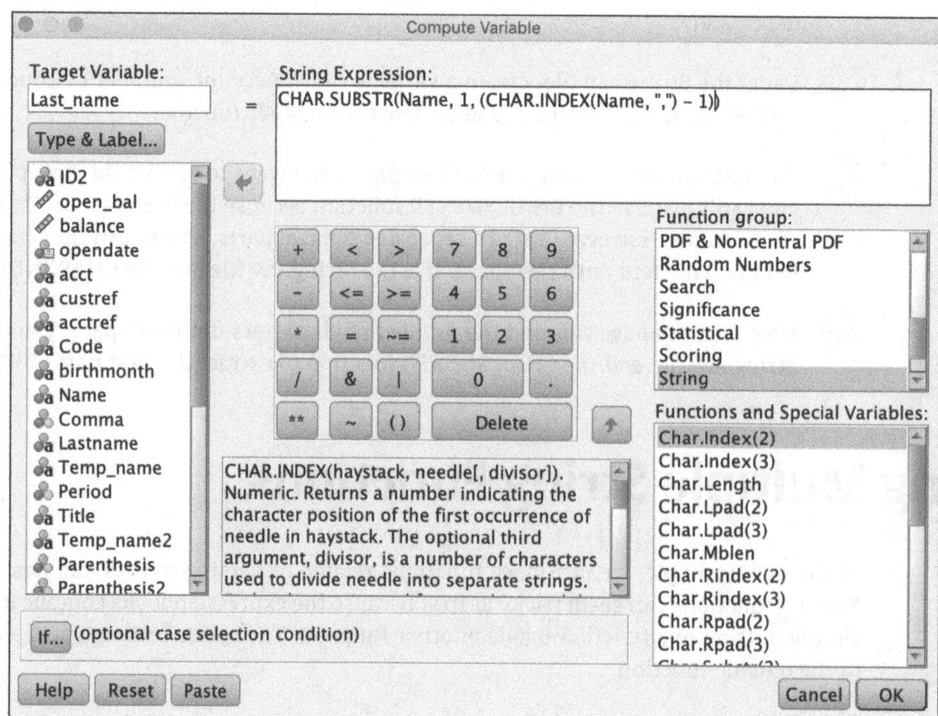

FIGURE 5-5: Creating a variable with all caps.

8. **Click OK.**

   The CHAR.INDEX function tells SPSS to use the Name variable and count the number of characters until the comma (","). The CHAR.SUBSTR function tells SPSS: Using the Name variable and a starting position of 1, extract the characters up to the comma. Then add -1 so that you don't extract the comma. You've created a variable that has the person's last name.

See the following for an example of using the compute variable transformation.

Q. In the preceding example, you extracted the person's last name. Now create a version of the Name variable without the last name, so you can use a cleaner version of the variable in the upcoming exercises. Use the String.sav file and the CHAR.INDEX and CHAR.SUBSTR functions.

A. Use the Compute Variable transformation to do the following:

a. Name the new variable Lastname_Removed.

b. Click the Type & Label button and change the variable type to String with a width of 50.

c. Type **(CHAR.SUBSTR(Name, CHAR.INDEX(Name, ",") + 2))** in the String Expression box.

   As before, the CHAR.INDEX function tells SPSS to use the Name variable and count the number of characters until the comma (","). The CHAR.SUBSTR function tells SPSS to use the Name variable and, starting at the position specified by the CHAR.INDEX function, extract the characters up to the comma. You add +2 to extract the extra spaces after the comma. You've created a cleaner version of the Name variable, which you'll use in the next examples.

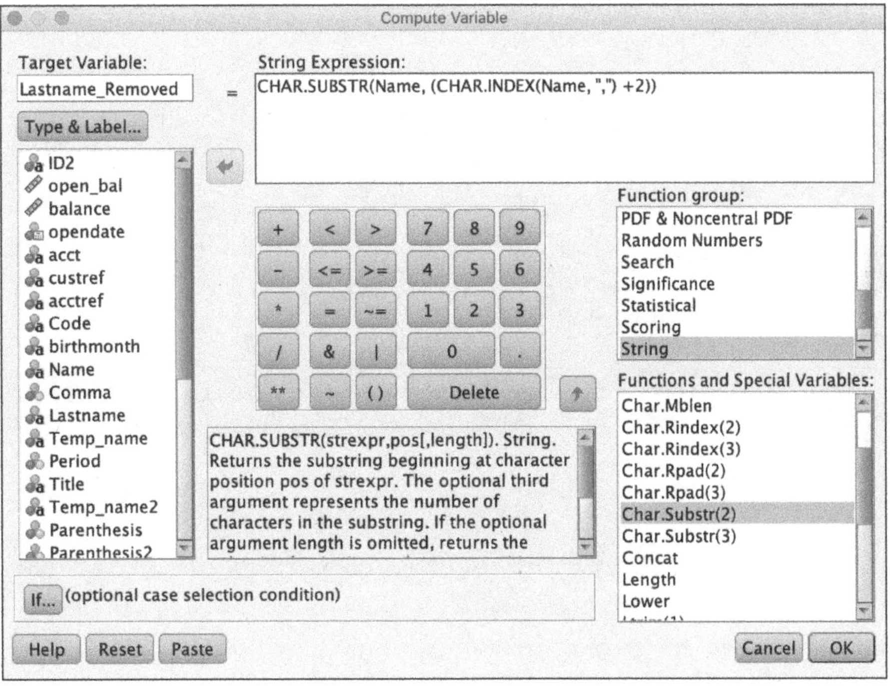

17 Using the String.sav file, extract the person's title from the new Lastname_Removed variable. Then create a cleaner version of the Lastname_Removed variable (one that no longer has the portion you just fixed), so you can use it in the next examples. Use the CHAR.INDEX and CHAR.SUBSTR functions.

18 Now that you've extracted the last name and title, find the location of the opening and closing parentheses so you can extract the maiden name. Use the String.sav file and the CHAR.INDEX function.

19 Now that you've identified the location of the opening and closing parentheses, extract the maiden name. Use the String.sav file and the CHAR.SUBSTR function.

20 Finally, create the person's full name by using their first name, last name, and, if applicable, maiden name instead of their husband's first name. Use the String.sav file, the CONCAT function, and the IF dialog.

Answers to Problems in Computing Variables

1 Do the following:

a. Use the Compute Variable transformation.

b. Type **ABS(CHILDS – CHLDIDEL)** in the Numeric Expression box.

c. Name the new variable.

You can now see the absolute difference between how many children each respondent has and their ideal number of children.

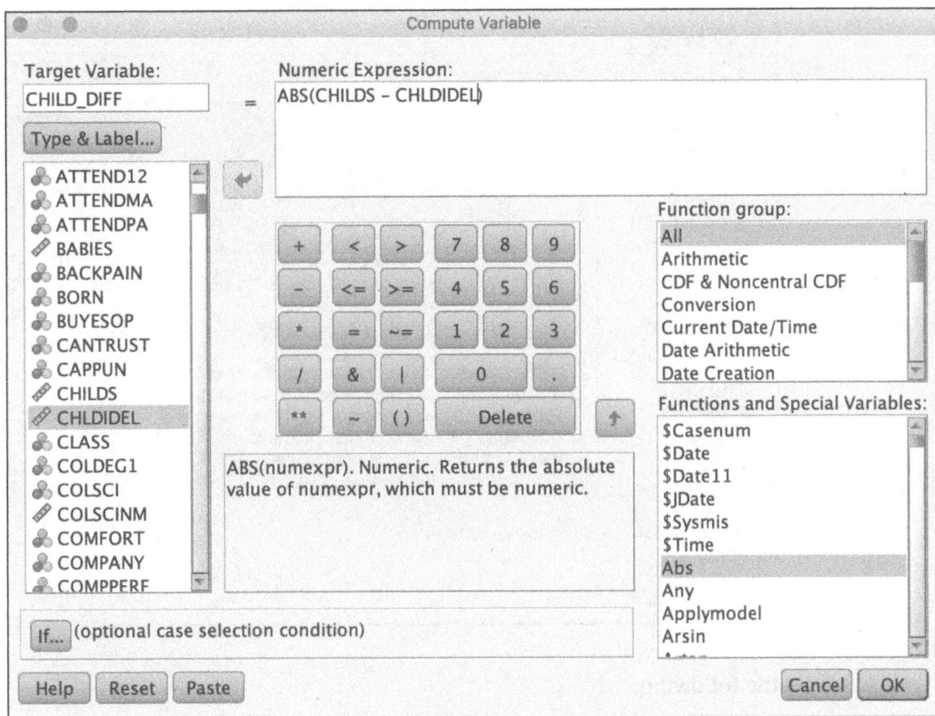

2 Do the following:

a. Use the Compute Variable transformation.

b. Type **ANY(MARITAL 2, 3, 4)** in the Numeric Expression box.

c. Name the new variable.

You've created a variable that identifies with a value of 1 anyone who is divorced, widowed, or separated, and identifies with a value of 0 everyone else.

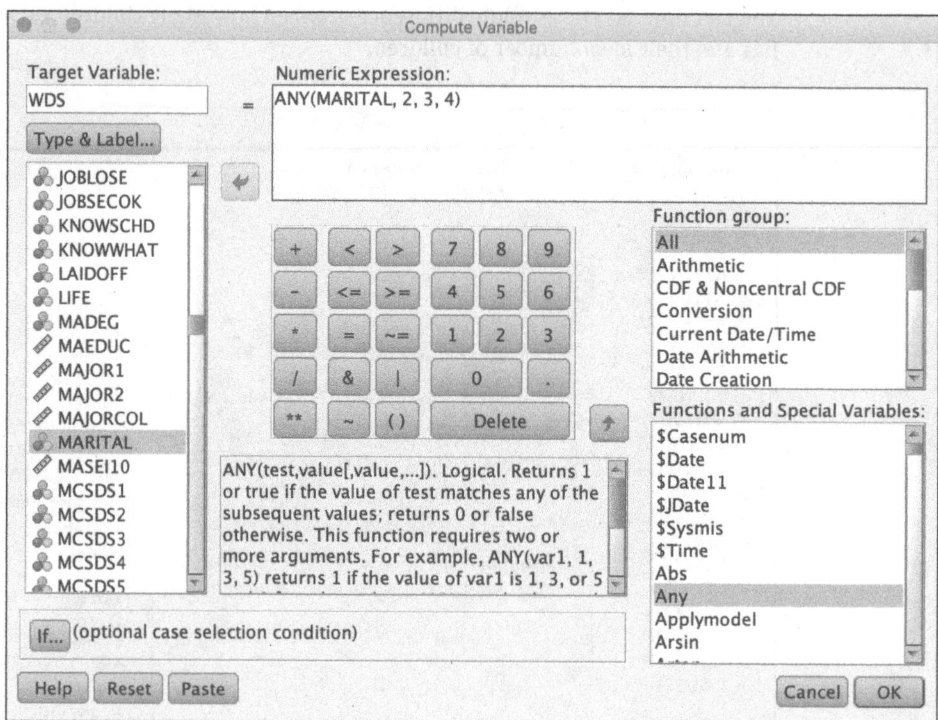

3 Do the following:

a. Use the Compute Variable transformation.

b. Type **RANGE(AGE 18, 65)** in the Numeric Expression box.

c. Name the new variable.

You've created a variable that identifies with a value of 1 for anyone between the ages of 18 and 65 and identifies with a value of 0 for everyone else.

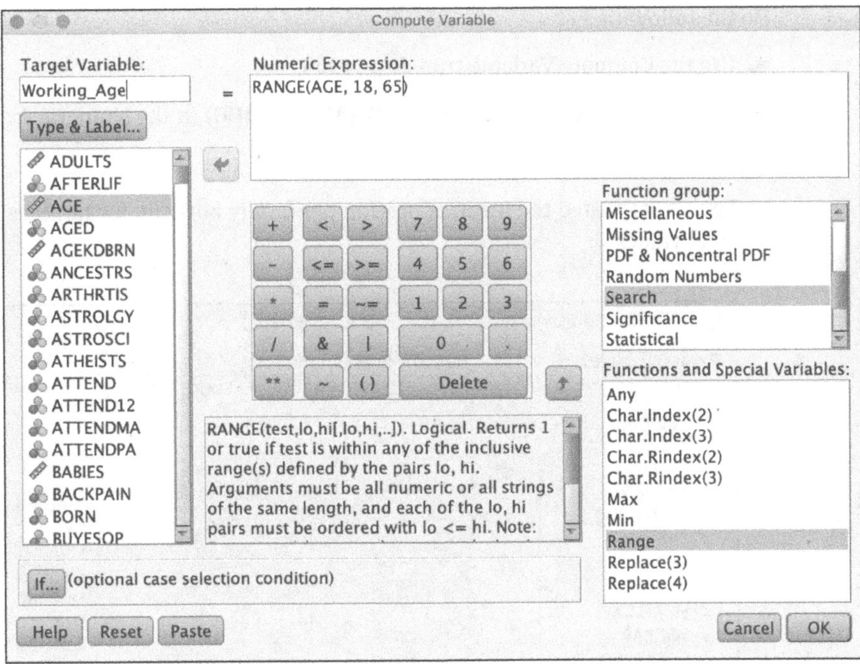

4. Do the following:

   a. Use the Compute Variable transformation.

   b. Type **MAX(EDUC, MAEDUC, PAEDUC, SPEDUC)** in the Numeric Expression box.

   c. Name the new variable.

   You've created a variable that identifies the maximum years of education in a family.

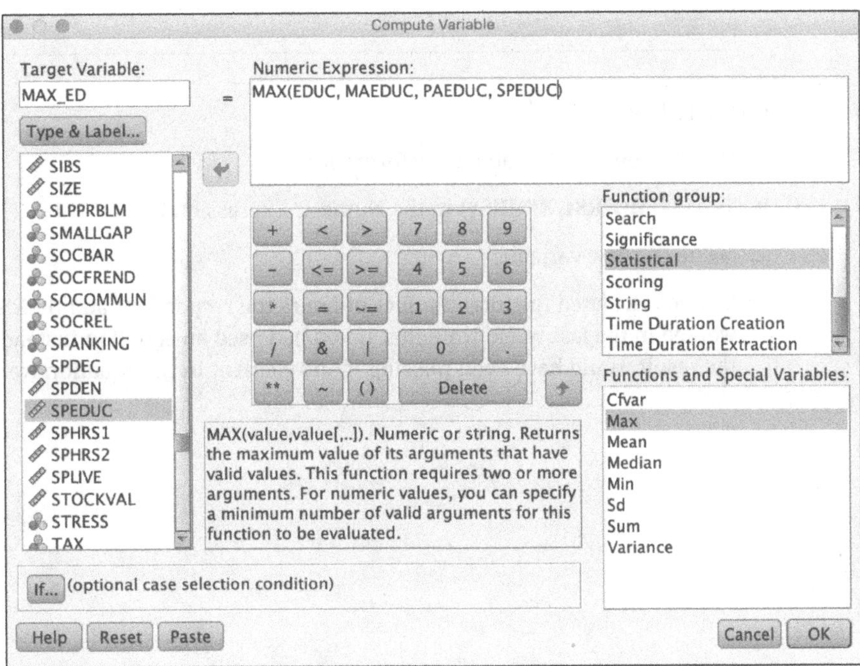

5. Do the following:

   a. Use the Compute Variable transformation.

   b. Type **MEAN.3(EDUC, MAEDUC, PAEDUC, SPEDUC)** in the Numeric Expression box.

   c. Name the new variable.

   You've calculated the mean education per family but only for families with at least three valid values.

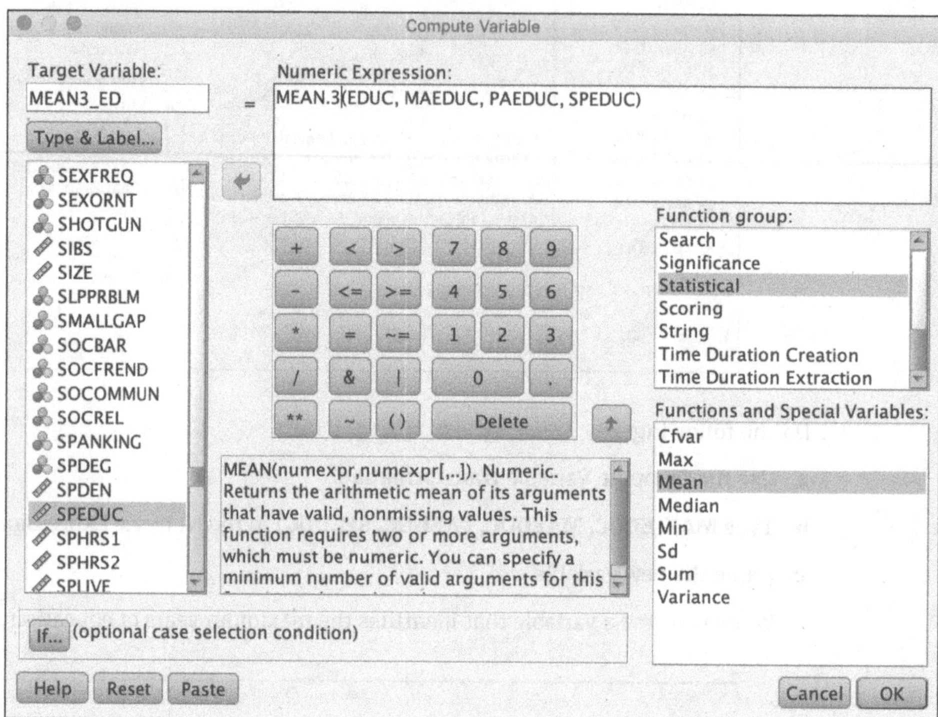

6. Do the following:

   a. Use the Compute Variable transformation.

   b. Type **SUM(HRS1, SPHRS1)** in the Numeric Expression box.

   c. Name the new variable.

   You've calculated the total number of hours the respondent and their spouse have worked in the last week. Note that if you had used an equation instead of a function, the result would have been missing if any element in the equation were missing.

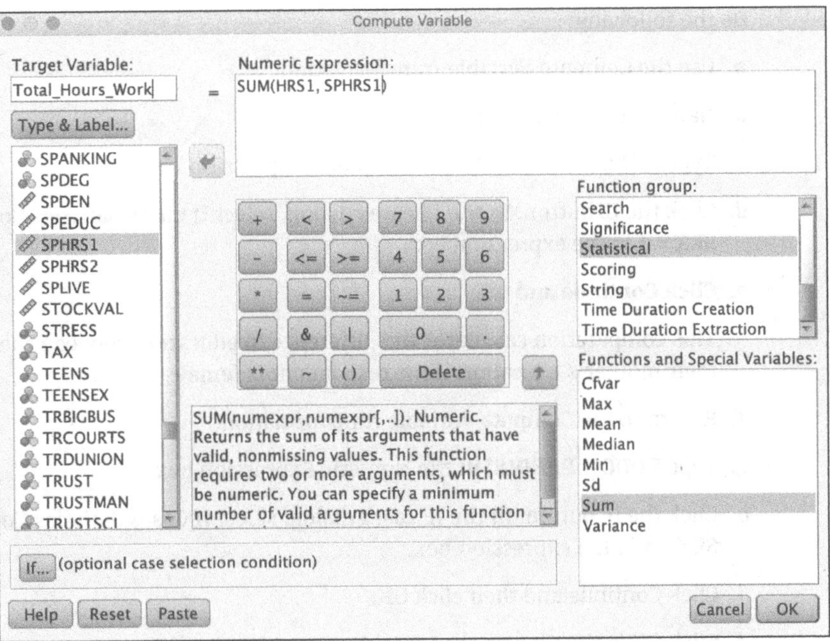

7. Do the following:

   a. Use the Compute Variable transformation.

   b. Type **NVALID(MCSDS1, MCSDS2, MCSDS3, MCSDS4, MCSDS5, MCSDS6, MCSDS7)** in the Numeric Expression box.

   c. Name the new variable.

   You've calculated the number of times respondents provided a valid value for the MCSDS variables.

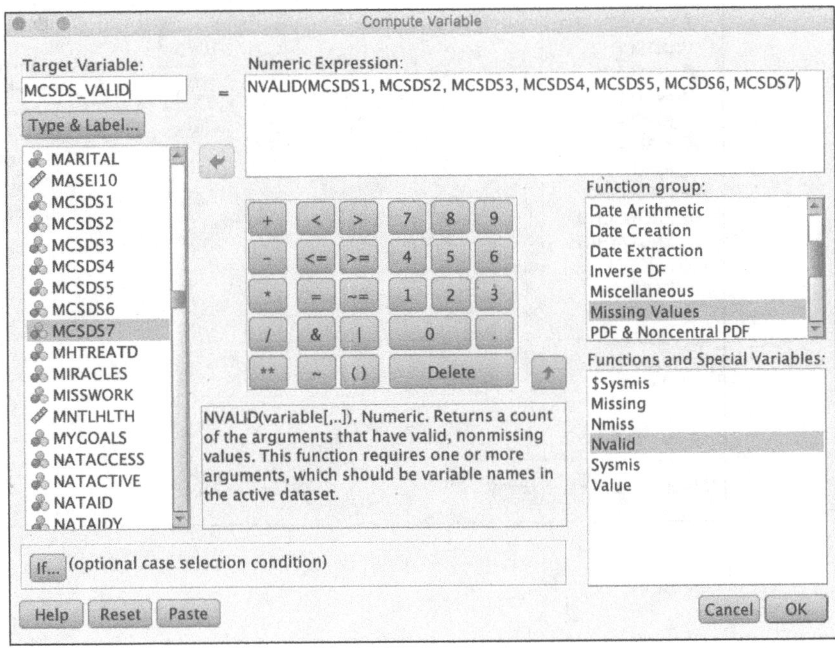

8 Do the following:

a. Use the Compute Variable transformation.

b. Name the new variable.

c. Type **EDUC – MAEDUC** in the Numeric Expression box.

d. Click the IF button. In the IF cases dialog, select If Cases Satisfies Condition and type **SEX = 2** in the expression box.

e. **Click Continue and OK.**

This computation calculates the difference in education between the respondent and their mother's education if the respondent is female.

f. Return to the Compute Variable transformation.

g. Type **EDUC – PAEDUC** in the Numeric Expression box.

h. Click the IF button. In the IF cases dialog, select If Cases Satisfies Condition and type **SEX = 1** in the expression box.

i. Click Continue and then click OK.

j. Click OK in the Change the Variable pop-up.

This computation calculates the difference in education between the respondent and their father's education if the respondent is male. You've calculated the difference in education between the respondent and their same-sex parent.

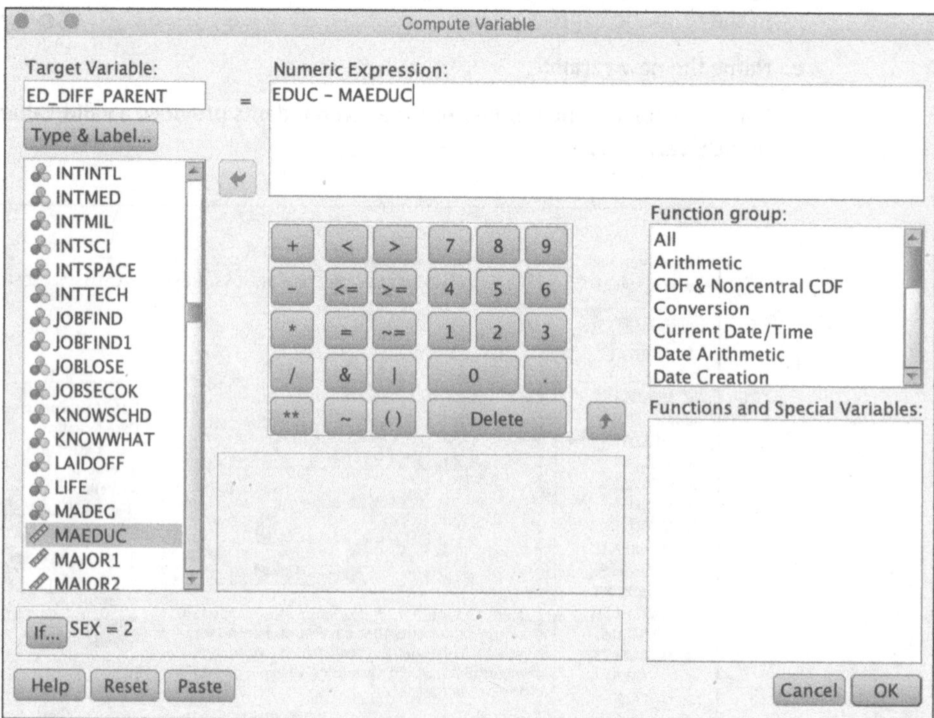

9 Do the following:

a. Use the Compute Variable transformation.

b. Name the new variable.

c. Type **$CASENUM** in the Numeric Expression box.

This computation creates an ID variable for a dataset that does not have a case identifier. Now you can manipulate the dataset and get the data back to the original order whenever you want.

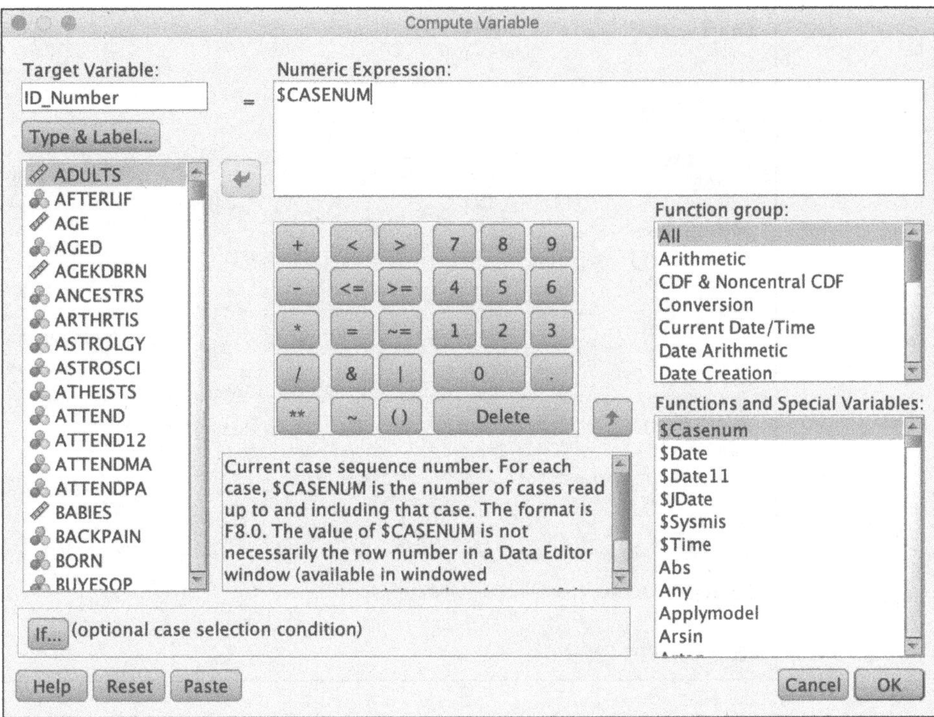

10. Do the following:

    a. Use the Compute Variable transformation.

    b. Name the new variable.

    c. Type **$SYSMIS** in the Numeric Expression box.

    This computation creates an empty variable for a dataset that you can use as a separator in the data file or as a starting point for a more complex variable that has IF conditions.

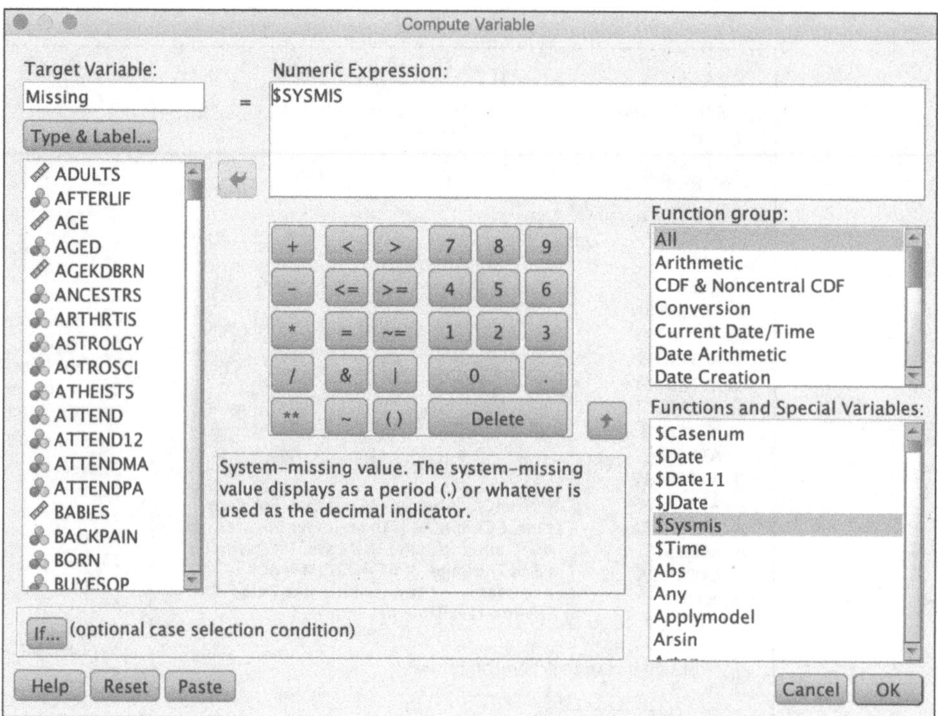

11. Do the following:

    a. Use the Compute Variable transformation.

    b. Name the new variable.

    c. Type **MISSING(HAPMAR)** in the Numeric Expression box.

    This variable transformation determines where you have a system or user-missing value for the HAPMAR variable. This new variable identifies with a value of 1 anyone who has a system or user-missing value for the HAPMAR variable with a value of 1, and identifies with a value of 0 everyone else.

The MISSING function is true if a value is either user-defined missing or system missing. The SYSMIS function returns true only when the value is system missing.

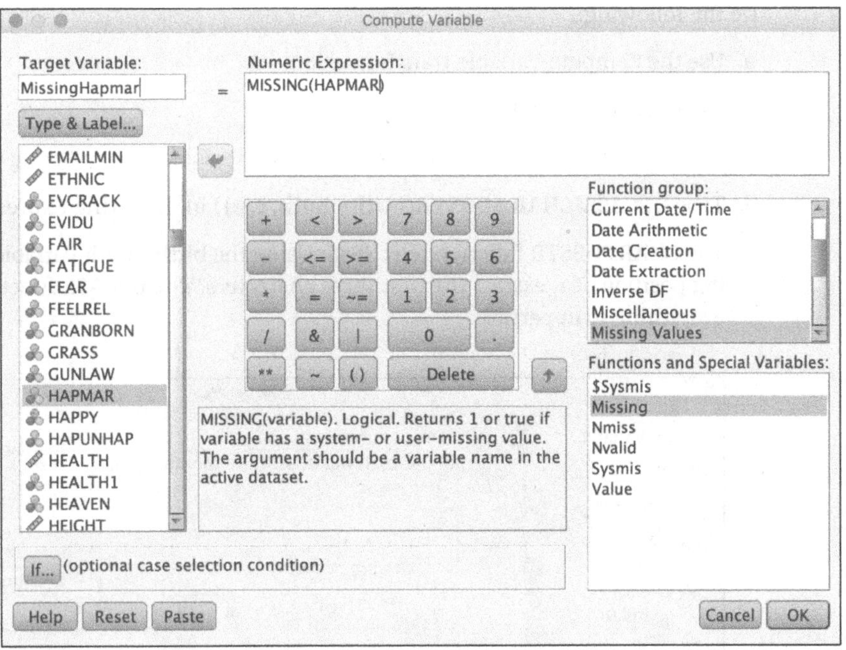

12 Do the following:

a. Use the Compute Variable transformation.

b. Name the new variable.

c. Type **VALUELABEL(DEGREE)** in the Numeric Expression box.

This new variable now has its value labels as the actual values, which can be useful when using some statistical algorithms.

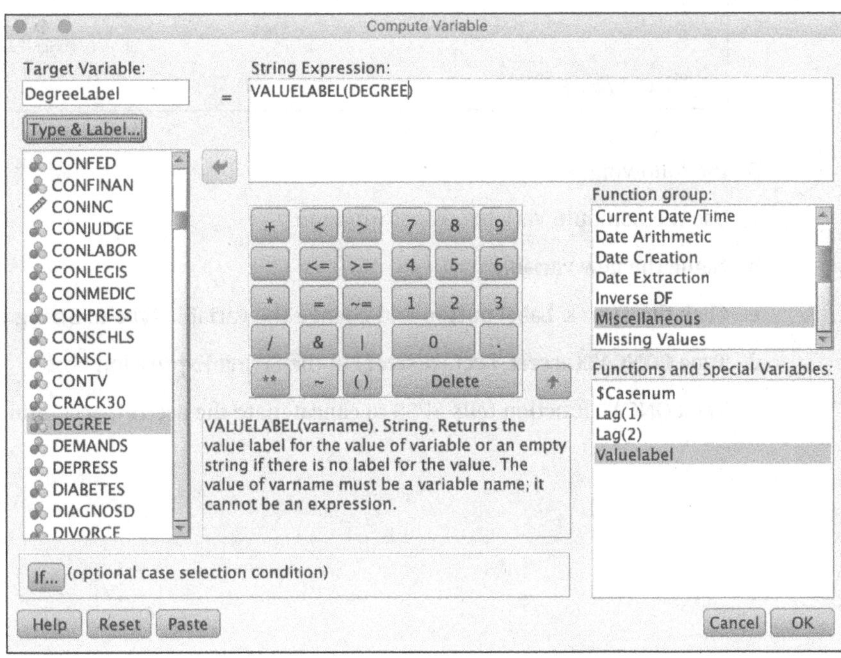

13 Do the following:

a. Use the Compute Variable transformation.

b. Name the new variable.

c. Click the Type & Label button and change the variable type to String with a width of 3.

d. Type **UPCASE(CHAR.SUBSTR(birthmonth, 1, 3))** in the String Expression box.

The CHAR.SUBSTR function tells SPSS: Using the birthmonth variable and a starting position of 1, extract the first three characters. The UPCASE function then converts everything to uppercase.

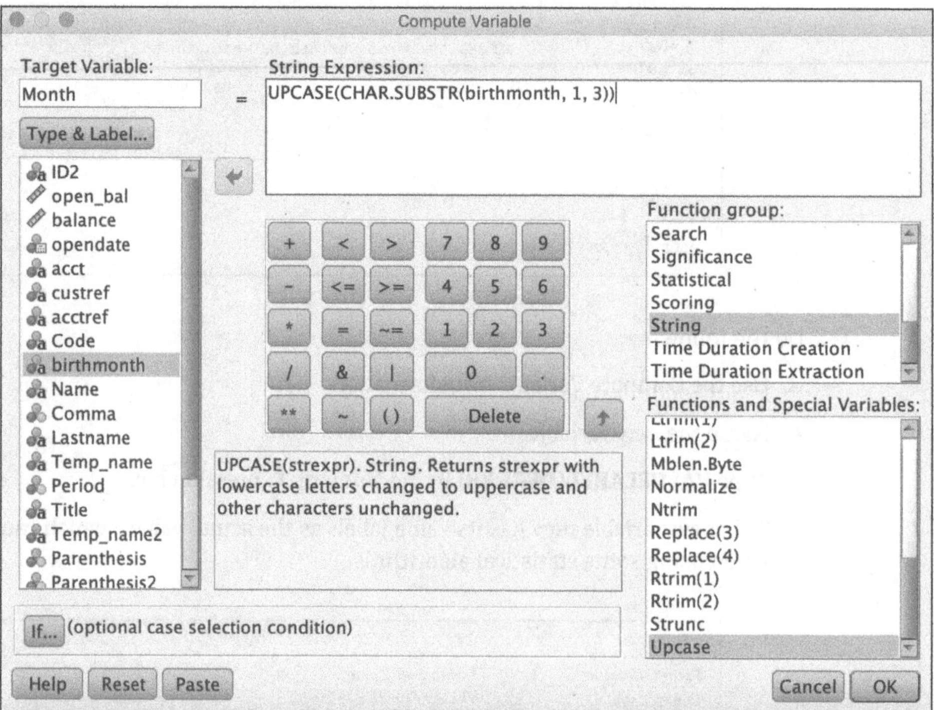

14 Do the following:

a. Use the Compute Variable transformation.

b. Name the new variable.

c. Click the Type & Label button and change the variable type to String with a width of 25.

d. Type **CONCAT(accref, acct, custref)** in the String Expression box.

The CONCAT function tells SPSS to concatenate the acctref, acct, and custref variables.

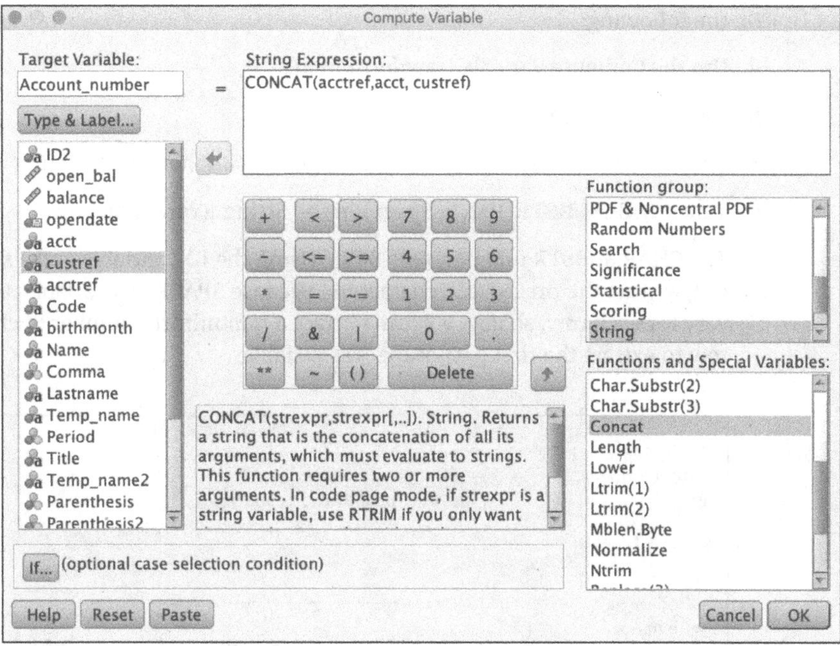

15. Do the following:

   a. Use the Compute Variable transformation.

   b. Name the new variable **Space**.

   c. Type **CHAR.INDEX(ID2, " ")** in the Numeric Expression box.

   The CHAR.INDEX function tells SPSS to use the ID2 variable and count the number of characters until the space (" ").

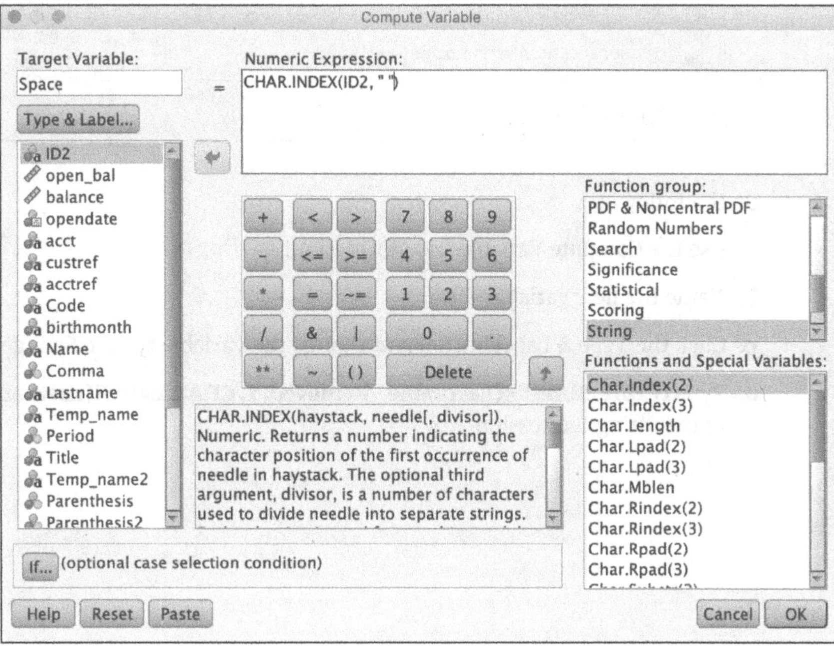

16 Do the following:

a. Use the Compute Variable transformation.

b. Name the new variable.

c. Click the Type & Label button and change the variable type to String with a width of 10.

d. Type **CHAR.SUBSTR(ID2, 1, Space)** in the String Expression box.

The CHAR.SUBSTR function tells SPSS: Using the ID2 variable and a starting position of 1, extract the first SPACE characters. Because SPACE is a variable, the values will vary for each row, so now you can extract a nonuniform number of characters in each row to extract the first part of the ID2 variable.

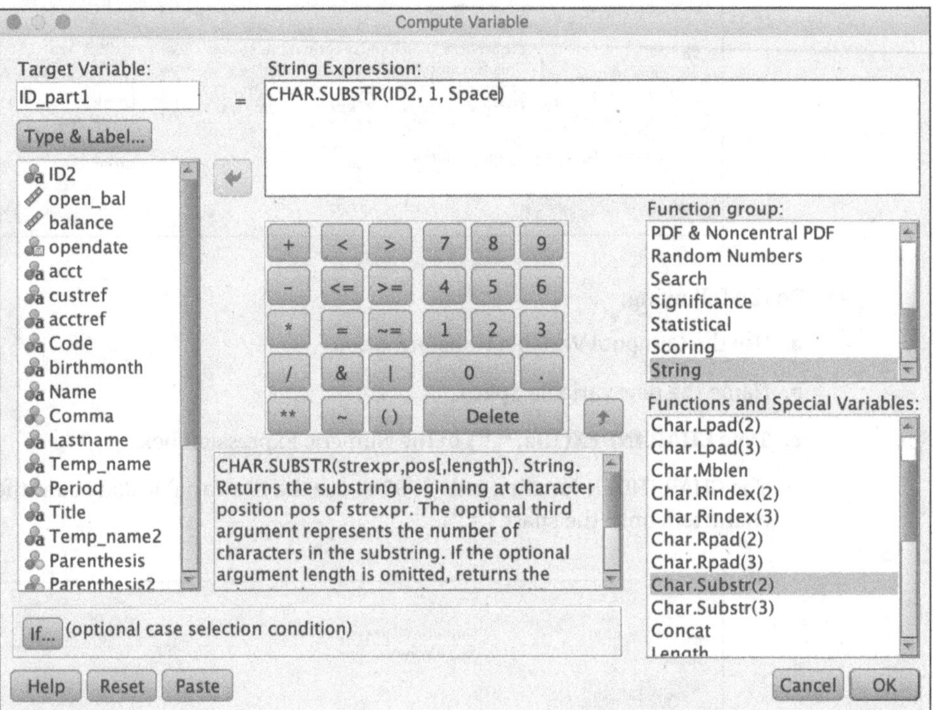

17 Do the following:

a. Use the Compute Variable transformation.

b. Name the new variable **Title.**

c. Click the Type & Label button and change the variable type to String with a width of 50.

d. Type **(CHAR.SUBSTR(Lastname_Removed, 1, CHAR.INDEX(Lastname_Removed, ".”))** in the String Expression box.

As before, the CHAR.INDEX function tells SPSS to use the Lastname_Removed variable and count the number of characters until the period ("."). The CHAR.SUBSTR function tells SPSS to use the Lastname_Removed variable and, starting at position 1, extract the characters up until the period. You've extracted the title from the name.

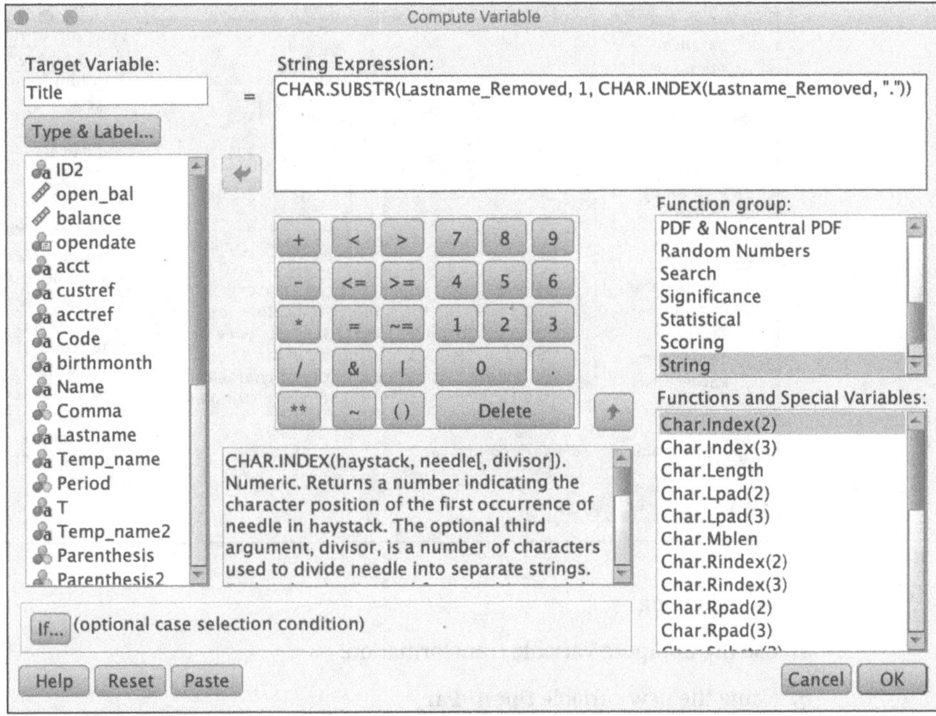

Next you use the Compute Variable transformation to create a cleaner version of the Lastname_Removed variable, which you use in the next examples.

e. Name a new variable **Title_removed.**

f. Click the Type & Label button and change the variable type to String with a width of 50.

g. Type **(CHAR.SUBSTR(Lastname_Removed, CHAR.INDEX(Lastname_Removed, ".") + 2)** in the String Expression box.

As before, the CHAR.INDEX function tells SPSS to use the Lastname_Removed variable and count the number of characters until the period ("."). The CHAR.SUBSTR function tells SPSS to use the Lastname_Removed variable, and starting at the position specified by the CHAR.INDEX function, extract the characters up until the period. You add + 2 so that you extract the extra spaces after the period. You've created a cleaner version of the Lastname_Removed variable, which you'll use in the next examples.

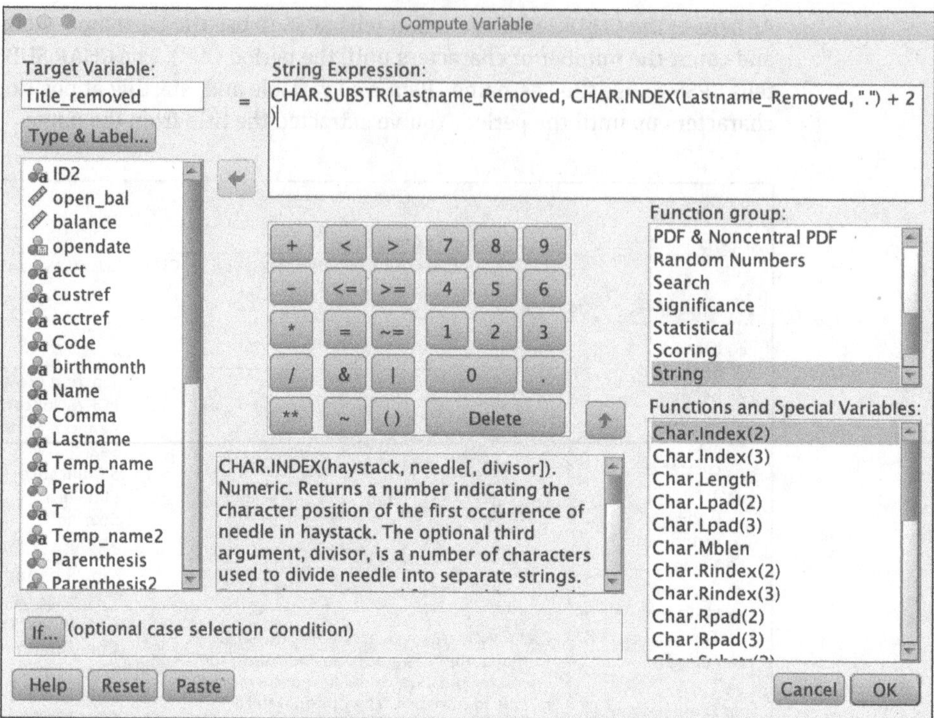

18 Do the following:

a. Use the Compute Variable transformation.

b. Name the new variable **Open_Par**.

c. Type **CHAR.INDEX(Title_Removed, "(")** in the Numeric Expression box.

As before, the CHAR.INDEX function tells SPSS to use the Title_Removed variable and count the number of characters until the opening parenthesis ("("). You've identified the location of the opening parenthesis.

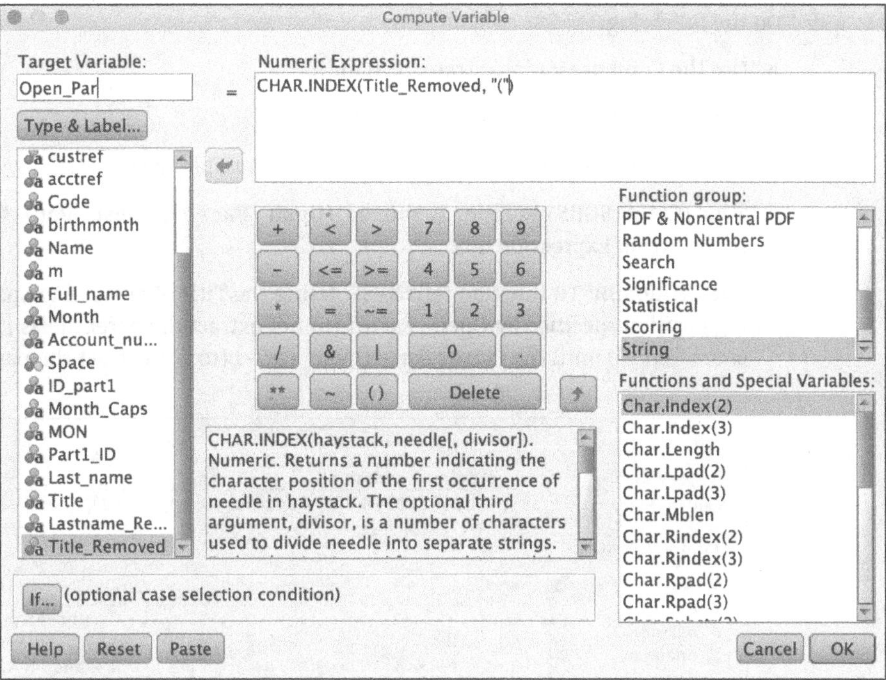

d. Name a new variable **Close_Par.**

e. Type **CHAR.INDEX(Title_Removed, ")")** in the Numeric Expression box.

As before, the CHAR.INDEX function tells SPSS to use the Title_Removed variable and count the number of characters until the closing parenthesis (")"). You've identified the location of the closing parenthesis.

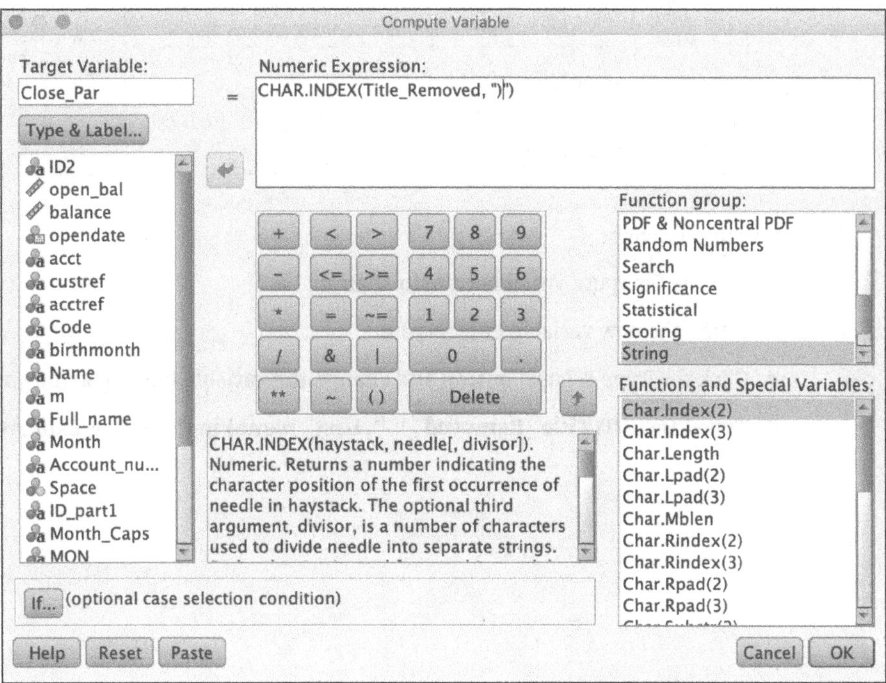

19. Do the following:

   a. Use the Compute Variable transformation.

   b. Name the new variable **Maiden_Name.**

   c. Click the Type & Label button and change the variable type to String with a width of 50.

   d. Type **(CHAR.SUBSTR(Title_Removed, (Open_Par + 1), (Close_Par – Open_Par – 1))** in the String Expression box.

   The CHAR.SUBSTR function tells SPSS: Using the Title_Removed variable and a starting position specified by Open_Par + 1 (to not extract the parenthesis itself), extract the characters up until the Close_Par - Open_Par -1 (to not extract the parenthesis).

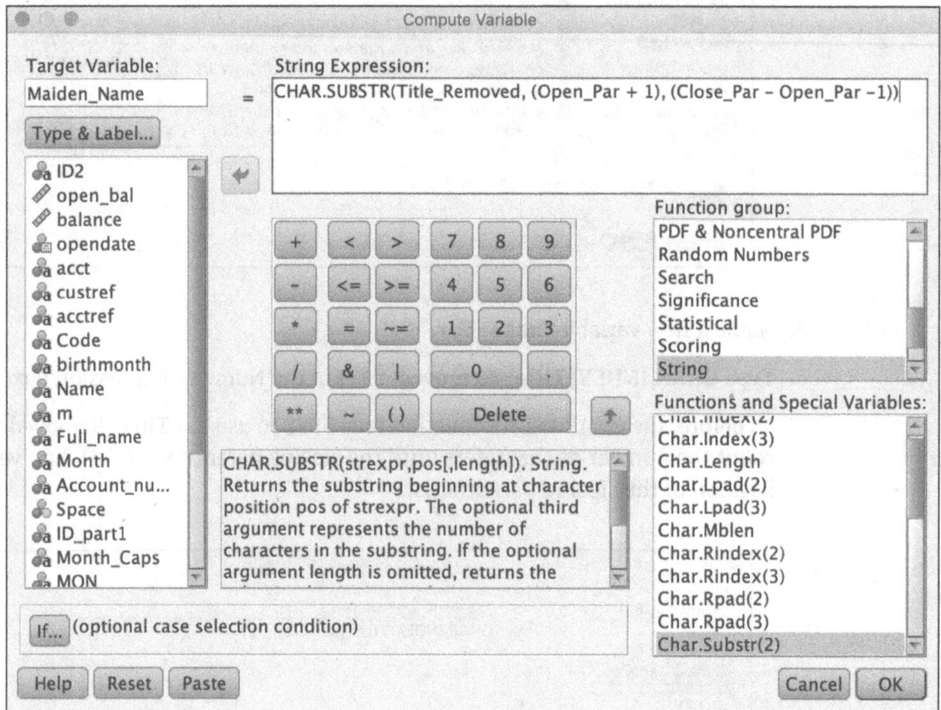

20. Do the following:

   a. Use the Compute Variable transformation.

   b. Name the new variable **Full_Name.**

   c. Click the Type & Label button and change the variable type to String with a width of 50.

   d. Type **CONCAT(Title_Removed , " ", Last_name)** in the String Expression box.

The CONCAT function tells SPSS to concatenate the Title_Removed (the person's first name) variable, a space, and the Last_name variable. This equation works for everyone without a maiden name.

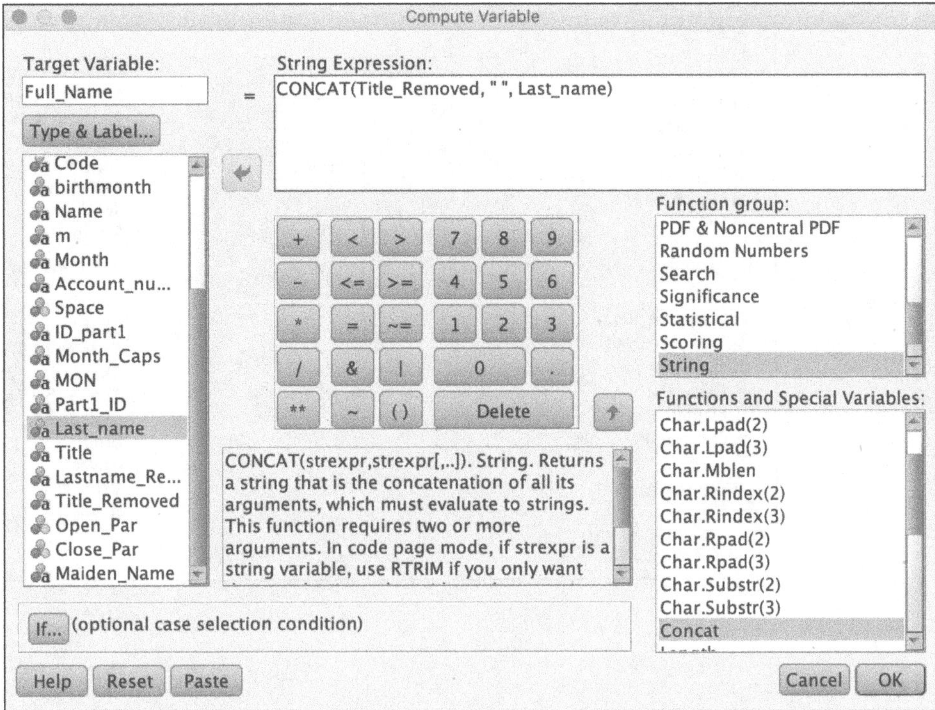

e. To fix the Full_name variable for those people with a maiden name, go back to the Compute Variable transformation.

f. Type **CONCAT(Maiden_Name , " ", Last_name)** in the String Expression box.

g. Click the IF button. In the IF cases dialog, select If Cases Satisfies Condition and type **Open_Par > 0** in the expression box.

h. Click Continue and then click OK.

i. Click OK in the Change the Variable pop-up.

The IF statement now uses the person's maiden name instead of the husband's first name whenever there's a maiden name.

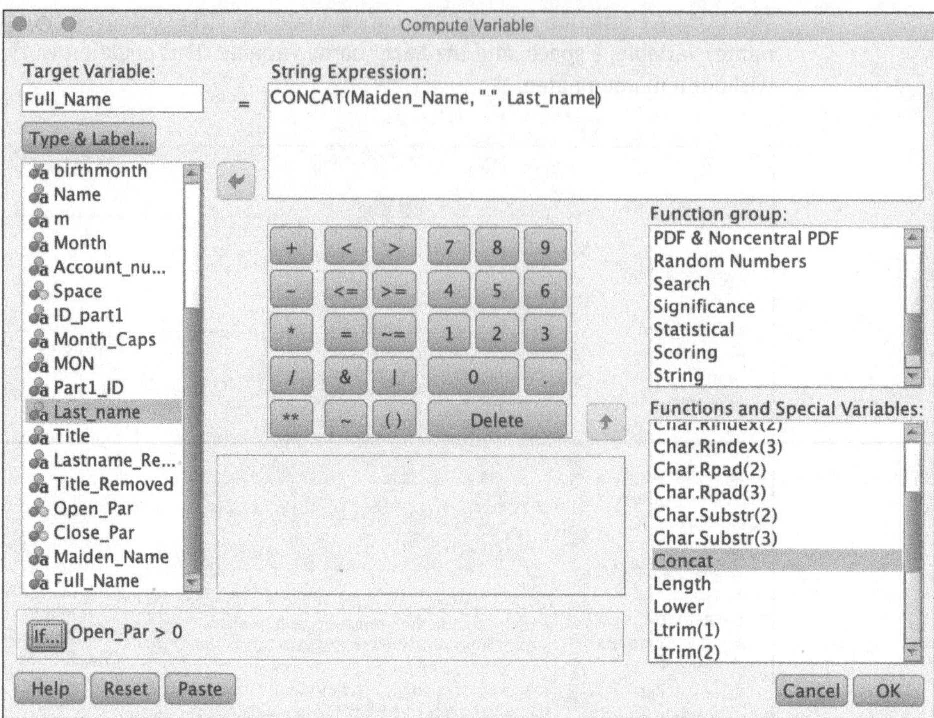
Compute Variable
Target Variable:
Full_Name
String Expression:
CONCAT(Maiden_Name, " ", Last_name)
Type & Label...
birthmonth
Name
m
Month
Account_nu...
Space
ID_part1
Month_Caps
MON
Part1_ID
Last_name
Title
Lastname_Re...
Title_Removed
Open_Par
Close_Par
Maiden_Name
Full_Name
Function group:
PDF & Noncentral PDF
Random Numbers
Search
Significance
Statistical
Scoring
String
Functions and Special Variables:
Char.Rindex(3)
Char.Rpad(2)
Char.Rpad(3)
Char.Substr(2)
Char.Substr(3)
Concat
Length
Lower
Ltrim(1)
Ltrim(2)
Delete
If... Open_Par > 0
Help
Reset
Paste
Cancel
OK

3 Analyzing Data

IN THIS PART . . .

Summarize variables.

Compare group proportions.

Observe how groups change from one time point to the next.

Compare group means.

Assess relationships.

Make predictions.

IN THIS CHAPTER

» Understanding level of measurement

» Summarizing data using frequencies

» Describing data with Descriptives

» Calculating z-scores

Chapter 6
Using Descriptive Statistics

After you have collected data and input it into SPSS Statistics, your next task is to perform preliminary analyses so that you can check the quality of your data, as well as describe the sample. Knowing the level of measurement of your variables is crucial to requesting the appropriate summary statistics. In this chapter, we discuss level of measurement, as well as three techniques for summarizing your data — frequencies, Descriptives, and z-scores — so that you can determine whether the data has discrepancies.

Level of Measurement

The level of measurement of a variable, or how data is coded, is important because it determines the appropriate summary statistics, tables, and graphs to describe data. *Statistical measures of central tendency* are used to summarize the distribution of a variable. Measures of central tendency are often referred to as the *average*. There are three main central tendency measures:

- **Mode:** The mode for any variable is merely the category or value that contains the most cases. This measure is typically used on nominal or ordinal data and can easily be determined by examining a frequency table.
- **Median:** If all the cases for a variable are arranged in order according to their value, the median is the value that splits the cases into two equally sized groups. The median is the same as the 50th percentile.

» **Mean:** The mean is the simple mathematical average of all the values in the distribution (that is, the sum of the values of all cases divided by the total number of cases). It's the most commonly reported measure of central tendency.

Measures of dispersion or variability describe the degree of spread, dispersion, or variability around the measure of central tendency. You might think of variability as a measure of the extent to which observations cluster within the distribution. A number of measures of dispersion are available, including simple measures such as maximum, minimum, and range, as well as standard deviation and variance:

» **Maximum:** The highest value for a variable.

» **Minimum:** The lowest value in the distribution.

» **Range:** The difference between the maximum and minimum values gives a general impression of how broad the distribution is.

» **Variance:** Both the variance and standard deviation provide information about the amount of spread around the mean value. They are overall measures of how clustered the data values are around the mean. The variance is calculated by summing the square of the difference between the value and the mean for each case and dividing this quantity by the number of cases minus 1. Because the variance measure is expressed in squared units, results are more difficult to interpret.

» **Standard deviation:** The standard deviation is the square root of the variance, which restores the value of variability to the units of measurement of the original variable. Standard deviation is easier to interpret because it uses the original measurement units, not squared units as in the variance.

REMEMBER

Choosing the appropriate summary statistic for each level of measurement is not the same as setting up the metadata in variable view. We review and rehearse declaring the level of measurement in Chapter 2.

As a first step in analyzing your data, you must gain a big-picture view of all the variables you're working with so you can detect any unusual or unexpected values. Take a look at the values and the number of responses in each of the categories of a variable. For some variables, you want to look at simple summary measures, including the central tendency of the distribution (arithmetic mean) and the dispersion (spread around the central point). These summaries can often answer important questions without requiring you to do any further analysis.

Summaries of individual variables provide the basis for conducting more complex analyses involving those variables. For instance, single variable analyses are useful for establishing base rates for the population sampled, such as determining the percentage of customers who are satisfied with the services they've received during the year.

In addition, if you have a frequency table with many values, you may be able to see ways to make the table more succinct by combining similar categories (or values) and then reporting statistics on the revised table. In addition, summaries can serve data-checking purposes because unusual values are easily apparent in tables.

Focusing on Frequencies

Statistical software applications report results using tables and graphs, and it's important to take those results and translate them into a form that's appropriate for your target audience. The most common technique for describing data is to request a *frequency table*, which provides a summary showing the number and percentage of cases falling into each category of a variable. Users can also request additional summary statistics.

In the following, you use the Frequencies procedure to create a frequency table and get summary statistics for variables:

1. **Choose File ⇨ Open ⇨ Data and load the GSS2018.sav file.**

   You can download the file from the book's companion website at `www.dummies.com/go/spssstatisticsworkbookfd`. This file contains data from the General Social Survey (GSS), a nationally representative survey of adults in the United States that collects data on contemporary opinions, attitudes, and behaviors.

2. **Choose Analyze ⇨ Descriptive Statistics ⇨ Frequencies.**

   The Frequencies dialog appears.

3. **Select the HAPMAR variable (happiness of marriage) and place it in the Variable(s) box.**

   If you were to run the Frequencies procedure now, you would get a table showing the distribution of the variable. However, it's customary to request additional summary statistics.

4. **Click the Statistics button.**

   The Frequencies: Statistics dialog appears.

5. **In the Central Tendency section, select the Mode check box.**

   WARNING

   This dialog provides many statistics, but it's critical that you request only those appropriate for the level of measurement of the variables you placed in the Variable(s) box. For nominal variables, the mode is the only suitable statistic.

6. **Click Continue.**

7. **Click the Charts button.**

   The Frequencies: Charts dialog appears.

8. **In the Chart Type section, click the Bar Charts radio button; in the Chart Values section, click the Percentages radio button.**

   This dialog has options for pie charts and bar charts. Either type of chart is acceptable for a nominal variable. Charts can be built using either counts or percentages, but percentages are usually a better choice.

9. **Click Continue, and then click OK.**

   SPSS runs the Frequencies procedure and calculates the summary statistics, frequency table, and bar chart you requested.

The frequency table, which is shown in Figure 6-1, displays the distribution of the happiness of marriage (hapmar) variable. The information in the frequency table is comprised of counts and percentages.

Happiness of marriage

		Frequency	Percent	Valid Percent	Cumulative Percent
Valid	VERY HAPPY	638	27.2	64.3	64.3
	PRETTY HAPPY	324	13.8	32.7	97.0
	NOT TOO HAPPY	30	1.3	3.0	100.0
	Total	992	42.2	100.0	
Missing	IAP	1348	57.4		
	DK	1	.0		
	NA	7	.3		
	Total	1356	57.8		
Total		2348	100.0		

FIGURE 6-1: The frequency table for the happiness of marriage variable.

The Frequency column contains *counts*, or the number of occurrences of each data value. So, for the happiness of marriage variable, it's easy to see why the Very Happy category is the mode — 638 people chose this response option.

The Percent column shows the percentage of cases in each category relative to the number of cases in the entire dataset, including those with missing values. In the example, 638 people who are Very Happy account for 27.2% of all cases.

The Valid Percent column contains the percentage of cases in each category relative to the number of valid (nonmissing) cases. In our example, the 638 people who are Very Happy make up 64.3% of all valid responses (those who are not currently married were not asked this question, which is why so much data is missing).

The Cumulative Percent column contains the percentage of cases whose values are less than or equal to the indicated value. Cumulative percent is useful only for ordinal or scale variables.

TIP

Depending on your research question, the Percent column or the Valid Percent column may be useful when you have a lot of missing data or a variable was not applicable to a large percentage of people.

Bar charts, like the one in Figure 6-2, summarize the distribution observed in the frequency table and enable you to see the distribution. For the happiness of marriage variable, more than half of the people are in the Very Happy category.

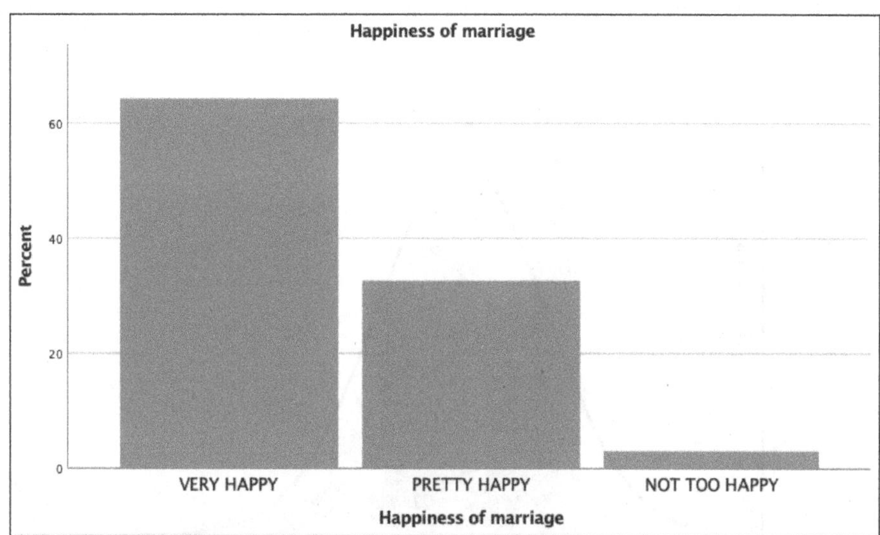

FIGURE 6-2: A bar chart for the happiness of marriage variable.

See the following for an example of running the Frequencies Procedure.

EXAMPLE

Q. Using the GSS2018.sav file, run the Frequencies procedure on the CHILDS (number of children) variable. Request the appropriate summary statistics and graph. Summarize the findings.

A. For continuous variables, all of the summary statistics are appropriate. More than 2000 people answered this question, and they have on average of 1.86 children. The number of children ranged from 0 to 8. In an ideal world, you would like the mean, median, and mode to be similar, because they're all measures of central tendency. In this example, these values are not too different, which is an indication that this variable is somewhat normally distributed. You can confirm the normal distribution by visually checking the distribution of the variable with a histogram.

Statistics

Number of children

N	Valid	2344
	Missing	4
Mean		1.86
Median		2.00
Std. Deviation		1.674
Variance		2.803
Range		8
Minimum		0
Maximum		8

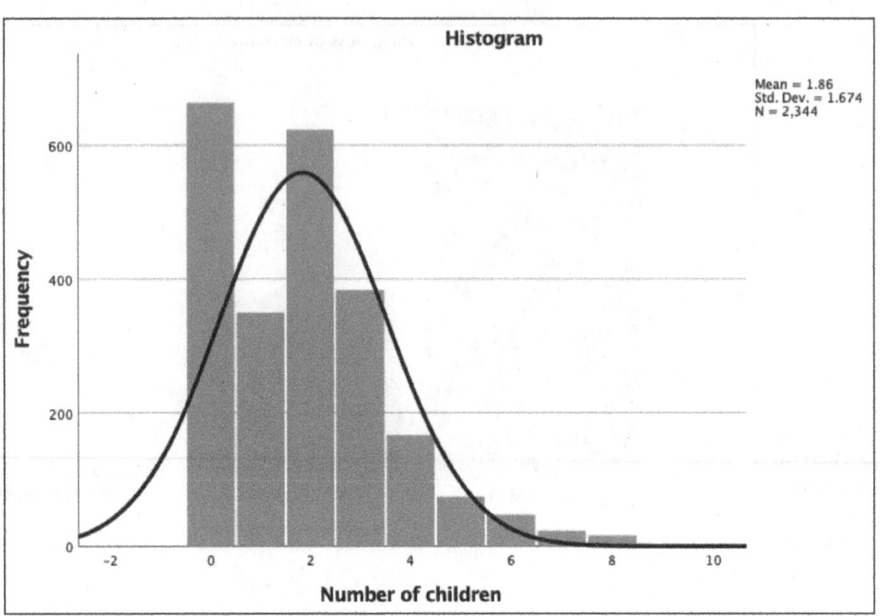

1. Using the GSS2018.sav file, run the Frequencies procedure on the MARITAL (marital status) variable. Request the appropriate summary statistics and graph. Summarize the findings.

2. Using the GSS2018.sav file, run the Frequencies procedure on the DEGREE (respondent's highest degree) variable. Request the appropriate summary statistics and graph. Summarize the findings.

3 Using the GSS2018.sav file, run the Frequencies procedure on the POLVIEWS (political views — higher numbers indicate that the person is more conservative) variable. Request the appropriate summary statistics and graph. Summarize the findings.

4 Using the GSS2018.sav file, run the Frequencies procedure on the TVHOURS (hours per day watching TV) variable. Request the appropriate summary statistics and graph. Summarize the findings.

Summarizing Variables with the Descriptives Procedure

When summarizing continuous variables, you can use the Descriptives procedure as an alternative to the Frequencies procedure. The *Descriptives procedure* provides an efficient summary of various statistics and the number of valid values for each variable so you can quickly determine if anything in the data is awry.

To use the Descriptives procedure, follow these steps:

1. **Choose File ➪ Open ➪ Data and load the GSS2018.sav data file.**

   Download the file at `www.dummies.com/go/spssstatisticsworkbookfd`.

2. **Choose Analyze ➪ Descriptive Statistics ➪ Descriptives.**

   The Descriptives dialog appears.

3. **Select the AGE variable and place it in the Variable(s) box.**
4. **Click OK.**

   SPSS runs the Descriptives procedure and calculates the summary statistics.

The minimum and maximum values provide an efficient way to check for values outside the expected range (see Figure 6-3). If you see a value that is too low or too high, you might have data errors or potential outliers. Likewise, it's always important to investigate the mean and standard deviation to determine whether the values make sense. Ask yourself, "Is this what I was expecting?" Sometimes a mean may be lower or higher than expected, which can be an indication that there is a problem with how the data was coded or collected. Also remember that if a standard deviation has a value of zero, this means that every person in the dataset provided the same value, which from a statistical perspective, isn't very useful.

Descriptive Statistics

	N	Minimum	Maximum	Mean	Std. Deviation
Age of respondent	2341	18	89	48.97	18.061
Valid N (listwise)	2341				

FIGURE 6-3: The Descriptive Statistics table.

See the following for an example of running the Descriptives procedure.

EXAMPLE

Q. Using the GSS2018.sav file, run the Descriptives procedure on the CHILDS (number of children) variable. Summarize the findings.

A. More than 2000 people answered this question, and they have on average 1.86 children. The number of children ranged from 0 to 8.

Descriptive Statistics

	N	Minimum	Maximum	Mean	Std. Deviation
Number of children	2344	0	8	1.86	1.674
Valid N (listwise)	2344				

5 Using the GSS2018.sav file, run the Descriptives procedure on the SIBS (number of brother and sisters) variable. Summarize the findings.

6 Using the GSS2018.sav file, run the Descriptives procedure on the TVHOURS (hours per day watching TV) variable. Summarize the findings.

7 Using the GSS2018.sav file, run the Descriptives procedure on the POLVIEWS (political views — higher numbers indicate that the person is more conservative) variable. Summarize the findings.

8 Using the GSS2018.sav file, run the Descriptives procedure on the EDUC (highest year of school completed) variable. Summarize the findings.

Working with Z-Scores

Standardized scores, or *z-scores,* can be used to calculate the relative position of each value in the distribution. They indicate the number of standard deviations a value is above or below the sample mean. For example, if you obtain a score of 70 out of 100 on a math test, that information is not enough to tell how well you did in relation to others taking the test. However, if you know the mean score is 60, the standard deviation is 10, and the scores are normally distributed, you can calculate the proportion of people who achieved a score at least as high as your own.

Z-scores are calculated by subtracting the mean from the value and then dividing by the standard deviation for the sample:

Z = (score – mean) / standard deviation

For the example, you can subtract the mean from the value (70–60=10) and divide the difference by the standard deviation (10/10=1.0).

So in this case, the score of 70 is 1.0 standard deviation above the mean. You can then use this information to determine what percentage of cases fall above or below this value, because you know that the mean of a standardized distribution is 0 and the standard deviation is 1. In the example, you'd find that 34% of cases are likely to have a score above 70. You can use this information also to identify *outliers* (unusual scores), which are typically defined as cases that are more than three standard deviations away from the mean, that is, values with z-scores greater than the absolute value of 3.0.

To use the Descriptives procedure to calculate z-scores, follow these steps:

1. **Choose File ⇨ Open ⇨ Data and load the GSS2018.sav data file.**

   Download the file at `www.dummies.com/go/spssstatisticsworkbookfd`.

2. **Choose Analyze ⇨ Descriptive Statistics ⇨ Descriptives.**
3. **Select the AGE variable and place it in the Variable(s) box.**
4. **Select the Save Standardized Values as Variables option.**
5. **Click OK.**

   SPSS runs the Descriptives procedure and calculates the z-scores.

6. **Switch over to the Data Editor window.**

Figure 6-4 shows the new standardized variable created at the end of the data file. By default, the new variable name is the old variable name prefixed with the letter Z. You can save this variable in the file and use it in any statistical procedure.

	ZAGE
1	-.33063
2	1.38579
3	-.38599
4	.77674
5	1.21969
6	.99821
7	.55527
8	-.33063
9	.72137
10	.33379

FIGURE 6-4: The Age variable has been standardized.

Focusing on the first row, note that the first person has a z-score of –.33, which indicates that this person is slightly younger (43) than the average age of the sample (48.97).

See the following for an example of calculating z-scores.

EXAMPLE

Q. Using the GSS2018.sav file, run the Descriptives procedure on the CHILDS (number of children) variable and calculate the z-scores. Describe your findings.

A. The person in the first row has a z-score of –.1.11, which indicates that this person has fewer children (0) than the average of the sample (1.86).

	ZCHILDS
1	-1.10816
2	.68365
3	.08638
4	.08638
5	-1.10816
6	.08638
7	2.47546
8	-1.10816
9	1.28092
10	.08638

9 Using the GSS2018.sav file, run the Descriptives procedure on the SIBS (number of brother and sisters) variable and calculate the z-scores. Summarize the findings.

10 Using the GSS2018.sav file, run the Descriptives procedure on the TVHOURS (hours per day watching TV) variable and calculate the z-scores. Summarize the findings.

11 Using the GSS2018.sav file, run the Descriptives procedure on the POLVIEWS (political views — higher numbers indicate that the person is more conservative) variable and calculate the z-scores. Summarize the findings.

12 Using the GSS2018.sav file, run the Descriptives procedure on the EDUC (highest year of school completed) variable and calculate the z-scores. Summarize the findings.

Answers to Problems in Using Descriptive Statistics

1. For the MARITAL (marital status) variable, the largest category, or mode, is the married group; 998 people chose this option. You could use either a bar chart or a pie chart to depict this variable. We used the pie chart option. The married group accounts for almost 50% of the sample.

Marital status

		Frequency	Percent	Valid Percent	Cumulative Percent
Valid	MARRIED	998	42.5	42.5	42.5
	WIDOWED	200	8.5	8.5	51.1
	DIVORCED	403	17.2	17.2	68.2
	SEPARATED	75	3.2	3.2	71.4
	NEVER MARRIED	670	28.5	28.6	100.0
	Total	2346	99.9	100.0	
Missing	NA	2	.1		
Total		2348	100.0		

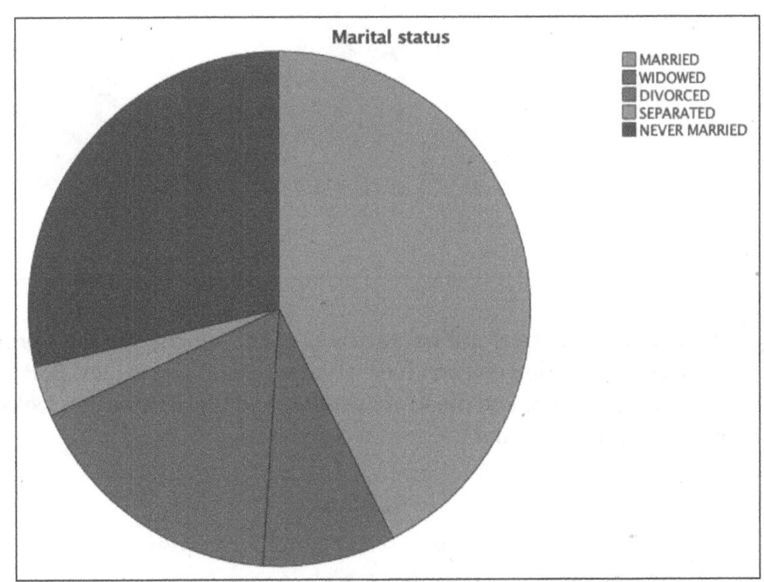

2 For the DEGREE variable, the largest category, or mode, is the high school group; 1178 people chose this option. Very little data is missing for this variable. We used a bar chart to view this distribution. The high school group is much larger than all other categories, and the junior college group has the least number of people.

R's highest degree

		Frequency	Percent	Valid Percent	Cumulative Percent
Valid	LT HIGH SCHOOL	262	11.2	11.2	11.2
	HIGH SCHOOL	1178	50.2	50.2	61.3
	JUNIOR COLLEGE	196	8.3	8.3	69.7
	BACHELOR	465	19.8	19.8	89.5
	GRADUATE	247	10.5	10.5	100.0
	Total	2348	100.0	100.0	

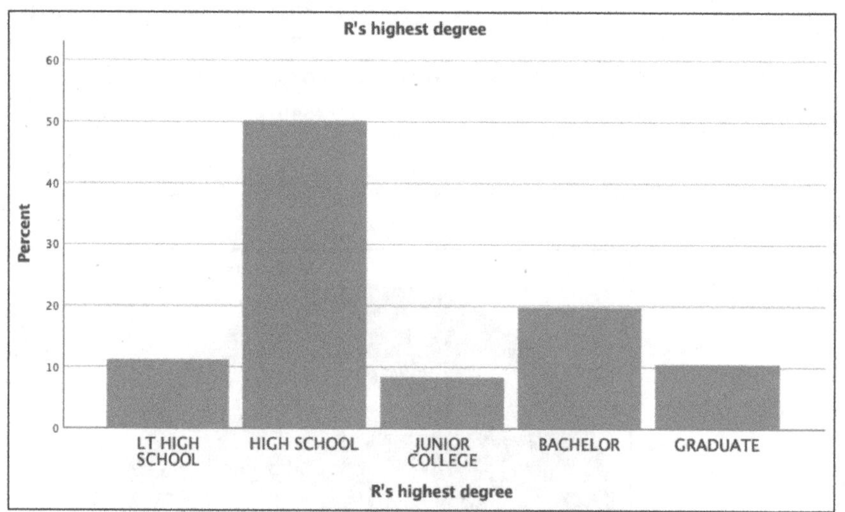

3 It's customary to treat ordinal variables as continuous variables provided you do not have a bimodal distribution and you have at least five unique values. Both conditions are met here. The mean, median, and mode are similar, and the distribution looks fairly normal with most values in the center and fewer in the extremes.

Statistics

Think of self as liberal or conservative

N	Valid	2247
	Missing	101
Mean		4.05
Median		4.00
Mode		4
Std. Deviation		1.500
Variance		2.249
Range		6
Minimum		1
Maximum		7

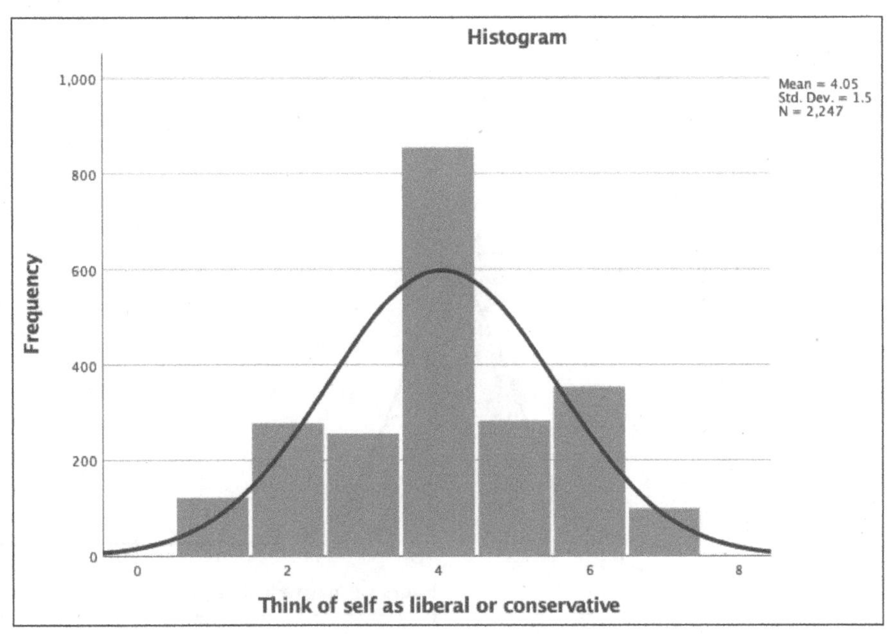

4. On average, people watch almost three hours of TV per day. The values have a considerable range from 0 to 24, which indicates outliers in the distribution. The histogram supports this indication because it is positively skewed and some extreme values are beyond watching more than 10 hours of TV per day.

Statistics

Hours per day watching TV

N	Valid	1555
	Missing	793
Mean		2.94
Median		2.00
Std. Deviation		2.837
Variance		8.049
Range		24
Minimum		0
Maximum		24

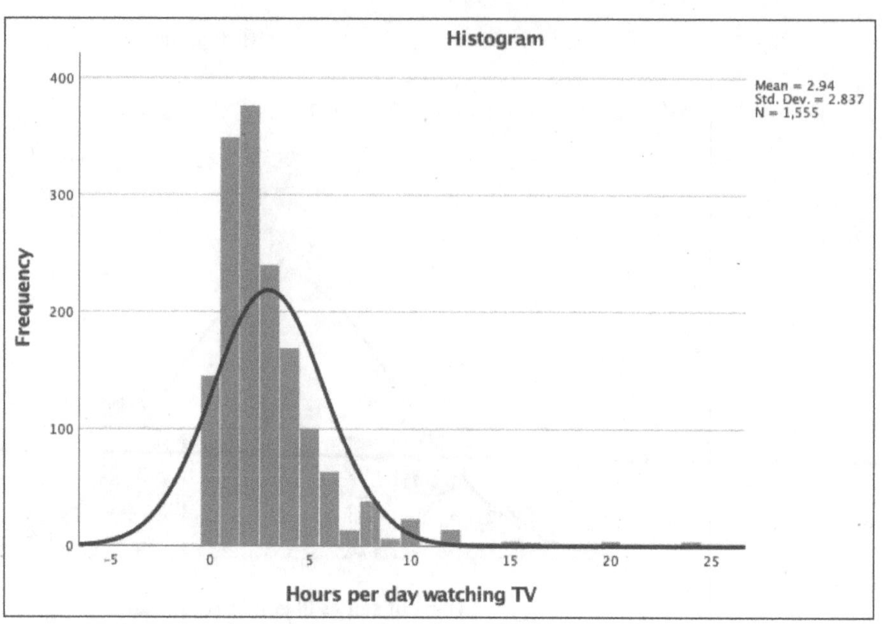

5. More than 2000 people answered this question, and they have on average 3.58 siblings. The number of siblings ranged from 0 to 25, so this distribution has outliers.

Descriptive Statistics

	N	Minimum	Maximum	Mean	Std. Deviation
Number of brothers and sisters	2343	0	25	3.58	2.863
Valid N (listwise)	2343				

6. More than 1500 people answered this question, and they watch on average 2.94 hours of TV per day. The number of hours of watching TV per day ranged from 0 to 24, so this distribution has outliers.

Descriptive Statistics

	N	Minimum	Maximum	Mean	Std. Deviation
Hours per day watching TV	1555	0	24	2.94	2.837
Valid N (listwise)	1555				

7 More than 2000 people answered this question. The political values ranged from 1 to 7, and the mean is right at the midpoint, with a value of 4.05.

Descriptive Statistics

	N	Minimum	Maximum	Mean	Std. Deviation
Think of self as liberal or conservative	2247	1	7	4.05	1.500
Valid N (listwise)	2247				

8 More than 2000 people answered this question, and they have on average 13.73 years of education. The number of years of education ranged from 0 to 20.

Descriptive Statistics

	N	Minimum	Maximum	Mean	Std. Deviation
Highest year of school completed	2345	0	20	13.73	2.974
Valid N (listwise)	2345				

9 The person in the first row has a z-score of .15, which indicates that this person has more siblings (4) than the average of the sample (3.58).

	ZSIBS
1	.14683
2	.14683
3	-.55170
4	-.20243
5	-.20243
6	-.90096
7	1.19463
8	-.55170
9	.14683
10	-.55170

10 The person in the first row has a z-score of .02, which indicates that this person watches more TV per day (3) than the sample's average (2.94).

	ZTVHOURS
1	.02176
2	.
3	-.68319
4	-.68319
5	.
6	2.48910
7	.
8	.
9	.37424
10	-.33072

11 The person in the first row has a z-score of 1.30, which indicates that this person is more conservative (6) than the sample's average (3.58).

	ZPOLVIEWS
1	1.30107
2	.
3	.63421
4	-.03265
5	1.96793
6	-.69950
7	-.03265
8	.63421
9	-.03265
10	.

12 The person in the first row has a z-score of .09, which indicates that this person has more education (14) than the sample's average (13.73).

	ZEDUC
1	.09018
2	-1.25467
3	.76261
4	.76261
5	1.43503
6	.76261
7	-.24603
8	-.58224
9	-1.92709
10	-.58224

IN THIS CHAPTER

» Assessing relationships with the chi-square test of independence

» Comparing column proportions

» Viewing clustered bar charts

Chapter 7
Testing Relationships with the Chi-Square Test of Independence

Cross tabulations display the joint distribution of two or more categorical variables, which allows researchers to focus on both descriptive and causal relationships between the variables. When you examine a table with categorical variables, you want to know whether an observed relationship is likely to exist in the target population or is due to random sampling variation. To determine whether a relationship between two or more variables is statistically significant in a cross tabulation, you use statistical tests. Otherwise, you might make decisions based on observed category percentage differences that are not likely to exist in a population.

Cross tabulations are commonly used to explore how demographic characteristics are related to attitudes and behaviors. They are used also to see how one attitude is related to another. For example, you might want to know whether

- Satisfaction with an instructor in a training workshop was related to satisfaction with the course material
- Eating more often at fast-food restaurants was related to more frequent shopping at convenience stores
- Certain types of people are more likely to buy laptops versus desktop computers

In this chapter, you perform a cross tabulation and then use the chi-square test of independence to see whether a statistically significant relationship exists between two categorical variables. Given that a statistically significant relationship exists, you then use the compare column proportions test to determine which groups differ from each other. Finally, to support your analysis, you learn how to display cross tabulations graphically.

Running the Chi-Square Test of Independence

The purpose of the *chi-square test of independence* is to study the relationship between two or more categorical variables to determine whether one category of a variable is more likely to be associated with a category of another variable. Going back to the idea of hypothesis testing, you can set up two hypotheses:

- **Null hypothesis:** Variables are not related to each other (variables are independent).
- **Alternative hypothesis:** Variables are related to each other (variables are associated).

Here's how to perform the chi-square test of independence:

1. **Choose File ⇨ Open ⇨ Data and load the GSS2018.sav file.**

   You can download the file from the book's companion website at `www.dummies.com/go/spssstatisticsworkbookfd`. This file contains data from the General Social Survey (GSS), a nationally representative survey of adults in the United States that collects data on contemporary opinions, attitudes, and behaviors.

2. **Choose Analyze ⇨ Descriptive Statistics ⇨ Crosstabs.**
3. **Select the AFTERLIF (belief in life after death) variable, and place it in the Row(s) box.**
4. **Select SEX, and place it in the Column(s) box.**

   TIP

   Although you can place the variables in either the Rows or Columns box, it's customary to place the independent variable in the column of the cross tabulation table.

5. **Click the Cells button.**
6. **In the Percentages area, click the Column check box, and then click Continue.**

   REMEMBER

   Although you can request row and column percentages, most researchers request percentages based on the independent variable, which as mentioned is typically placed in the column dimension.

7. **Click the Statistics button.**
8. **Click the Chi-Square check box, and then click Continue.**
9. **Click OK.**

The cross tabulation table shows the relationship between the variables (see Figure 7-1). Each cell in the table represents a unique combination of the variables' values. For example, the first cell in the cross tabulation table shows the number of males that definitely believe in life after death (233).

Belief in life after death * Respondents sex Crosstabulation

			Respondents sex MALE	FEMALE	Total
Belief in life after death	YES, DEFINITELY	Count	233	429	662
		% within Respondents sex	50.3%	64.4%	58.6%
	YES, PROBABLY	Count	106	136	242
		% within Respondents sex	22.9%	20.4%	21.4%
	NO, PROBABLY NOT	Count	66	60	126
		% within Respondents sex	14.3%	9.0%	11.2%
	NO, DEFINITELY NOT	Count	58	41	99
		% within Respondents sex	12.5%	6.2%	8.8%
Total		Count	463	666	1129
		% within Respondents sex	100.0%	100.0%	100.0%

FIGURE 7-1: The cross tabulation table.

TIP

Although looking at counts is useful, it's usually much easier to detect patterns by examining percentages. This is why you clicked the Column check box in the Cell Display dialog.

Looking at the first column in the cross tabulation table, you can see that 50.3% of males and 64.4% of females definitely believe in life after death. It seems as though females are more likely than males to definitely believe in life after death. And 12.5% of males but only 6.2% of females definitely do not believe in life after death.

These differences in percentages would certainly lead you to conclude that sex is related to belief in life after death. But how do you know if these differences in percentages are real differences or due to chance? To answer this question, you need to perform the chi-square test of independence.

The Chi-Square Tests table shown in Figure 7-2 provides three chi-square values. Concentrate on the Pearson chi-square statistic, which is adequate for almost all purposes. The Pearson chi-square statistic is calculated by testing the difference between the *observed counts* (the number of cases observed in each cross tabulation cell) and the *expected counts* (the number of cases that should have been observed in each cross tabulation cell if no relationship existed between the variables). So, the Pearson chi-square statistic is an indication of misfit between observed minus expected counts.

Chi-Square Tests

	Value	df	Asymptotic Significance (2-sided)
Pearson Chi-Square	29.404[a]	3	.000
Likelihood Ratio	29.158	3	.000
Linear-by-Linear Association	29.277	1	.000
N of Valid Cases	1129		

a. 0 cells (0.0%) have expected count less than 5. The minimum expected count is 40.60.

FIGURE 7-2: The Chi-Square Tests table.

The number of degrees of freedom (df) is determined by the number of cells in the table. The actual chi-square value (here, 29.404) is used with df to calculate the significance for the chi-square statistic. You can see the probability value for the chi-square statistic in the Asymptotic Significance (2-sided) column.

The significance value provides the probability of the null hypothesis being true; the lower the number, the less likely that the variables are unrelated. Analysts often use a cutoff value of 0.05 or lower to determine whether the results are statistically significant. For example, with a cutoff value of 0.05, if the significance value is smaller than 0.05, the null hypothesis is rejected. In this case, the probability of the null hypothesis being true is very small — in fact, it's less than 0.05, so you can reject the null hypothesis and have no choice but to say that you found support for the research hypothesis. Therefore, you can conclude that a relationship exists between the sex and belief in life after death.

REMEMBER

Every statistical test has assumptions. The better you meet these assumptions, the more you can trust the results of the test. The chi-square test of independence assumes the following:

- The variables are categorical (nominal or ordinal).
- Each case is assessed only once (hence, this test is not used in a test-retest scenario).
- The levels of the variables are mutually exclusive (each person can be in only one category of a variable).
- The sample is large enough.

The example meets all these assumptions. The last assumption is assessed by the footnote to the Chi-Square Tests table (refer to Figure 7-2), which notes the number of cells with expected counts less than 5. In the example, you have 0 cells (0%) with expected counts less than 5. Ideally, this is what you want to see because the chi-square is not as reliable when sample sizes are very small (expected counts less than 5).

If more than 20% of the cells have expected counts less than 5, consider increasing your sample size if you can or reducing the number of cells in your cross tabulation table (by combining or removing categories). Otherwise, you don't have enough data to trust the result of the analysis.

See the following for an example of running a chi-square test of independence.

EXAMPLE

Q. Using the GSS2018.sav file, run a chi-square test of independence assessing the relationship between HAPMAR (happiness of marriage) and SEX. Is one of the genders more likely than the other to be happy in their marriage? Did you meet the assumptions?

A. The percentages of males and females are very similar in the happiness categories. In fact, the chi-square test of independence shows that we do not have significant differences between the genders with regard to happiness of marriage, therefore both genders are equally happy in their marriage. We also did meet the assumptions, as the expected counts are greater than 5.

Happiness of marriage * Respondents sex Crosstabulation

			Respondents sex		
			MALE	FEMALE	Total
Happiness of marriage	VERY HAPPY	Count	307	331	638
		% within Respondents sex	65.6%	63.2%	64.3%
	PRETTY HAPPY	Count	151	173	324
		% within Respondents sex	32.3%	33.0%	32.7%
	NOT TOO HAPPY	Count	10	20	30
		% within Respondents sex	2.1%	3.8%	3.0%
Total		Count	468	524	992
		% within Respondents sex	100.0%	100.0%	100.0%

Chi-Square Tests

	Value	df	Asymptotic Significance (2-sided)
Pearson Chi-Square	2.577[a]	2	.276
Likelihood Ratio	2.633	2	.268
Linear-by-Linear Association	1.401	1	.236
N of Valid Cases	992		

a. 0 cells (0.0%) have expected count less than 5. The minimum expected count is 14.15.

1. Using the GSS2018.sav file, run a chi-square test of independence assessing the relationship between COMPUSE (does respondent use computer) and BORN (was respondent born in this country). Are there differences in computer usage based on where respondents were born? Did you meet the assumptions?

2. Using the GSS2018.sav file, run a chi-square test of independence assessing the relationship between CLASS (subjective class identification) and BORN (was respondent born in this country). Are there differences in social class based on where respondents were born? Did you meet the assumptions?

3 Using the GSS2018.sav file, run a chi-square test of independence assessing the relationship between AFTERLIF (belief in life after death) and DEGREE (respondent's highest degree). Are there differences in belief in life after death based on respondents' highest degree? Did you meet the assumptions?

4 Using the GSS2018.sav file, run a chi-square test of independence assessing the relationship between CLASS (subjective class identification) and DEGREE (respondent's highest degree). Are there differences in social class based on respondents' highest degree? Did you meet the assumptions?

Comparing Column Proportions

After determining that a relationship exists between two variables, the next step is to determine the nature of the relationship — that is, which groups differ from each other. You can determine which groups differ from each other by comparing column proportions as follows:

1. **Choose File ⇨ Open ⇨ Data and load the GSS2018.sav file.**

   You can download the file from the book's companion website at `www.dummies.com/go/spssstatisticsworkbookfd`.

2. **Choose Analyze ⇨ Descriptive Statistics ⇨ Crosstabs.**
3. **Select the AFTERLIF (belief in life after death) variable, and place it in the Row(s) box.**
4. **Select SEX, and place it in the Column(s) box.**
5. **Click the Cells button.**
6. **In the Percentages area, click the Column check box.**

7. **Select the Compare Column Proportions check box.**
8. **Select the Adjust P-Values (Bonferroni Method) check box, and then click Continue.**

   The compare column proportions test (also called the z-test) determines which groups differ from each other. The Adjust P-Values (Bonferroni Method) option is a technical adjustment that controls for Type I error rates.

9. **Click the Statistics button.**
10. **Select the Chi-Square check box, and then click Continue.**
11. **Click OK.**

A modified version of the cross tabulation table now includes the column proportions test notations. Subscript letters are assigned to the categories of the column variable. For each pair of columns, the column proportions are compared using a z-test. If a pair of values is significantly different, different subscript letters are displayed in each cell.

As you can see in Figure 7-3, the different subscript letters indicate that the proportion of males who definitely believe in life after death (50.3%) is smaller and significantly different, according to the z-test, than the proportion of females who definitely believe in life after death (64.4%). In addition, the proportion of females who probably do not believe in life after death (9.0%) is significantly smaller than the proportion of males who probably do not believe in life after death (14.3%). Finally, the proportion of females who definitely do not believe in life after death (6.2%) is significantly smaller than the proportion of males who definitely do not believe in life after death (12.5%). In other words, because the groups have different subscripts, they are significantly different from each other.

Belief in life after death * Respondents sex Crosstabulation

			Respondents sex		
			MALE	FEMALE	Total
Belief in life after death	YES, DEFINITELY	Count	233_a	429_b	662
		% within Respondents sex	50.3%	64.4%	58.6%
	YES, PROBABLY	Count	106_a	136_a	242
		% within Respondents sex	22.9%	20.4%	21.4%
	NO, PROBABLY NOT	Count	66_a	60_b	126
		% within Respondents sex	14.3%	9.0%	11.2%
	NO, DEFINITELY NOT	Count	58_a	41_b	99
		% within Respondents sex	12.5%	6.2%	8.8%
Total		Count	463	666	1129
		% within Respondents sex	100.0%	100.0%	100.0%

Each subscript letter denotes a subset of Respondents sex categories whose column proportions do not differ significantly from each other at the .05 level.

FIGURE 7-3: The cross tabulation table with the compare column proportions test.

See the following for an example of comparing column proportions.

EXAMPLE

Q. Using the GSS2018.sav file, rerun the chi-square test depicting the relationship between HAPMAR (happiness of marriage) and SEX. If you have a statistically significant result, compare column proportions to determine which groups differ from each other. Summarize the findings.

A. We don't have a statistically significant result. Therefore, we do not need to compare column proportions because the sexes do not differ with regard to martial happiness.

Happiness of marriage * Respondents sex Crosstabulation

			Respondents sex		
			MALE	FEMALE	Total
Happiness of marriage	VERY HAPPY	Count	307a	331a	638
		% within Respondents sex	65.6%	63.2%	64.3%
	PRETTY HAPPY	Count	151a	173a	324
		% within Respondents sex	32.3%	33.0%	32.7%
	NOT TOO HAPPY	Count	10a	20a	30
		% within Respondents sex	2.1%	3.8%	3.0%
Total		Count	468	524	992
		% within Respondents sex	100.0%	100.0%	100.0%

Each subscript letter denotes a subset of Respondents sex categories whose column proportions do not differ significantly from each other at the .05 level.

5 Using the GSS2018.sav file, rerun the chi-square test depicting the relationship between COMPUSE (does respondent use computer) and BORN (was respondent born in this country). If you have a statistically significant result, compare column proportions to determine which groups differ from each other. Summarize the findings.

6 Using the GSS2018.sav file, rerun the chi-square test depicting the relationship between CLASS (subjective class identification) and BORN (was respondent born in this country). If you have a statistically significant result, compare column proportions to determine which groups differ from each other. Summarize the findings.

7 Using the GSS2018.sav file, rerun the chi-square test depicting the relationship between AFTERLIF (belief in life after death) and DEGREE (respondent's highest degree). If you have a statistically significant result, compare column proportions to determine which groups differ from each other. Summarize the findings.

8 Using the GSS2018.sav file, rerun the chi-square test depicting the relationship between CLASS (subjective class identification) and DEGREE (respondent's highest degree). If you have a statistically significant result, compare column proportions to determine which groups differ from each other. Summarize the findings.

Creating a Clustered Bar Chart

For presentations, a graph showing the relationship between two categorical variables is often useful, and a *clustered bar chart* is the most effective graph for displaying the results of a cross tabulation.

In this section, you create a clustered bar chart of the cross tabulation table of belief in life after death and sex. You want to see how belief in life after death varies across sex, so you use sex as the clustering variable. This setup is equivalent to the percentage you requested in the cross tabulation table.

Follow these steps to create a clustered bar chart:

1. **Choose File ⇨ Open ⇨ Data and load the GSS2018.sav file.**

   You can download the file from the book's companion website at `www.dummies.com/go/spssstatisticsworkbookfd`.

2. **Choose Graphs ⇨ Chart Builder.**
3. **In the Choose From list, select Bar.**

4. **Select the second graph image (with the Clustered Bar tooltip) and drag it to the panel at the top of the window.**
5. **Select the SEX variable, and place it in the Cluster on X: Set Color box.**
6. **Select the AFTERLIF variable, and place it in the X-Axis box.**
7. **In the Statistics Area of the Element Properties tab, change Statistic to Percentage().**
8. **Click the Set Parameters button and select Total for Each Legend Variable Category (Same Fill Color).**
9. **Click Continue, and then click OK.**

   The graph in Figure 7-4 appears.

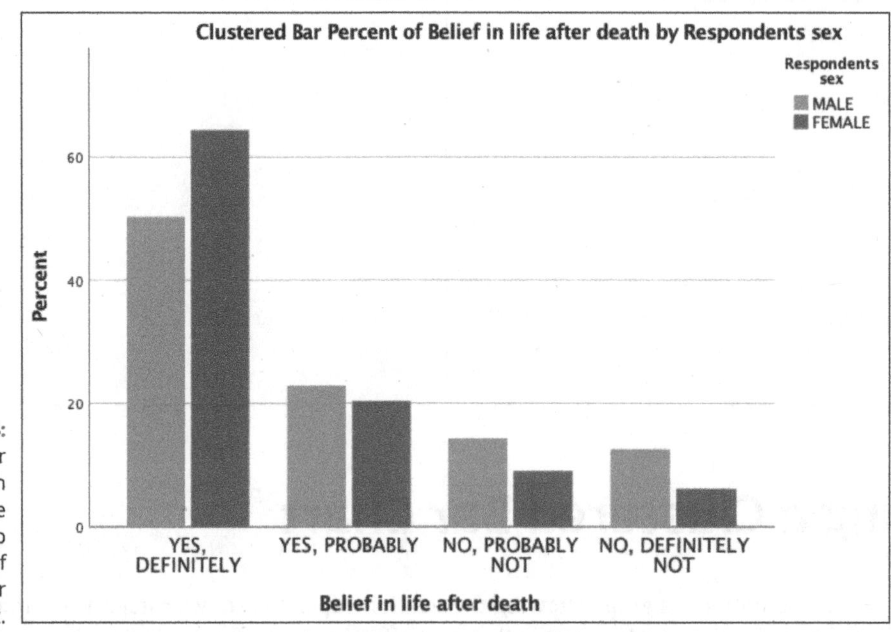

FIGURE 7-4: A cluster bar graph displaying the relationship between belief in life after death and sex.

A cross tabulation table with percentages based on sex lets you compare percentages of that variable within categories of belief in life after death, which is mirrored in the clustered bar chart. The height of the bars is significantly different for three of the four categories of the AFTERLIF (belief in life after death) variable.

See the following for an example of making a clustered bar chart.

EXAMPLE

Q. Using the GSS2018.sav file, create a clustered bar chart depicting the relationship between HAPMAR (happiness of marriage) and SEX. Describe your findings.

A. When comparing the sexes, the bars are quite similar for every category marital happiness, again showing that the genders are equally happy in their marriage.

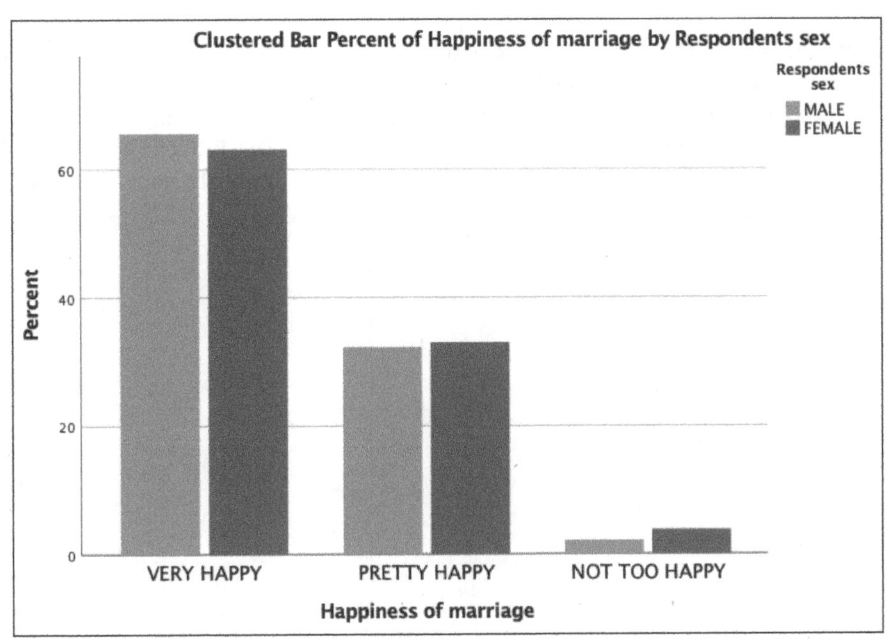

9. Using the GSS2018.sav file, create a clustered bar chart depicting the relationship between COMPUSE (does respondent use computer) and BORN (was respondent born in this country). Describe your findings.

10. Using the GSS2018.sav file, create a clustered bar chart depicting the relationship between CLASS (subjective class identification) and BORN (was respondent born in this country). Describe your findings.

11. Using the GSS2018.sav file, create a clustered bar chart depicting the relationship between AFTERLIF (belief in life after death) and DEGREE (respondent's highest degree). Describe your findings.

12. Using the GSS2018.sav file, create a clustered bar chart depicting the relationship between CLASS (subjective class identification) and DEGREE (respondent's highest degree). Describe your findings.

Answers to Problems in Testing Relationships with the Chi-Square Test of Independence

1 The percentages of those born and not born in this country are different with regard to computer usage. The chi-square test of independence shows that we do have statistically significant differences between where respondents are born and using a computer because the significance value is less than .05. We also did meet the assumptions because the expected counts are greater than 5.

R use computer * Was R born in this country Crosstabulation

			Was R born in this country		
			YES	NO	Total
R use computer	YES	Count	1098	148	1246
		% within Was R born in this country	81.6%	69.8%	80.0%
	NO	Count	248	64	312
		% within Was R born in this country	18.4%	30.2%	20.0%
Total		Count	1346	212	1558
		% within Was R born in this country	100.0%	100.0%	100.0%

Chi-Square Tests

	Value	df	Asymptotic Significance (2-sided)	Exact Sig. (2-sided)	Exact Sig. (1-sided)
Pearson Chi-Square	15.826[a]	1	.000		
Continuity Correction[b]	15.100	1	.000		
Likelihood Ratio	14.507	1	.000		
Fisher's Exact Test				.000	.000
Linear-by-Linear Association	15.816	1	.000		
N of Valid Cases	1558				

a. 0 cells (0.0%) have expected count less than 5. The minimum expected count is 42.45.

b. Computed only for a 2x2 table

2 The percentages of those born and not born in this country are different with regard to social class. The chi-square test of independence shows that we do have statistically significant differences between where respondents were born and social class because the significance value is less than .05. We also did meet the assumptions because the expected counts are greater than 5.

Subjective class identification * Was R born in this country Crosstabulation

			Was R born in this country		
			YES	NO	Total
Subjective class identification	LOWER CLASS	Count	187	24	211
		% within Was R born in this country	9.2%	8.0%	9.0%
	WORKING CLASS	Count	861	159	1020
		% within Was R born in this country	42.4%	53.0%	43.7%
	MIDDLE CLASS	Count	906	113	1019
		% within Was R born in this country	44.6%	37.7%	43.7%
	UPPER CLASS	Count	79	4	83
		% within Was R born in this country	3.9%	1.3%	3.6%
Total		Count	2033	300	2333
		% within Was R born in this country	100.0%	100.0%	100.0%

Chi-Square Tests

	Value	df	Asymptotic Significance (2-sided)
Pearson Chi-Square	14.831[a]	3	.002
Likelihood Ratio	15.901	3	.001
Linear-by-Linear Association	6.160	1	.013
N of Valid Cases	2333		

a. 0 cells (0.0%) have expected count less than 5. The minimum expected count is 10.67.

(3) The percentages of the degree groups are similar in the belief in life after death categories. The chi-square test of independence shows that we do not have statistically significant differences between the degree groups with regard to belief in life after death. We also did meet the assumptions because the expected counts are greater than 5.

Belief in life after death * R's highest degree Crosstabulation

			R's highest degree					
			LT HIGH SCHOOL	HIGH SCHOOL	JUNIOR COLLEGE	BACHELOR	GRADUATE	Total
Belief in life after death	YES, DEFINITELY	Count	74	344	50	133	61	662
		% within R's highest degree	57.8%	58.8%	59.5%	59.6%	56.0%	58.6%
	YES, PROBABLY	Count	27	124	19	53	19	242
		% within R's highest degree	21.1%	21.2%	22.6%	23.8%	17.4%	21.4%
	NO, PROBABLY NOT	Count	14	65	11	22	14	126
		% within R's highest degree	10.9%	11.1%	13.1%	9.9%	12.8%	11.2%
	NO, DEFINITELY NOT	Count	13	52	4	15	15	99
		% within R's highest degree	10.2%	8.9%	4.8%	6.7%	13.8%	8.8%
Total		Count	128	585	84	223	109	1129
		% within R's highest degree	100.0%	100.0%	100.0%	100.0%	100.0%	100.0%

Chi-Square Tests

	Value	df	Asymptotic Significance (2-sided)
Pearson Chi-Square	8.547[a]	12	.741
Likelihood Ratio	8.497	12	.745
Linear-by-Linear Association	.044	1	.834
N of Valid Cases	1129		

a. 0 cells (0.0%) have expected count less than 5. The minimum expected count is 7.37.

(4) The percentages of the degree groups are very different with regard to social class. In fact, the chi-square test of independence shows that we do have statistically significant differences between degree groups and social class, as the significance value is less than .05. We also did meet the assumptions, as the expected counts are greater than 5.

Subjective class identification * R's highest degree Crosstabulation

			R's highest degree					
			LT HIGH SCHOOL	HIGH SCHOOL	JUNIOR COLLEGE	BACHELOR	GRADUATE	Total
Subjective class identification	LOWER CLASS	Count	48	134	14	10	5	211
		% within R's highest degree	18.7%	11.4%	7.2%	2.2%	2.0%	9.0%
	WORKING CLASS	Count	124	602	107	141	46	1020
		% within R's highest degree	48.2%	51.4%	54.9%	30.4%	18.8%	43.7%
	MIDDLE CLASS	Count	78	411	72	288	170	1019
		% within R's highest degree	30.4%	35.1%	36.9%	62.1%	69.4%	43.7%
	UPPER CLASS	Count	7	25	2	25	24	83
		% within R's highest degree	2.7%	2.1%	1.0%	5.4%	9.8%	3.6%
Total		Count	257	1172	195	464	245	2333
		% within R's highest degree	100.0%	100.0%	100.0%	100.0%	100.0%	100.0%

Chi-Square Tests

	Value	df	Asymptotic Significance (2-sided)
Pearson Chi-Square	295.560[a]	12	.000
Likelihood Ratio	302.557	12	.000
Linear-by-Linear Association	245.633	1	.000
N of Valid Cases	2333		

a. 0 cells (0.0%) have expected count less than 5. The minimum expected count is 6.94.

(5) **We have a statistically significant result. The percentages of those born in this country (81.6%) and not born in this country (69.8%) are different with regard to computer usage. However, in this case you technically do not need to use the compare column proportions test because there are only two categories for the independent and dependent variables. When you have this 2 x 2 cross tabulation, you know that both groups differ from each other if you have a statistically significant result.**

R use computer * Was R born in this country Crosstabulation

			Was R born in this country		
			YES	NO	Total
R use computer	YES	Count	1098a	148b	1246
		% within Was R born in this country	81.6%	69.8%	80.0%
	NO	Count	248a	64b	312
		% within Was R born in this country	18.4%	30.2%	20.0%
Total		Count	1346	212	1558
		% within Was R born in this country	100.0%	100.0%	100.0%

Each subscript letter denotes a subset of Was R born in this country categories whose column proportions do not differ significantly from each other at the .05 level.

(6) **We have a statistically significant result. The percentages of those born and not born in this country are different with regard to social class. In general, those born in this country are significant less likely to be in the working class than those not born in this country. In addition, those born in this country are significantly more likely to be in the middle and upper class than those not born in this country.**

Subjective class identification * Was R born in this country Crosstabulation

			Was R born in this country		
			YES	NO	Total
Subjective class identification	LOWER CLASS	Count	187_a	24_a	211
		% within Was R born in this country	9.2%	8.0%	9.0%
	WORKING CLASS	Count	861_a	159_b	1020
		% within Was R born in this country	42.4%	53.0%	43.7%
	MIDDLE CLASS	Count	906_a	113_b	1019
		% within Was R born in this country	44.6%	37.7%	43.7%
	UPPER CLASS	Count	79_a	4_b	83
		% within Was R born in this country	3.9%	1.3%	3.6%
Total		Count	2033	300	2333
		% within Was R born in this country	100.0%	100.0%	100.0%

Each subscript letter denotes a subset of Was R born in this country categories whose column proportions do not differ significantly from each other at the .05 level.

7 We don't have a statistically significant result. Therefore, we do not need to compare column proportions because the degree groups do not differ with regard to belief in life after death. That is, the percentages of the degree groups are very similar in the belief in life after death categories.

Belief in life after death * R's highest degree Crosstabulation

			R's highest degree					
			LT HIGH SCHOOL	HIGH SCHOOL	JUNIOR COLLEGE	BACHELOR	GRADUATE	Total
Belief in life after death	YES, DEFINITELY	Count	74_a	344_a	50_a	133_a	61_a	662
		% within R's highest degree	57.8%	58.8%	59.5%	59.6%	56.0%	58.6%
	YES, PROBABLY	Count	27_a	124_a	19_a	53_a	19_a	242
		% within R's highest degree	21.1%	21.2%	22.6%	23.8%	17.4%	21.4%
	NO, PROBABLY NOT	Count	14_a	65_a	11_a	22_a	14_a	126
		% within R's highest degree	10.9%	11.1%	13.1%	9.9%	12.8%	11.2%
	NO, DEFINITELY NOT	Count	13_a	52_a	4_a	15_a	15_a	99
		% within R's highest degree	10.2%	8.9%	4.8%	6.7%	13.8%	8.8%
Total		Count	128	585	84	223	109	1129
		% within R's highest degree	100.0%	100.0%	100.0%	100.0%	100.0%	100.0%

Each subscript letter denotes a subset of R's highest degree categories whose column proportions do not differ significantly from each other at the .05 level.

(8) We have a statistically significant result. The percentages of degree groups are different with regard to social class. In general, those with more education are significant less likely to be in the lower- and working-class categories, and those with more education are significantly more likely to be in the middle- and upper-class categories.

Subjective class identification * R's highest degree Crosstabulation

			R's highest degree					
			LT HIGH SCHOOL	HIGH SCHOOL	JUNIOR COLLEGE	BACHELOR	GRADUATE	Total
Subjective class identification	LOWER CLASS	Count	48_{a}	134_{b}	$14_{b,c}$	10_{d}	$5_{c,d}$	211
		% within R's highest degree	18.7%	11.4%	7.2%	2.2%	2.0%	9.0%
	WORKING CLASS	Count	124_{a}	602_{a}	107_{a}	141_{b}	46_{c}	1020
		% within R's highest degree	48.2%	51.4%	54.9%	30.4%	18.8%	43.7%
	MIDDLE CLASS	Count	78_{a}	411_{a}	72_{a}	288_{b}	170_{b}	1019
		% within R's highest degree	30.4%	35.1%	36.9%	62.1%	69.4%	43.7%
	UPPER CLASS	Count	$7_{a,b}$	25_{b}	$2_{a,b}$	$25_{a,c}$	24_{c}	83
		% within R's highest degree	2.7%	2.1%	1.0%	5.4%	9.8%	3.6%
Total		Count	257	1172	195	464	245	2333
		% within R's highest degree	100.0%	100.0%	100.0%	100.0%	100.0%	100.0%

Each subscript letter denotes a subset of R's highest degree categories whose column proportions do not differ significantly from each other at the .05 level.

(9) We have a statistically significant result. These differences in percentages are shown in the clustered bar chart, with the bars for birth location quite different with regard to computer usage.

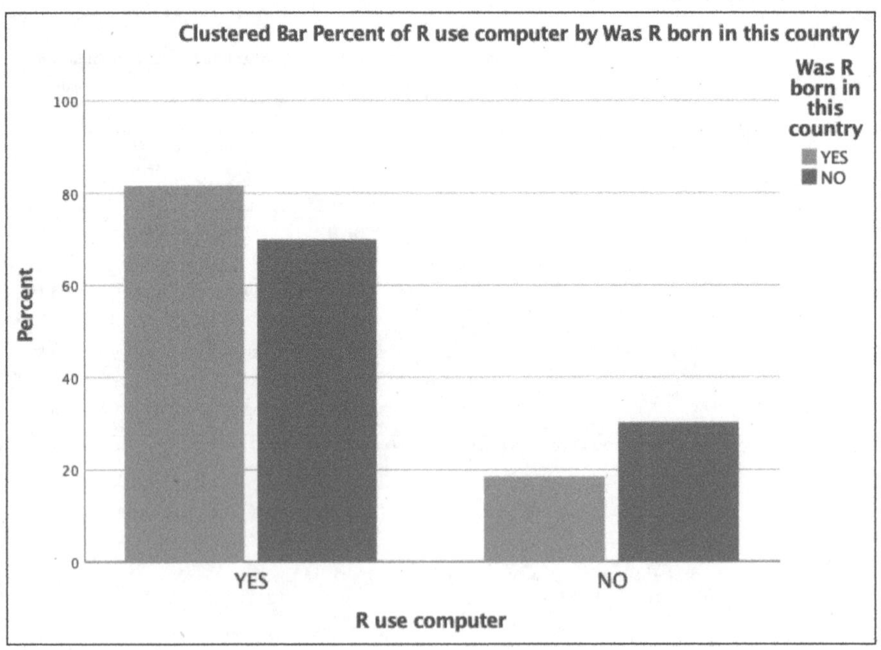

10 We have a statistically significant result. These differences in percentages are shown in the clustered bar chart, with the bars for birth location quite different with regard to social class.

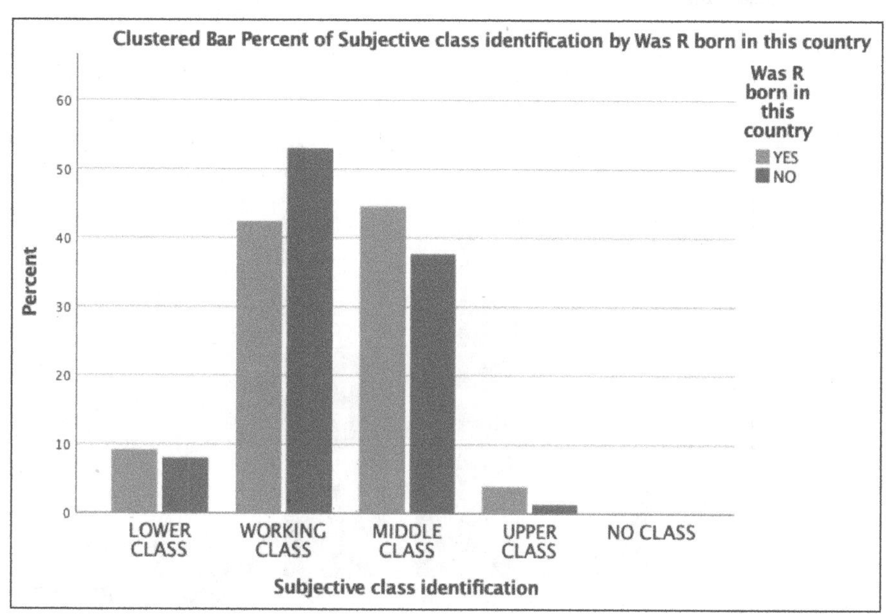

11 When comparing the degree groups, the bars are similar for every category of belief in life after death.

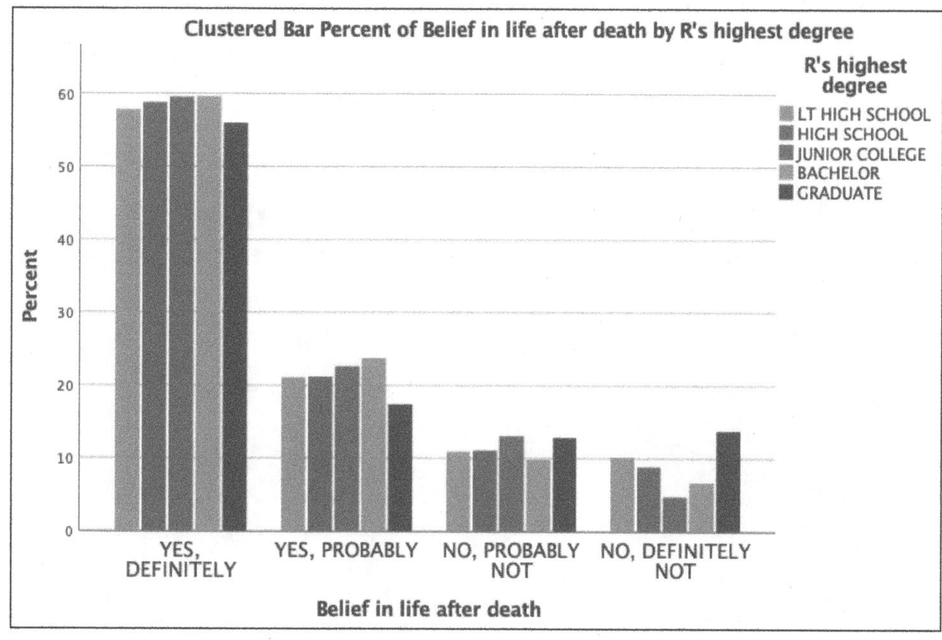

12 We have a statistically significant result. These differences in percentages are shown in the clustered bar chart, with the bars for the degree groups quite different with regard to social class.

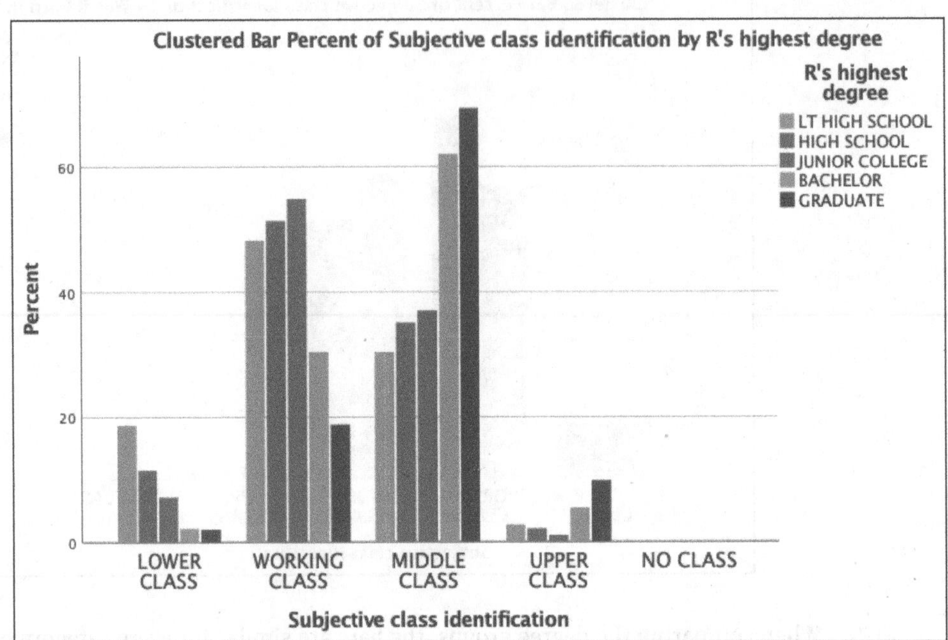

IN THIS CHAPTER

» Comparing groups with the independent-samples t-test

» Comparing time points or conditions with the paired-samples t-test

» Viewing error bar charts

Chapter 8

Comparing Two Groups with T-Tests

In this chapter, we explain how to compare two different groups on a continuous outcome variable. The independent-samples t-test determines if a significant difference exists between two different groups. For example, you might want to know whether one group of customers purchases more items, on average, than a second group of customers; or whether drug A reduces depression levels more than drug B; or whether student test scores in one class are higher than in a second class.

You also learn how to compare a group of people who have been assessed at two different points in time or under two different conditions, such as before and after an intervention. The paired-samples t-test determines if there's a significant difference or change from one point in time or condition to another. For example, you might want to know whether measures administered before and after some type of treatment differ; or whether differences in ratings exist between two competing products; or whether satisfaction changes after a special customer care program is implemented.

Finally, you use error bar charts to graphically display the results of both types of t-tests.

Running the Independent-Samples T-Test Procedure

The purpose of the independent-samples t-test is to examine whether the means of two separate groups differ from each other on a continuous dependent variable (for example, females and males on income). For example, you might compare two different marketing campaigns to determine their effect on sales.

You can set up two hypotheses:

- **Null hypothesis:** The means of the two groups will be the same.
- **Alternative hypothesis:** The means of the two groups will differ from each other.

Here's how to perform an independent-samples T-test:

1. **Choose File ⇨ Open ⇨ Data and load the GSS2018.sav file.**

   You can download the file from the book's companion website at `www.dummies.com/go/spssstatisticsworkbookfd`. This file contains data from the General Social Survey (GSS), a nationally representative survey of adults in the United States that collects data on contemporary opinions, attitudes, and behaviors.

2. **Choose Analyze ⇨ Compare Means ⇨ Independent-Samples T Test.**

   The Independent-Samples T Test dialog appears.

3. **Select the AGEKDBRN variable (age when respondents had their first child) and place it in the Test Variable(s) box.**

   Continuous variables are placed in the Test Variable(s) box.

4. **Select the BORN variable (whether respondents were born in the US or another country) and place it in the Grouping Variable box.**

   Categorical independent variables are placed in the Grouping Variable box. Note that you can have only one independent variable and that SPSS requires you to indicate which groups are to be compared.

5. **Click the Define Groups button.**

   The Define Groups dialog appears.

6. **In the Group 1 box, type** 1; **in the Group 2 box, type** 2.

TIP

   If the independent variable is continuous, you can specify a cut-off value to define the two groups. Cases less than or equal to the cut point go into the first group, and cases greater than the cut point are in the second group. Also, if the independent variable is categorical but has more than two categories, you can still use it by specifying only two categories to compare in an analysis.

7. **Click Continue.**

   You're returned to the Independent-Samples T Test dialog.

8. **Click OK.**

The Group Statistics table shown in Figure 8-1 provides sample sizes, means, standard deviations, and standard errors for the two groups on each of the dependent variables.

Group Statistics

	Was R born in this country	N	Mean	Std. Deviation	Std. Error Mean
R's age when 1st child born	YES	1435	24.25	5.727	.151
	NO	231	24.61	5.810	.382

FIGURE 8-1: The Group Statistics table.

The average age that people have their first child is about the same for people born and not born in this country. Assessing mean differences is precisely what the independent-samples t-test assesses — whether the differences between the means are significantly different or due to chance.

The Independent Samples Test table, shown in Figure 8-2, displays the result of the independent-samples t-test. To understand how to work with the Independent Samples Test table, you'll need to review the assumptions for conducting an independent-samples t-test.

Independent Samples Test

		Levene's Test for Equality of Variances		t-test for Equality of Means					95% Confidence Interval of the Difference	
		F	Sig.	t	df	Sig. (2-tailed)	Mean Difference	Std. Error Difference	Lower	Upper
R's age when 1st child born	Equal variances assumed	.027	.869	-.891	1664	.373	-.362	.407	-1.160	.436
	Equal variances not assumed			-.881	306.370	.379	-.362	.411	-1.171	.447

FIGURE 8-2: The Independent Samples Test table.

REMEMBER

Every statistical test has assumptions. The better you meet these assumptions, the more you can trust the results of the test. The Independent-Samples T-Test has four assumptions:

- The dependent variable is continuous.
- Only two different groups are compared.
- The dependent variable is normally distributed within each category of the independent variable (normality).
- Similar variation exists within each category of the independent variable (homogeneity of variance).

REMEMBER

The *assumption of homogeneity of variance* says that similar variation exists within each category of the independent variable — in other words, the standard deviation of each group is similar. Violating the assumption of homogeneity of variance is more critical than violating the assumption of normality because when the former occurs, the significance or probability value reported by SPSS is incorrect and the test statistics must be adjusted.

Levene's test for equality of variances assesses the assumption of homogeneity of variance by evaluating the null hypothesis that the dependent variable's variance is the same in the two groups. When Levene's test is not statistically significant (that is, the assumption of equal variance was met), you can continue with the regular independent-samples t-test and use the results from the Equal Variances Assumed row.

When Levene's test is statistically significant (that is, the assumption of equal variance was not met), differences in variation exist between the groups, so you have to make an adjustment to the independent-samples t-test. In this situation, you would use the results from the Equal Variances Not Assumed row. Note that if the assumption of homogeneity of variance is not met, you can still do the test but must apply a correction.

In the left section of the Independent Samples Test table, Levene's test for equality of variances is displayed. The F column displays the actual test result, which is used to calculate the significance level (the Sig. column).

In the example, the assumption of homogeneity of variance was met because the value in the Sig. column is greater than 0.05 (no difference exists in the variation of the groups). Therefore, you can look at the row that specifies that equal variances are assumed.

Now that you've determined whether the assumption of homogeneity of variance was met, you're ready to see if the differences between the means are significantly different or due to random variation. The t column displays the result of the t-test and the df column tells SPSS Statistics how to determine the probability of the t-statistic. The Sig. (2-tailed) column tells you the probability of the null hypothesis being correct. If the probability value is very low (less than 0.05), you can conclude that the means are significantly different.

In Figure 8-2, you can see that no significant differences exist between people born and not born in this country with regard to how old they were when they had their first child. This makes sense given that the means of the two groups are fairly similar.

See the following for an example of running an independent-samples t-test.

EXAMPLE

Q. Using the GSS2018.sav file, run an independent-samples t-test depicting the relationship between HRS1 (hours worked last week) and SEX. Did you meet the assumption of homogeneity of variance? Are there any significant differences between the sexes?

A. The assumption of homogeneity of variance was not met, so you have to look at the Equal variances not assumed row. A statistically significant difference between the sexes was found, with men (44.60) working significantly more hours than women (38.15).

Group Statistics

	Respondents sex	N	Mean	Std. Deviation	Std. Error Mean
Number of hours worked last week	MALE	670	44.60	14.969	.578
	FEMALE	711	38.15	13.276	.498

Independent Samples Test

		Levene's Test for Equality of Variances		t-test for Equality of Means						
									95% Confidence Interval of the Difference	
		F	Sig.	t	df	Sig. (2-tailed)	Mean Difference	Std. Error Difference	Lower	Upper
Number of hours worked last week	Equal variances assumed	9.774	.002	8.476	1379	.000	6.445	.760	4.954	7.937
	Equal variances not assumed			8.446	1336.464	.000	6.445	.763	4.948	7.942

1. Using the GSS2018.sav file, run an independent-samples t-test depicting the relationship between HRSRELAX (hours to relax per day) and SEX. Did you meet the assumption of homogeneity of variance? Are there any significant differences between the sexes?

2. Using the GSS2018.sav file, run an independent-samples t-test depicting the relationship between SEI10 (2010 socioeconomic index) and SEX. Did you meet the assumption of homogeneity of variance? Are there any significant differences between the sexes?

3. Using the GSS2018.sav file, run an independent-samples t-test depicting the relationship between AGE and SEX. Did you meet the assumption of homogeneity of variance? Are there any significant differences between the sexes?

4. Using the GSS2018.sav file, run an independent-samples t-test depicting the relationship between POLVIEWS (political views—higher numbers indicate that the person is more conservative) and DEGREE (compare those with less than a high school degree and those with a graduate degree). Did you meet the assumption of homogeneity of variance? Are there any significant differences between the sexes?

Running the Paired-Samples T-Test Procedure

The purpose of the *paired-samples t-test* is to assess whether means of the same or similar group differ from each other under two separate conditions (for example, before and after an intervention). This test determines if there is a significant difference or change from one point in time or condition to another. For example, you might compare a participant's weight before and after participating in a weight-loss program to see if the program had an effect.

To the extent that an individual's outcomes across the two conditions are related, the paired-samples t-test provides a more powerful statistical analysis (that is, a greater probability of finding true effects if they exist) than the independent-samples t-test because each person serves as their own control.

You can set up two hypotheses:

- **Null hypothesis:** The mean of the difference or change variable will be 0.
- **Alternative hypothesis**: The mean of the difference or change variable will not be 0.

Follow these steps to perform a paired-samples t-test:

1. **Choose File ⇨ Open ⇨ Data and load the GSS2018.sav file.**

   Download the file at `www.dummies.com/go/spssstatisticsworkbookfd`.

2. **Choose Analyze ⇨ Compare Means ⇨ Paired-Samples T Test.**

   The Paired-Samples T Test dialog appears.

3. **Select the CHILDS (number of children) variable and the CHLDIDEL (ideal number of children) variable, and place them in the Paired Variables box.**

   These variables are now in the same row, so they will be compared.

   REMEMBER

   Technically, the order in which you select the pair of variables does not matter; the calculations will be the same. However, SPSS subtracts the second variable from the first. So in terms of presentation, you might want to be careful how SPSS displays the results so as to facilitate reader understanding.

4. **Click OK.**

REMEMBER

Every statistical test has assumptions. The better you meet these assumptions, the more you can trust the results of the test. The paired-samples t-test has three assumptions:

- The dependent variable is continuous.
- The two points in time or conditions that are compared are on the same continuous dependent variable. In other words, the dependent variable should be measured in a consistent format, not, for example, on a seven-point scale the first time and on a ten-point scale the second time.
- The difference scores are normally distributed (normality).

The Paired Samples Statistics table, shown in Figure 8-3, provides sample sizes, means, standard deviations, and standard errors for the two variables. You can see that 1555 people were in this analysis, with an average number of children of 1.94 and an average ideal number of children of 3.30.

FIGURE 8-3: The Paired Samples Statistics table.

Paired Samples Statistics

		Mean	N	Std. Deviation	Std. Error Mean
Pair 1	Number of children	1.94	1555	1.685	.043
	Ideal number of children	3.30	1555	2.019	.051

The null hypothesis is that the two means are equal. The mean difference between actual and ideal number of children is about 1.35. The Paired Samples Test table in Figure 8-4 reports this information along with the sample standard deviation and standard error.

FIGURE 8-4: The Paired Samples Test table.

Paired Samples Test

		Paired Differences							
					95% Confidence Interval of the Difference				
		Mean	Std. Deviation	Std. Error Mean	Lower	Upper	t	df	Sig. (2-tailed)
Pair 1	Number of children - Ideal number of children	-1.351	2.437	.062	-1.472	-1.230	-21.861	1554	.000

The t column displays the result of the t-test, and the df column tells SPSS Statistics how to determine the probability of the t-statistic. The Sig. (2-tailed) column tells you the probability of the null hypothesis being correct. If the probability value is very low (less than 0.05), you can conclude that the means are significantly different. Significant differences exist between the means, so you can conclude that a difference exists between actual and ideal number of children.

See the following for an example of running a paired-samples t-test.

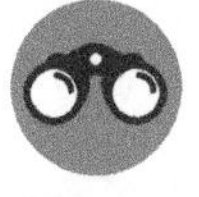
EXAMPLE

Q. Using the GSS2018.sav file, run a paired-samples t-test to determine if differences exist between MAEDUC (number of years of mother's education) and PAEDUC (number of years of father's education). Are there any significant differences?

A. The mean education for fathers (11.90) and mothers (11.91) are practically identical, so there are no significant differences. This result suggests that couples tend to marry partners with similar education.

Paired Samples Statistics

		Mean	N	Std. Deviation	Std. Error Mean
Pair 1	Highest year school completed, mother	11.91	1561	3.810	.096
	Highest year school completed, father	11.90	1561	4.174	.106

Paired Samples Test

		Paired Differences					t	df	Sig. (2-tailed)
		Mean	Std. Deviation	Std. Error Mean	95% Confidence Interval of the Difference				
					Lower	Upper			
Pair 1	Highest year school completed, mother - Highest year school completed, father	.010	3.168	.080	-.148	.167	.120	1560	.905

5 Using the GSS2018.sav file, run a paired-samples t-test to determine if differences exist between MNTHLTH (number of days of poor mental health) and PHYSHLTH (number of days of poor physical health). Are there any significant differences?

6 Using the GSS2018.sav file, run a paired-samples t-test to determine whether differences exist between EMAILHR (number of hours of email per week) and WWWHR (number of hours online per week). Are there any significant differences?

7 The RepMeas1.sav file, which you can download at `www.dummies.com/go/spssstatisticsworkbookfd`, contains data for ten adults who were asked to recall nonsense syllables under four different drugs. Using the RepMeas1.sav file, run a paired-samples t-test to determine whether differences exist between the number of nonsense syllables recalled under drug1 and drug2. Are there any significant differences?

8 Using the RepMeas1.sav file, run a paired-samples t-test to determine whether differences exist between the number of nonsense syllables recalled under drug1 and drug3. Are there any significant differences?

Comparing the Means Graphically

For presentations, it's often useful to show a graph of the results of a t-test. A typical chart for displaying the group sample means is a bar chart. However, many people prefer to use an error bar chart instead because it focuses more on the precision of the estimated mean for each group than the mean itself.

The error bar chart generates a graph depicting the relationship between a continuous variable and a categorical variable, providing a visual sense of how far the groups are separated.

Follow these steps to create a simple error bar chart:

1. **Choose File ➪ Open ➪ Data and load the GSS2018.sav data file.**

   Download the file at `www.dummies.com/go/spssstatisticsworkbookfd`.

2. **Choose Graphs ➪ Chart Builder.**
3. **In the Choose From list, select Bar.**
4. **Select the seventh graph image (the one with the Simple Error Bar tooltip) and drag it to the panel at the top of the window.**
5. **Select the AGEKDBRN (age when respondents had their first child) variable, and place IT in the Y-Axis box.**
6. **Select the born (whether respondents were born in the US or another country) variable, and place it in the X-Axis box.**
7. **Click OK.**

   The graph in Figure 8-5 appears.

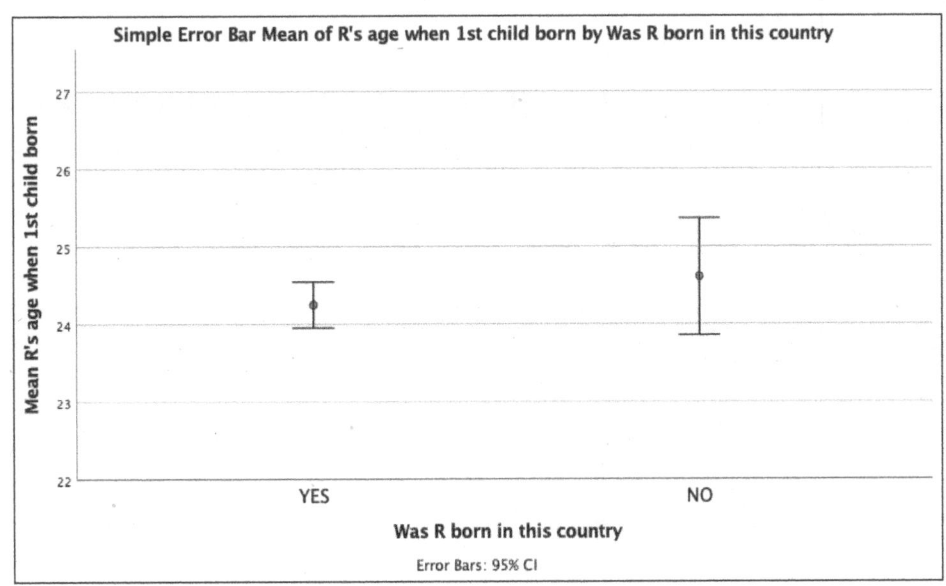

FIGURE 8-5: Error bar graph displaying the age when respondents had their first child and where respondents were born.

This chart represents the mean age when participants had their first child for each location along with 95% confidence intervals. The confidence intervals for the two locations overlap, which is consistent with the result from the t-test and indicates that the groups are not significantly different from each other.

See the following for an example of making an error bar chart.

EXAMPLE

Q. Using the GSS2018.sav file, create an error bar chart depicting the relationship between CHILDS (number of children) and CHLDIDEL (ideal number of children). Describe your findings.

TIP

When creating an error bar chart to depict the findings of a paired-samples t-test, you need to add both continuous variables to the y-axis.

A. No overlap exists between actual and ideal number of children. This result suggests that parents tend to have fewer children than they ideally would like to have.

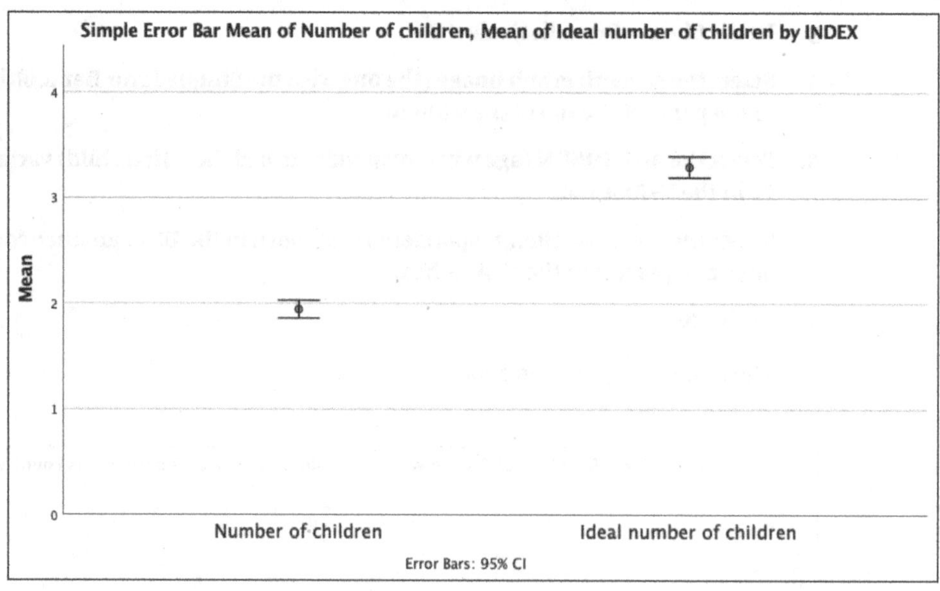

9 Using the GSS2018.sav file, create an error bar chart depicting the relationship between HRS1 (hours worked last week) and SEX. Describe your findings.

10 Using the GSS2018.sav file, create an error bar chart depicting the relationship between SEI10 (2010 socioeconomic index) and SEX. Describe your findings.

11 Using the GSS2018.sav file, create an error bar chart depicting the relationship between MAEDUC (number of years of mother's education) and PAEDUC (number of years of father's education). Describe your findings.

12 Using the GSS2018.sav file, create an error bar chart depicting the relationship between MNTHLTH (number of days of poor mental health) and PHYSHLTH (number of days of poor physical health). Describe your findings.

Answers to Problems in Comparing Two Groups with T-Tests

1. The assumption of homogeneity of variance was met, so you have to look at the Equal variances assumed row. A statistically significant difference was found between the sexes, where men (3.97) had significantly more time to relax per day than women (3.50).

Group Statistics

	Respondents sex	N	Mean	Std. Deviation	Std. Error Mean
Hours per day R have to relax	MALE	676	3.97	2.836	.109
	FEMALE	729	3.50	2.714	.101

Independent Samples Test

		Levene's Test for Equality of Variances		t-test for Equality of Means						
									95% Confidence Interval of the Difference	
		F	Sig.	t	df	Sig. (2-tailed)	Mean Difference	Std. Error Difference	Lower	Upper
Hours per day R have to relax	Equal variances assumed	.069	.792	3.200	1403	.001	.474	.148	.183	.764
	Equal variances not assumed			3.195	1383.327	.001	.474	.148	.183	.765

2. The assumption of homogeneity of variance was not met, so you have to look at the Equal variances not assumed row. A statistically significant difference was not found between the sexes, so the mean SEI for men (47.38) and women (46.58) are similar.

Group Statistics

	Respondents sex	N	Mean	Std. Deviation	Std. Error Mean
R's socioeconomic index (2010)	MALE	1022	47.384	22.3824	.7001
	FEMALE	1226	46.578	23.5561	.6728

Independent Samples Test

		Levene's Test for Equality of Variances		t-test for Equality of Means						
									95% Confidence Interval of the Difference	
		F	Sig.	t	df	Sig. (2-tailed)	Mean Difference	Std. Error Difference	Lower	Upper
R's socioeconomic index (2010)	Equal variances assumed	6.544	.011	.827	2246	.408	.8065	.9755	-1.1064	2.7195
	Equal variances not assumed			.831	2207.972	.406	.8065	.9710	-1.0976	2.7106

3. The assumption of homogeneity of variance was met, so you have to look at the Equal variances assumed row. A statistically significant difference was not found between the sexes, so the mean ages for men (49.26) and women (48.74) are similar.

Group Statistics

	Respondents sex	N	Mean	Std. Deviation	Std. Error Mean
Age of respondent	MALE	1049	49.26	18.066	.558
	FEMALE	1292	48.74	18.060	.502

Independent Samples Test

		Levene's Test for Equality of Variances		t-test for Equality of Means						
									95% Confidence Interval of the Difference	
		F	Sig.	t	df	Sig. (2-tailed)	Mean Difference	Std. Error Difference	Lower	Upper
Age of respondent	Equal variances assumed	.048	.827	.686	2339	.493	.515	.751	-.957	1.987
	Equal variances not assumed			.686	2240.997	.493	.515	.751	-.957	1.987

4 The assumption of homogeneity of variance was not met, so you have to look at the Equal variances not assumed row. A statistically significant difference was not found between graduate (3.77) and less than high school (3.98) degrees.

Group Statistics

	R's highest degree	N	Mean	Std. Deviation	Std. Error Mean
Think of self as liberal or conservative	LT HIGH SCHOOL	232	3.98	1.586	.104
	GRADUATE	243	3.77	1.680	.108

Independent Samples Test

		Levene's Test for Equality of Variances		t-test for Equality of Means						
									95% Confidence Interval of the Difference	
		F	Sig.	t	df	Sig. (2-tailed)	Mean Difference	Std. Error Difference	Lower	Upper
Think of self as liberal or conservative	Equal variances assumed	7.148	.008	1.392	473	.165	.209	.150	-.086	.504
	Equal variances not assumed			1.394	472.939	.164	.209	.150	-.086	.503

5 The mean number of poor mental health days (3.53) and the mean number of poor physical health days (2.62) have statistically significant differences. This result suggests that participants did not have the same number of poor mental and physical health days; they had more days of poor mental health than physical health.

Paired Samples Statistics

		Mean	N	Std. Deviation	Std. Error Mean
Pair 1	Days of poor mental health past 30 days	3.53	1404	6.836	.182
	Days of poor physical health past 30 days	2.62	1404	6.093	.163

Paired Samples Test

		Paired Differences							
					95% Confidence Interval of the Difference				
		Mean	Std. Deviation	Std. Error Mean	Lower	Upper	t	df	Sig. (2-tailed)
Pair 1	Days of poor mental health past 30 days - Days of poor physical health past 30 days	.905	7.548	.201	.510	1.300	4.494	1403	.000

6 The mean number of hours of email per week (7.43) and the mean number of hours online per week (13.91) have statistically significant differences. This result suggests that participants spend more time online than on email.

Paired Samples Statistics

		Mean	N	Std. Deviation	Std. Error Mean
Pair 1	Email hours per week	7.43	1361	11.895	.322
	Www hours per week	13.91	1361	17.143	.465

Paired Samples Test

		Paired Differences					t	df	Sig. (2-tailed)
					95% Confidence Interval of the Difference				
		Mean	Std. Deviation	Std. Error Mean	Lower	Upper			
Pair 1	Email hours per week - Www hours per week	-6.475	17.775	.482	-7.421	-5.530	-13.440	1360	.000

7 The mean number of nonsense syllables recalled under drug1 (26.90) and drug2 (25.70) are practically identical, so there are no significant differences. This result suggests that both drugs are equally effective on memory.

Paired Samples Statistics

		Mean	N	Std. Deviation	Std. Error Mean
Pair 1	drug1	26.90	10	8.279	2.618
	drug2	25.70	10	6.219	1.967

Paired Samples Test

		Paired Differences					t	df	Sig. (2-tailed)
					95% Confidence Interval of the Difference				
		Mean	Std. Deviation	Std. Error Mean	Lower	Upper			
Pair 1	drug1 - drug2	1.200	3.425	1.083	-1.250	3.650	1.108	9	.297

8 The mean number of nonsense syllables recalled under drug1 (26.90) and drug3 (16.10) have statistically significant differences. This result suggests that participants recalled more nonsense syllables when taking drug1 than when taking drug3.

Paired Samples Statistics

		Mean	N	Std. Deviation	Std. Error Mean
Pair 1	drug1	26.90	10	8.279	2.618
	drug3	16.10	10	3.665	1.159

Paired Samples Test

		Paired Differences					t	df	Sig. (2-tailed)
					95% Confidence Interval of the Difference				
		Mean	Std. Deviation	Std. Error Mean	Lower	Upper			
Pair 1	drug1 - drug3	10.800	5.432	1.718	6.914	14.686	6.287	9	.000

9 No overlap exists between the genders with regard to the number of hours worked in the last week. This result suggests that significant differences exist between men and women with regard to the number of hours worked, with men working more hours than women.

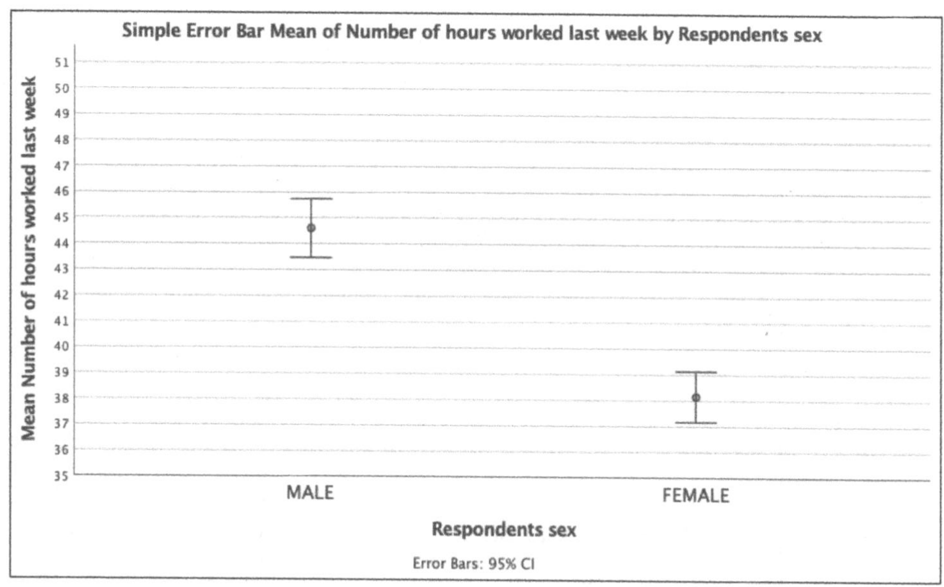

10 An overlap exists between the genders with regard to socioeconomic index. This result suggests that no significant differences exist between men and women regarding the socioeconomic index; in other words, they are equivalent.

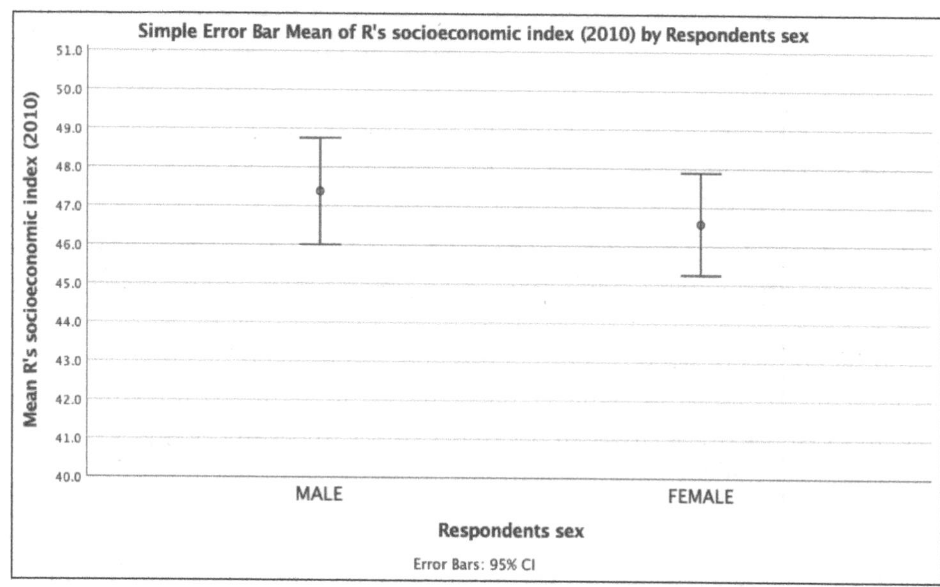

11 Overlap exists between the education of mothers and fathers. This result suggests that no significant differences exist between mothers and fathers with regard to education; in other words, they are equivalent.

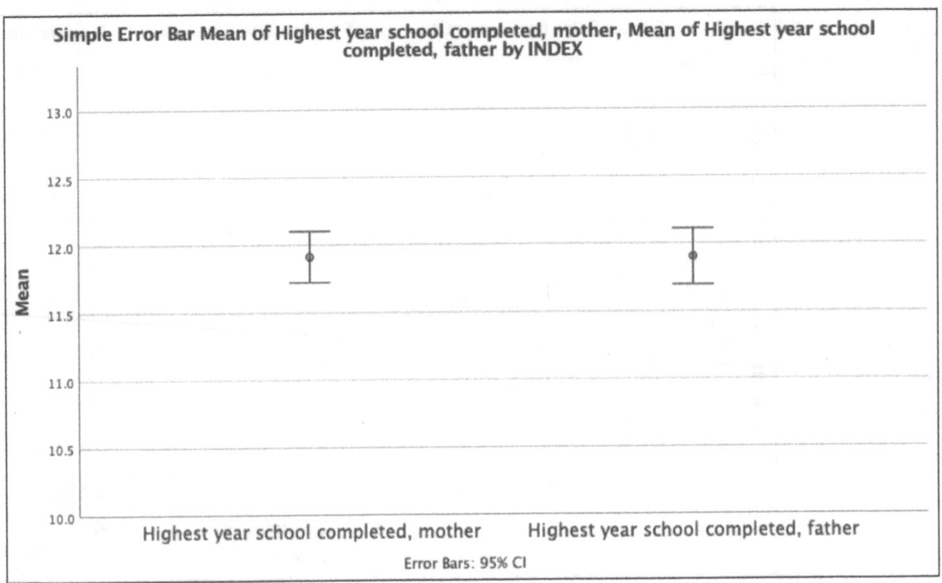

12 No overlap exists between MNTHLTH (number of days of poor mental health) and PHYSHLTH (number of days of poor physical health). This result suggests that significant differences exist because there were higher values for mental health.

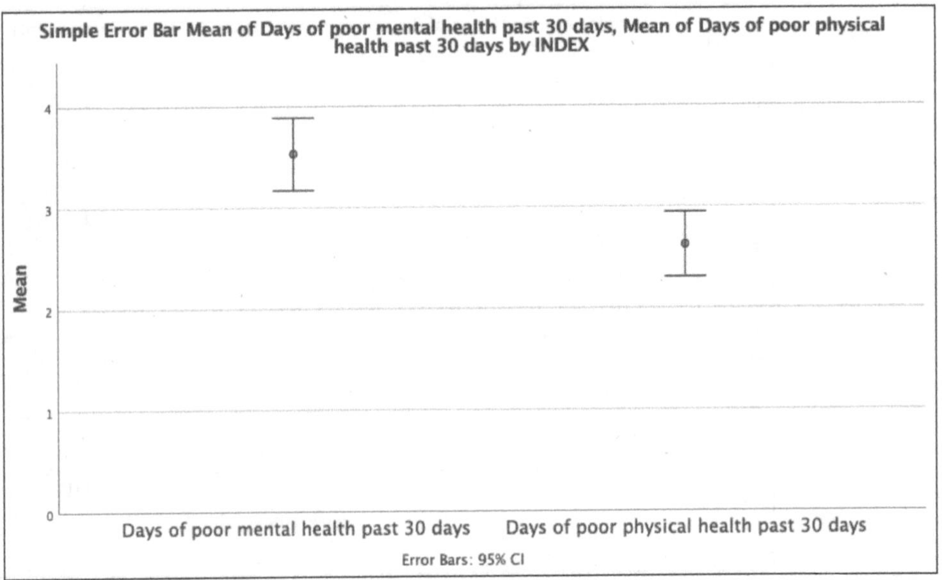

IN THIS CHAPTER

» Comparing groups using One-Way ANOVA

» Using post hoc tests

» Viewing error bar charts

Chapter 9 Comparing More Than Two Groups with ANOVA

This chapter is an extension of Chapter 8, where we discuss t-tests, which are used when the dependent variable is continuous and the independent variable has two categories. In this chapter, we talk about the One-Way Analysis of Variance (ANOVA) procedure, which is a direct extension of the independent samples t-test. One-Way ANOVA is used when you have a continuous dependent variable and the independent variable has two or more categories (for example, control group, treatment group 1, treatment group 2, and so on), thus letting you know which groups score higher or lower than the others. For example, you might want to use ANOVA to know whether customer groups differ in attitude toward a product or service; or different drugs more effectively reduce depression levels; or which promotional campaigns produce the largest sales.

Running the One-Way ANOVA Procedure

ANOVA is a general method of drawing conclusions regarding differences in population means when two or more comparison groups are involved. The independent-samples t-test applies only to the simplest instance (two groups), whereas the One-Way ANOVA procedure can accommodate more complex situations (three or more groups). The purpose of the One-Way ANOVA is to test whether the means of two or more separate groups differ from each other on a continuous dependent variable.

The basic logic underlying significance testing for comparing group means across more than two groups is the same as that for comparing the means of two groups. To make this assessment, the amount of variation among group means *(between-group variation)* is compared to the amount of variation among observations within each group *(within-group variation)*. Assuming that the group means are identical in the population (the null hypothesis), the only source of variation among sample means is the fact that the groups are composed of different individual observations. Thus a ratio of the two sources of variation (between-group/within-group) should be about 1.0 when population differences do not exist. (Note that when the null hypothesis is rejected, it does not follow that all group means differ significantly; the only thing that can be said is that not all group means are the same.)

You can set up two hypotheses:

- **Null hypothesis:** The means of the groups will be the same. Stating the null hypothesis as an equation, we can say that the null hypothesis is true when there is similar variation between and within samples, which results in an F statistic of 1.
- **Alternative hypothesis:** The means of the groups will differ from each other. Stating the alternative hypothesis as an equation, we can say that the null hypothesis is not true when the variation between samples is much larger than the variation within samples, which results in an F statistic larger than 1.

Follow these steps to perform a one-way ANOVA:

1. **Choose File ⇨ Open ⇨ Data and load the GSS2018.sav file.**

   You can download the file from the book's companion website at `www.dummies.com/go/spssstatisticsworkbookfd`. This file contains data from the General Social Survey (GSS), a nationally representative survey of adults in the United States that collects data on contemporary opinions, attitudes, and behaviors.

2. **Choose Analyze ⇨ Compare Means ⇨ One-Way ANOVA.**

   The One-Way ANOVA dialog appears.

3. **Select the AGEKDBRN (age when respondents had their first child) variable, and place it in the Dependent List box.**

   Continuous variables are placed in the Dependent List box.

4. **Select the DEGREE (respondent's highest degree) variable, and place it in the Factor box.**

   The categorical independent variable is placed in the Factor box.

5. **Select the Options button.**

   The Options dialog appears.

6. **Select the Descriptive, Homogeneity of Variance Test, and Welch Test options.**

   Descriptive provides group means and standard deviations. The homogeneity of variance test assesses the assumption of homogeneity of variance. Welch is a robust test that does not assume homogeneity of variance and thus can be used when this assumption is not met.

7. **Click Continue.**
8. **Click OK.**

It's important to begin by determining the number of cases in each group and calculating the means and standard deviations. The first table, which is shown in Figure 9-1, provides this information. The size of the groups ranges from 153 to 829 people. The mean ages of when respondents had their first child varies from 21.10 to 28.73; the One-Way ANOVA procedure will assess if these means differ. The standard deviations vary from 4.67 to 5.53; the test of homogeneity of variance will assess if these standard deviations differ.

Descriptives

R's age when 1st child born

	N	Mean	Std. Deviation	Std. Error	95% Confidence Interval for Mean		Minimum	Maximum
					Lower Bound	Upper Bound		
LT HIGH SCHOOL	209	21.10	4.674	.323	20.46	21.73	14	39
HIGH SCHOOL	829	23.07	5.232	.182	22.72	23.43	12	47
JUNIOR COLLEGE	153	24.67	5.118	.414	23.85	25.48	16	41
BACHELOR	313	27.21	5.532	.313	26.59	27.82	15	44
GRADUATE	162	28.73	5.475	.430	27.89	29.58	14	51
Total	1666	24.30	5.738	.141	24.02	24.57	12	51

FIGURE 9-1: The Descriptives table.

As education increases, the average age at which respondents have their first child also increases. Comparing these means is precisely what the One-Way ANOVA assesses — whether the differences between the means are significantly different from each other or if the differences are due to chance.

REMEMBER

Every statistical test has assumptions. The better you meet these assumptions, the more you can trust the results of the test. The One-Way ANOVA has four assumptions:

- The dependent variable is continuous.
- Two or more different groups are compared.
- The dependent variable is normally distributed within each category of the independent variable (normality).
- Similar variation exists within each category of the independent variable (homogeneity of variance).

The test of homogeneity of variance is shown in Figure 9-2. The null hypothesis here is that the variances are equal, so if the significance level is low enough (less than .05), you reject the null hypothesis and conclude that the variances are not equal and you need to use a correction, such as the Welch test.

Tests of Homogeneity of Variances

		Levene Statistic	df1	df2	Sig.
R's age when 1st child born	Based on Mean	1.708	4	1661	.146
	Based on Median	2.229	4	1661	.064
	Based on Median and with adjusted df	2.229	4	1642.591	.064
	Based on trimmed mean	1.931	4	1661	.103

FIGURE 9-2: The Tests of Homogeneity of Variances table.

In a One-Way ANOVA, similar to the independent-samples t-test, violation of the assumption of homogeneity of variance is more serious than violation of the assumption of normality. And like the independent-samples t-test, the One-Way ANOVA applies a two-step strategy for testing:

- Test the homogeneity of variance assumption.
- If the assumption holds, proceed with the standard test (ANOVA F-test) to test equality of means. If the null hypothesis of equal variances is rejected, use an adjusted F-test to test equality of means.

In our case, we have met the assumption of homogeneity of variance, so we can use the traditional ANOVA table. Most of the information in the ANOVA table (see Figure 9-3) is technical and not directly interpreted. Rather, the summaries are used to obtain the F statistic and, more importantly, the probability value used in evaluating the population differences.

ANOVA

R's age when 1st child born

	Sum of Squares	df	Mean Square	F	Sig.
Between Groups	9247.904	4	2311.976	84.276	.000
Within Groups	45566.832	1661	27.433		
Total	54814.735	1665			

FIGURE 9-3: The ANOVA table.

The standard ANOVA table provides the following information:

- **Sum of squares:** Intermediate summary numbers used in calculating the between-groups (deviations of individual group means around the total sample mean) and within-groups (deviations of individual observations around their respective sample group mean) variances
- **Degrees of freedom (df):** Information about the degrees of freedom, which is related to the number of groups and the number of individual observations within each group
- **Mean square:** Measures of the between-group and within-group variations (sum of squares divided by their respective degrees of freedom)

» **F statistic:** The ratio of between-group variation to within-group variation (close to 1.0 if the null hypothesis is true)

» **Significance (Sig.):** The probability of obtaining the sample F ratio (taking into account the number of groups and sample size), if the null hypothesis is true

Most researchers typically focus on the significance values. And as shown in Figure 9-3, the low significance value leads us to reject the null hypothesis of equal means.

When the condition of equal variances is not met, an adjusted F-test has to be used. SPSS Statistics provides two such tests: Welch and Brown-Forsythe. Both corrections make adjustments by reducing the sample size in the analysis so that the final result is more conservative. You selected Welch, but you can use either test. The Robust Tests of Equality of Means table provides the technical details to compute the significance value.

See the following for an example of running an ANOVA.

EXAMPLE

Q. Using the GSS2018.sav file, run an ANOVA depicting the relationship between CHILDS (number of children) and DEGREE (respondent's highest degree). Did you meet the assumption of homogeneity of variance? Do any significant differences exist among the degree groups?

A. We did not meet the assumption of homogeneity of variance, so you have to use the Welch test. However, we did find a statistically significant result, so differences exist among the degree groups with regard to number of children.

Tests of Homogeneity of Variances

		Levene Statistic	df1	df2	Sig.
Number of children	Based on Mean	13.013	4	2339	.000
	Based on Median	13.032	4	2339	.000
	Based on Median and with adjusted df	13.032	4	2308.325	.000
	Based on trimmed mean	12.297	4	2339	.000

Robust Tests of Equality of Means

Number of children

	Statistic[a]	df1	df2	Sig.
Welch	14.715	4	656.710	.000

a. Asymptotically F distributed.

1. Using the GSS2018.sav file, run an ANOVA depicting the relationship between HRS1 (hours worked last week) and DEGREE (respondent's highest degree). Did you meet the assumption of homogeneity of variance? Do any significant differences exist among the degree groups?

2. Using the GSS2018.sav file, run an ANOVA depicting the relationship between HRSRELAX (hours to relax per day) and DEGREE (respondent's highest degree). Did you meet the assumption of homogeneity of variance? Do any significant differences exist among the degree groups?

3. Using the GSS2018.sav file, run an ANOVA depicting the relationship between POLVIEWS (political views — higher numbers indicate the person is more conservative) and DEGREE (respondent's highest degree). Did you meet the assumption of homogeneity of variance? Do any significant differences exist among the degree groups?

4. Using the GSS2018.sav file, run an ANOVA depicting the relationship between TVHOURS (hours per day watching TV) and DEGREE (respondent's highest degree). Did you meet the assumption of homogeneity of variance? Do any significant differences exist among the degree groups?

Conducting Post Hoc Tests

Post hoc tests are performed only after the overall ANOVA indicates that significant differences exist among the groups so that you can determine specifically which groups differ from each other. A post hoc test compares every possible pair of groups.

However, as more tests are performed, the probability of obtaining at least one false-positive result increases. As an extreme example, with ten groups, you can make 45 pairwise group comparisons (n*(n-1)/2). If you assess each comparison at the .05 level, you would expect to obtain, on average, about 2 (.05*45) false-positive tests.

In an attempt to reduce the false-positive rate when multiple tests of this type are done, statisticians have developed a number of methods such as Bonferroni, Tukey, and Scheffe.

Follow these steps to perform post hoc tests for a One-Way ANOVA:

1. **Choose File ➪ Open ➪ Data and load the GSS2018.sav file.**

   Download the file at `www.dummies.com/go/spssstatisticsworkbookfd`.

2. **Choose Analyze ➪ Compare Means ➪ One-Way ANOVA.**
3. **Select the AGEKDBRN (age when respondents had their first child) variable, and place it in the Dependent List box.**
4. **Select the DEGREE (respondent's highest degree) variable, and place it in the Factor box.**
5. **Select the Options button.**
6. **Select the Descriptive, Homogeneity of Variance Test, and Welch Test options.**
7. **Click Continue.**
8. **Click the Post Hoc button.**

   The Post Hoc Multiple Comparisons dialog appears. Many tests are available due to the trade-offs in the control of Type I error (false positive), statistical power, assumption violations, and statistical distributions. Statisticians don't agree on a single test that is optimal for all situations, but the Bonferroni test is currently the most popular procedure. This procedure is easy to understand because the criterion level for each pairwise test is obtained by dividing the original criterion level (say .05) by the number of pairwise comparisons made. Thus, with five means, and therefore ten pairwise comparisons, each Bonferroni test will be performed at the .05/10, or .005, level.

9. **Select Bonferroni.**
10. **Click Continue.**
11. **Click OK.**

    The One-Way ANOVA is now rerun, along with the post hoc tests.

The Multiple Comparisons table in Figure 9-4 provides all pairwise comparisons. The rows are formed by every possible combination of groups. The Mean Difference (I-J) column contains the sample mean difference between each pair of groups. If this difference is statistically significant at the specified level after applying the post hoc adjustments, an asterisk (*) appears beside the mean difference. The actual significance value for the test appears in the Sig. column. Note that each pairwise comparison appears twice. For each such duplicate pair, the significance value is the same, but the signs are reversed for the mean difference. In our example, all the groups are significantly different from each other at the specified level after applying the post hoc adjustments.

Multiple Comparisons

Dependent Variable: R's age when 1st child born
Bonferroni

(I) R's highest degree	(J) R's highest degree	Mean Difference (I-J)	Std. Error	Sig.	95% Confidence Interval Lower Bound	95% Confidence Interval Upper Bound
LT HIGH SCHOOL	HIGH SCHOOL	-1.977*	.405	.000	-3.12	-.84
	JUNIOR COLLEGE	-3.571*	.557	.000	-5.14	-2.00
	BACHELOR	-6.112*	.468	.000	-7.43	-4.80
	GRADUATE	-7.639*	.548	.000	-9.18	-6.10
HIGH SCHOOL	LT HIGH SCHOOL	1.977*	.405	.000	.84	3.12
	JUNIOR COLLEGE	-1.594*	.461	.006	-2.89	-.30
	BACHELOR	-4.135*	.347	.000	-5.11	-3.16
	GRADUATE	-5.662*	.450	.000	-6.93	-4.40
JUNIOR COLLEGE	LT HIGH SCHOOL	3.571*	.557	.000	2.00	5.14
	HIGH SCHOOL	1.594*	.461	.006	.30	2.89
	BACHELOR	-2.541*	.517	.000	-3.99	-1.09
	GRADUATE	-4.068*	.590	.000	-5.73	-2.41
BACHELOR	LT HIGH SCHOOL	6.112*	.468	.000	4.80	7.43
	HIGH SCHOOL	4.135*	.347	.000	3.16	5.11
	JUNIOR COLLEGE	2.541*	.517	.000	1.09	3.99
	GRADUATE	-1.527*	.507	.026	-2.95	-.10
GRADUATE	LT HIGH SCHOOL	7.639*	.548	.000	6.10	9.18
	HIGH SCHOOL	5.662*	.450	.000	4.40	6.93
	JUNIOR COLLEGE	4.068*	.590	.000	2.41	5.73
	BACHELOR	1.527*	.507	.026	.10	2.95

*. The mean difference is significant at the 0.05 level.

FIGURE 9-4: The Multiple Comparisons table.

To summarize the entire table, we would say that all groups differed significantly from each other in terms of respondent's average age when they had their first child — as education increased, the respondent's age of when their first child was born increased.

See the following for an example of running a multiple comparisons.

EXAMPLE

Q. Using the GSS2018.sav file, rerun an ANOVA depicting the relationship between CHILDS (number of children) and DEGREE (respondent's highest degree). If you have a statistically significant result, perform the Bonferroni multiple comparison to determine which groups differ from each other. Summarize the findings.

A. As previously seen, we have a statistically significant result. The Multiple Comparisons table allows us to determine that the less than high school degree group has significantly more children than any of the other degree groups.

Descriptives

Number of children

	N	Mean	Std. Deviation	Std. Error	95% Confidence Interval for Mean Lower Bound	95% Confidence Interval for Mean Upper Bound	Minimum	Maximum
LT HIGH SCHOOL	260	2.68	2.059	.128	2.43	2.93	0	8
HIGH SCHOOL	1177	1.81	1.651	.048	1.71	1.90	0	8
JUNIOR COLLEGE	196	1.89	1.491	.107	1.68	2.10	0	8
BACHELOR	465	1.67	1.502	.070	1.53	1.80	0	8
GRADUATE	246	1.54	1.513	.096	1.35	1.73	0	8
Total	2344	1.86	1.674	.035	1.79	1.92	0	8

Multiple Comparisons

Dependent Variable: Number of children

Bonferroni

(I) R's highest degree	(J) R's highest degree	Mean Difference (I-J)	Std. Error	Sig.	95% Confidence Interval Lower Bound	95% Confidence Interval Upper Bound
LT HIGH SCHOOL	HIGH SCHOOL	.873*	.113	.000	.56	1.19
	JUNIOR COLLEGE	.793*	.156	.000	.36	1.23
	BACHELOR	1.014*	.128	.000	.66	1.37
	GRADUATE	1.140*	.147	.000	.73	1.55
HIGH SCHOOL	LT HIGH SCHOOL	-.873*	.113	.000	-1.19	-.56
	JUNIOR COLLEGE	-.080	.127	1.000	-.44	.28
	BACHELOR	.141	.090	1.000	-.11	.39
	GRADUATE	.267	.115	.207	-.06	.59
JUNIOR COLLEGE	LT HIGH SCHOOL	-.793*	.156	.000	-1.23	-.36
	HIGH SCHOOL	.080	.127	1.000	-.28	.44
	BACHELOR	.221	.140	1.000	-.17	.62
	GRADUATE	.347	.158	.279	-.10	.79
BACHELOR	LT HIGH SCHOOL	-1.014*	.128	.000	-1.37	-.66
	HIGH SCHOOL	-.141	.090	1.000	-.39	.11
	JUNIOR COLLEGE	-.221	.140	1.000	-.62	.17
	GRADUATE	.126	.130	1.000	-.24	.49
GRADUATE	LT HIGH SCHOOL	-1.140*	.147	.000	-1.55	-.73
	HIGH SCHOOL	-.267	.115	.207	-.59	.06
	JUNIOR COLLEGE	-.347	.158	.279	-.79	.10
	BACHELOR	-.126	.130	1.000	-.49	.24

*. The mean difference is significant at the 0.05 level.

5. Using the GSS2018.sav file, rerun an ANOVA depicting the relationship between HRS1 (hours worked last week) and DEGREE (respondent's highest degree). If you have a statistically significant result, perform the Bonferroni multiple comparison to determine which groups differ from each other. Summarize the findings.

6. Using the GSS2018.sav file, rerun an ANOVA depicting the relationship between HRSRELAX (hours to relax per day) and DEGREE (respondent's highest degree). If you have a statistically significant result, perform the Bonferroni multiple comparison to determine which groups differ from each other. Summarize the findings.

7. Using the GSS2018.sav file, rerun an ANOVA depicting the relationship between POLVIEWS (political views — higher numbers indicate the person is more conservative) and DEGREE (respondent's highest degree). If you have a statistically significant result, perform the Bonferroni multiple comparison to determine which groups differ from each other. Summarize the findings.

8. Using the GSS2018.sav file, rerun an ANOVA depicting the relationship between TVHOURS (hours per day watching TV) and DEGREE (respondent's highest degree). If you have a statistically significant result, perform the Bonferroni multiple comparison to determine which groups differ from each other. Summarize the findings.

Comparing the Means Graphically

As with t-tests, a simple error bar chart is the most effective way of comparing means. Follow these steps to create a simple error bar chart:

1. **Choose File ⇨ Open ⇨ Data and load the GSS2018.sav data file.**

   Download the file at `www.dummies.com/go/spssstatisticsworkbookfd`.

2. **Choose Graphs ⇨ Chart Builder.**
3. **In the Choose From list, select Bar.**
4. **Select the seventh graph image (the one with the Simple Error Bar tooltip) and drag it to the panel at the top of the window.**
5. **Select the AGEKDBRN (age when respondents had their first child) variable, and place it in the Y-axis box.**
6. **Select the DEGREE (respondent's highest degree) variable, and place it in the X-axis box.**
7. **Click OK.**

   The graph shown in Figure 9-5 appears.

This chart represents the mean age when participants had their first child for each degree along with 95% confidence intervals. The confidence intervals for the degree groups do not overlap, which is consistent with the result from the ANOVA and indicates that all the groups are significantly different from each other.

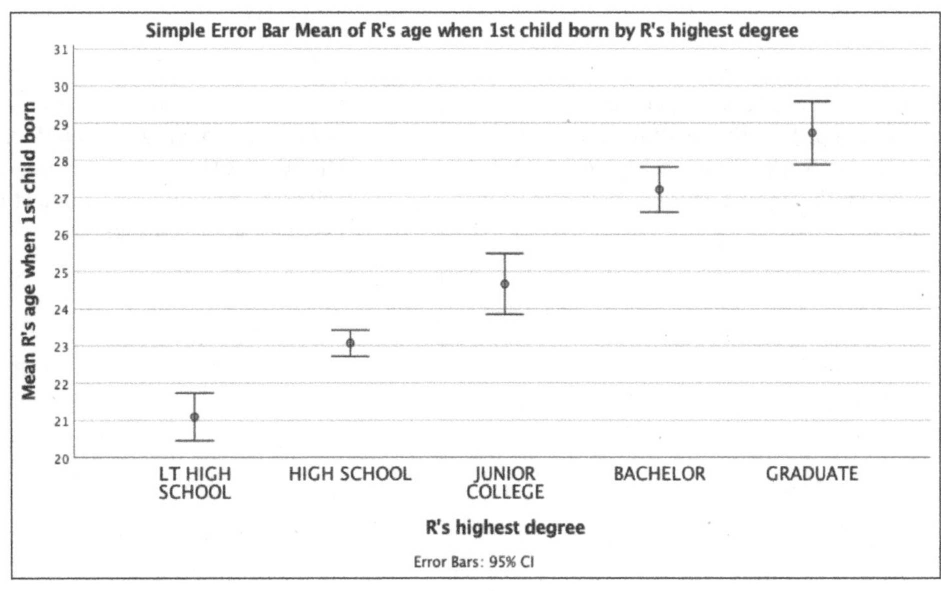

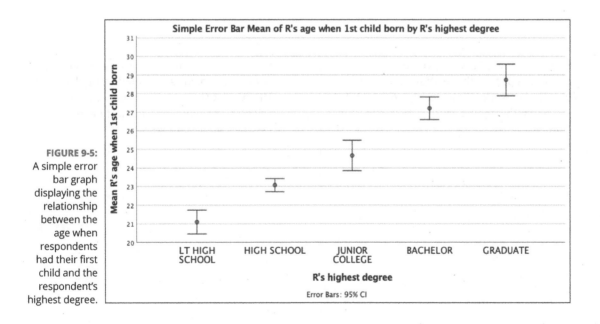

FIGURE 9-5: A simple error bar graph displaying the relationship between the age when respondents had their first child and the respondent's highest degree.

See the following for an example of making an error bar chart.

EXAMPLE

Q. Using the GSS2018.sav file, create an error bar chart depicting the relationship between CHILDS (number of children) and DEGREE (respondent's highest degree). Describe your findings.

A. No overlap exists between the less than high school degree group and all the other degree groups, whereas overlap does exist among the other degree groups. This result suggests that parents with less than a high school degree tend to have more children than all other degree groups.

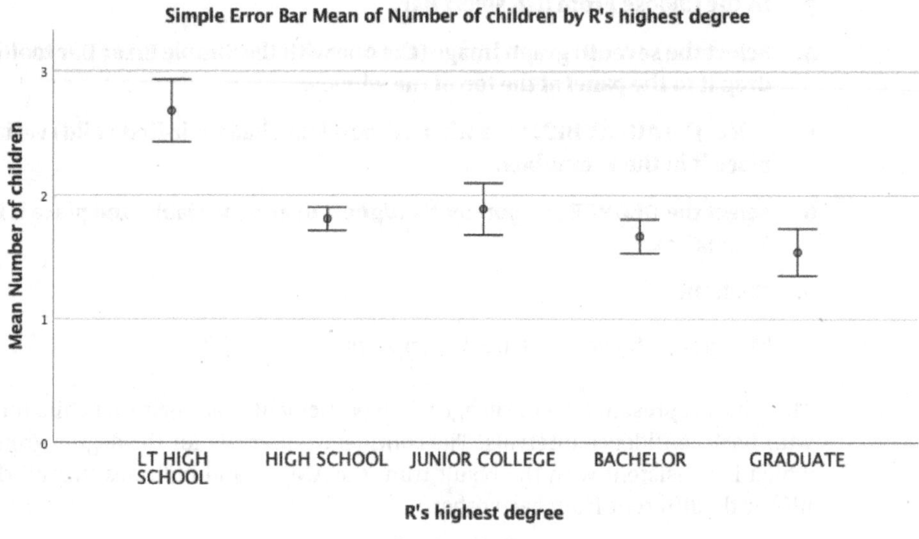

9 Using the GSS2018.sav file, create an error bar chart depicting the relationship between HRS1 (hours worked last week) and DEGREE (respondent's highest degree). Describe your findings.

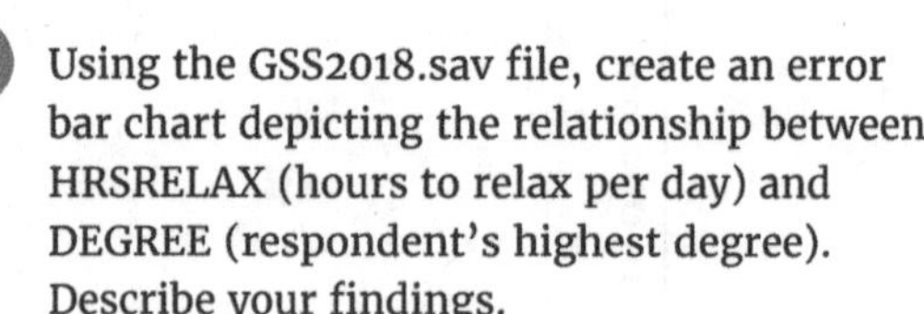

10 Using the GSS2018.sav file, create an error bar chart depicting the relationship between HRSRELAX (hours to relax per day) and DEGREE (respondent's highest degree). Describe your findings.

11 Using the GSS2018.sav file, create an error bar chart depicting the relationship between POLVIEWS (political views – higher numbers indicate the person is more conservative) and DEGREE (respondent's highest degree). Describe your findings.

12 Using the GSS2018.sav file, create an error bar chart depicting the relationship between TVHOURS (hours per day watching TV) and DEGREE (respondent's highest degree). Describe your findings.

Answers to Problems in Comparing More Than Two Groups with ANOVA

1. We met the assumption of homogeneity of variance, so we have to use the standard ANOVA table. We did not find a statistically significant difference between the degree groups — regardless of degree, respondents work the same number of hours.

Tests of Homogeneity of Variances

		Levene Statistic	df1	df2	Sig.
Number of hours worked last week	Based on Mean	1.241	4	1376	.291
	Based on Median	1.308	4	1376	.265
	Based on Median and with adjusted df	1.308	4	1364.864	.265
	Based on trimmed mean	1.218	4	1376	.301

ANOVA

Number of hours worked last week

	Sum of Squares	df	Mean Square	F	Sig.
Between Groups	1069.654	4	267.414	1.276	.277
Within Groups	288291.772	1376	209.514		
Total	289361.427	1380			

2. We did meet the assumption of homogeneity of variance, so we have to use the standard ANOVA table. We did find a statistically significant difference between the degree groups in terms of the number of hours to relax.

Tests of Homogeneity of Variances

		Levene Statistic	df1	df2	Sig.
Hours per day R have to relax	Based on Mean	2.344	4	1400	.053
	Based on Median	2.545	4	1400	.038
	Based on Median and with adjusted df	2.545	4	1358.834	.038
	Based on trimmed mean	2.581	4	1400	.036

ANOVA

Hours per day R have to relax

	Sum of Squares	df	Mean Square	F	Sig.
Between Groups	90.993	4	22.748	2.955	.019
Within Groups	10777.410	1400	7.698		
Total	10868.403	1404			

3 We did not meet the assumption of homogeneity of variance, so we have to use the Welch test. We did find a statistically significant result, so differences exist among degree groups with regard to political views.

Tests of Homogeneity of Variances

		Levene Statistic	df1	df2	Sig.
Think of self as liberal or conservative	Based on Mean	6.806	4	2242	.000
	Based on Median	6.829	4	2242	.000
	Based on Median and with adjusted df	6.829	4	2231.757	.000
	Based on trimmed mean	6.743	4	2242	.000

Robust Tests of Equality of Means

Think of self as liberal or conservative

	Statistic[a]	df1	df2	Sig.
Welch	6.628	4	617.372	.000

a. Asymptotically F distributed.

4 We did not meet the assumption of homogeneity of variance, so we have to use the Welch test. We did find a statistically significant result, so differences exist among degree groups with regard to the number of hours they watch TV.

Tests of Homogeneity of Variances

		Levene Statistic	df1	df2	Sig.
Hours per day watching TV	Based on Mean	15.995	4	1550	.000
	Based on Median	12.515	4	1550	.000
	Based on Median and with adjusted df	12.515	4	1334.078	.000
	Based on trimmed mean	13.526	4	1550	.000

Robust Tests of Equality of Means

Hours per day watching TV

	Statistic[a]	df1	df2	Sig.
Welch	24.694	4	430.198	.000

a. Asymptotically F distributed.

5 We do not have a statistically significant result, so we do not need to perform multiple comparisons — that is, the groups do not differ from each other so the means are similar.

Descriptives

Number of hours worked last week

	N	Mean	Std. Deviation	Std. Error	95% Confidence Interval for Mean		Minimum	Maximum
					Lower Bound	Upper Bound		
LT HIGH SCHOOL	118	39.67	15.974	1.471	36.76	42.58	3	89
HIGH SCHOOL	662	41.26	14.257	.554	40.18	42.35	1	89
JUNIOR COLLEGE	123	41.08	15.200	1.370	38.37	43.79	1	89
BACHELOR	318	40.93	13.953	.782	39.39	42.47	1	89
GRADUATE	160	43.40	14.660	1.159	41.11	45.69	1	89
Total	1381	41.28	14.480	.390	40.52	42.05	1	89

Multiple Comparisons

Dependent Variable: Number of hours worked last week

Bonferroni

(I) R's highest degree	(J) R's highest degree	Mean Difference (I-J)	Std. Error	Sig.	95% Confidence Interval	
					Lower Bound	Upper Bound
LT HIGH SCHOOL	HIGH SCHOOL	-1.595	1.446	1.000	-5.66	2.47
	JUNIOR COLLEGE	-1.412	1.865	1.000	-6.66	3.83
	BACHELOR	-1.258	1.560	1.000	-5.64	3.13
	GRADUATE	-3.731	1.756	.339	-8.67	1.21
HIGH SCHOOL	LT HIGH SCHOOL	1.595	1.446	1.000	-2.47	5.66
	JUNIOR COLLEGE	.183	1.421	1.000	-3.81	4.18
	BACHELOR	.337	.988	1.000	-2.44	3.11
	GRADUATE	-2.136	1.275	.942	-5.72	1.45
JUNIOR COLLEGE	LT HIGH SCHOOL	1.412	1.865	1.000	-3.83	6.66
	HIGH SCHOOL	-.183	1.421	1.000	-4.18	3.81
	BACHELOR	.154	1.537	1.000	-4.17	4.47
	GRADUATE	-2.319	1.736	1.000	-7.20	2.56
BACHELOR	LT HIGH SCHOOL	1.258	1.560	1.000	-3.13	5.64
	HIGH SCHOOL	-.337	.988	1.000	-3.11	2.44
	JUNIOR COLLEGE	-.154	1.537	1.000	-4.47	4.17
	GRADUATE	-2.472	1.403	.783	-6.42	1.47
GRADUATE	LT HIGH SCHOOL	3.731	1.756	.339	-1.21	8.67
	HIGH SCHOOL	2.136	1.275	.942	-1.45	5.72
	JUNIOR COLLEGE	2.319	1.736	1.000	-2.56	7.20
	BACHELOR	2.472	1.403	.783	-1.47	6.42

6 We have a statistically significant result. The Multiple Comparisons table enables us to determine that the only significant difference is between the high school degree group, which has significantly more time to relax than the graduate degree group.

Descriptives

Hours per day R have to relax

	N	Mean	Std. Deviation	Std. Error	95% Confidence Interval for Mean		Minimum	Maximum
					Lower Bound	Upper Bound		
LT HIGH SCHOOL	117	3.69	2.743	.254	3.19	4.19	0	14
HIGH SCHOOL	670	3.97	3.004	.116	3.74	4.20	0	21
JUNIOR COLLEGE	127	3.50	2.853	.253	3.00	4.00	0	20
BACHELOR	322	3.56	2.353	.131	3.30	3.82	0	24
GRADUATE	169	3.27	2.518	.194	2.88	3.65	0	16
Total	1405	3.72	2.782	.074	3.58	3.87	0	24

Multiple Comparisons

Dependent Variable: Hours per day R have to relax
Bonferroni

(I) R's highest degree	(J) R's highest degree	Mean Difference (I-J)	Std. Error	Sig.	95% Confidence Interval	
					Lower Bound	Upper Bound
LT HIGH SCHOOL	HIGH SCHOOL	-.276	.278	1.000	-1.06	.51
	JUNIOR COLLEGE	.196	.356	1.000	-.80	1.20
	BACHELOR	.133	.300	1.000	-.71	.98
	GRADUATE	.426	.334	1.000	-.51	1.36
HIGH SCHOOL	LT HIGH SCHOOL	.276	.278	1.000	-.51	1.06
	JUNIOR COLLEGE	.473	.269	.786	-.28	1.23
	BACHELOR	.410	.188	.296	-.12	.94
	GRADUATE	.702*	.239	.033	.03	1.37
JUNIOR COLLEGE	LT HIGH SCHOOL	-.196	.356	1.000	-1.20	.80
	HIGH SCHOOL	-.473	.269	.786	-1.23	.28
	BACHELOR	-.063	.291	1.000	-.88	.75
	GRADUATE	.230	.326	1.000	-.69	1.15
BACHELOR	LT HIGH SCHOOL	-.133	.300	1.000	-.98	.71
	HIGH SCHOOL	-.410	.188	.296	-.94	.12
	JUNIOR COLLEGE	.063	.291	1.000	-.75	.88
	GRADUATE	.293	.264	1.000	-.45	1.03
GRADUATE	LT HIGH SCHOOL	-.426	.334	1.000	-1.36	.51
	HIGH SCHOOL	-.702*	.239	.033	-1.37	-.03
	JUNIOR COLLEGE	-.230	.326	1.000	-1.15	.69
	BACHELOR	-.293	.264	1.000	-1.03	.45

*. The mean difference is significant at the 0.05 level.

7 We have a statistically significant result. The Multiple Comparisons table enables us to determine that the high school degree group is significantly more conservative (higher numbers for the political views variable indicate that the person is more conservative) than the bachelor and graduate degree groups.

Descriptives

Think of self as liberal or conservative

	N	Mean	Std. Deviation	Std. Error	95% Confidence Interval for Mean		Minimum	Maximum
					Lower Bound	Upper Bound		
LT HIGH SCHOOL	232	3.98	1.586	.104	3.77	4.18	1	7
HIGH SCHOOL	1123	4.19	1.423	.042	4.10	4.27	1	7
JUNIOR COLLEGE	193	4.17	1.420	.102	3.97	4.37	1	7
BACHELOR	456	3.85	1.530	.072	3.71	3.99	1	7
GRADUATE	243	3.77	1.680	.108	3.56	3.98	1	7
Total	2247	4.05	1.500	.032	3.99	4.11	1	7

Multiple Comparisons

Dependent Variable: Think of self as liberal or conservative
Bonferroni

(I) R's highest degree	(J) R's highest degree	Mean Difference (I-J)	Std. Error	Sig.	95% Confidence Interval	
					Lower Bound	Upper Bound
LT HIGH SCHOOL	HIGH SCHOOL	-.207	.108	.547	-.51	.10
	JUNIOR COLLEGE	-.193	.145	1.000	-.60	.22
	BACHELOR	.132	.120	1.000	-.21	.47
	GRADUATE	.209	.137	1.000	-.18	.59
HIGH SCHOOL	LT HIGH SCHOOL	.207	.108	.547	-.10	.51
	JUNIOR COLLEGE	.014	.116	1.000	-.31	.34
	BACHELOR	.339*	.083	.000	.11	.57
	GRADUATE	.416*	.106	.001	.12	.71
JUNIOR COLLEGE	LT HIGH SCHOOL	.193	.145	1.000	-.22	.60
	HIGH SCHOOL	-.014	.116	1.000	-.34	.31
	BACHELOR	.324	.128	.114	-.04	.68
	GRADUATE	.401	.144	.053	.00	.81
BACHELOR	LT HIGH SCHOOL	-.132	.120	1.000	-.47	.21
	HIGH SCHOOL	-.339*	.083	.000	-.57	-.11
	JUNIOR COLLEGE	-.324	.128	.114	-.68	.04
	GRADUATE	.077	.118	1.000	-.26	.41
GRADUATE	LT HIGH SCHOOL	-.209	.137	1.000	-.59	.18
	HIGH SCHOOL	-.416*	.106	.001	-.71	-.12
	JUNIOR COLLEGE	-.401	.144	.053	-.81	.00
	BACHELOR	-.077	.118	1.000	-.41	.26

*. The mean difference is significant at the 0.05 level.

8 We have a statistically significant result. The Multiple Comparisons table enables us to determine that the less than high school, high school, and junior college degree groups spend more time watching TV than the bachelor and graduate degree groups.

Descriptives

Hours per day watching TV

	N	Mean	Std. Deviation	Std. Error	95% Confidence Interval for Mean		Minimum	Maximum
					Lower Bound	Upper Bound		
LT HIGH SCHOOL	183	3.70	3.312	.245	3.22	4.18	0	24
HIGH SCHOOL	787	3.24	3.026	.108	3.03	3.45	0	24
JUNIOR COLLEGE	124	3.15	3.121	.280	2.60	3.71	0	20
BACHELOR	302	2.12	1.670	.096	1.93	2.31	0	12
GRADUATE	159	1.94	2.153	.171	1.61	2.28	0	20
Total	1555	2.94	2.837	.072	2.80	3.08	0	24

Multiple Comparisons

Dependent Variable: Hours per day watching TV
Bonferroni

(I) R's highest degree	(J) R's highest degree	Mean Difference (I-J)	Std. Error	Sig.	95% Confidence Interval	
					Lower Bound	Upper Bound
LT HIGH SCHOOL	HIGH SCHOOL	.457	.228	.453	-.18	1.10
	JUNIOR COLLEGE	.546	.323	.912	-.36	1.45
	BACHELOR	1.580*	.260	.000	.85	2.31
	GRADUATE	1.756*	.301	.000	.91	2.60
HIGH SCHOOL	LT HIGH SCHOOL	-.457	.228	.453	-1.10	.18
	JUNIOR COLLEGE	.089	.268	1.000	-.67	.84
	BACHELOR	1.123*	.188	.000	.59	1.65
	GRADUATE	1.299*	.242	.000	.62	1.98
JUNIOR COLLEGE	LT HIGH SCHOOL	-.546	.323	.912	-1.45	.36
	HIGH SCHOOL	-.089	.268	1.000	-.84	.67
	BACHELOR	1.034*	.296	.005	.20	1.87
	GRADUATE	1.210*	.333	.003	.27	2.15
BACHELOR	LT HIGH SCHOOL	-1.580*	.260	.000	-2.31	-.85
	HIGH SCHOOL	-1.123*	.188	.000	-1.65	-.59
	JUNIOR COLLEGE	-1.034*	.296	.005	-1.87	-.20
	GRADUATE	.176	.272	1.000	-.59	.94
GRADUATE	LT HIGH SCHOOL	-1.756*	.301	.000	-2.60	-.91
	HIGH SCHOOL	-1.299*	.242	.000	-1.98	-.62
	JUNIOR COLLEGE	-1.210*	.333	.003	-2.15	-.27
	BACHELOR	-.176	.272	1.000	-.94	.59

*. The mean difference is significant at the 0.05 level.

9 All degree groups overlap. This result suggests that no significant differences exist among the degree groups — regardless of degree, all groups work a similar number of hours.

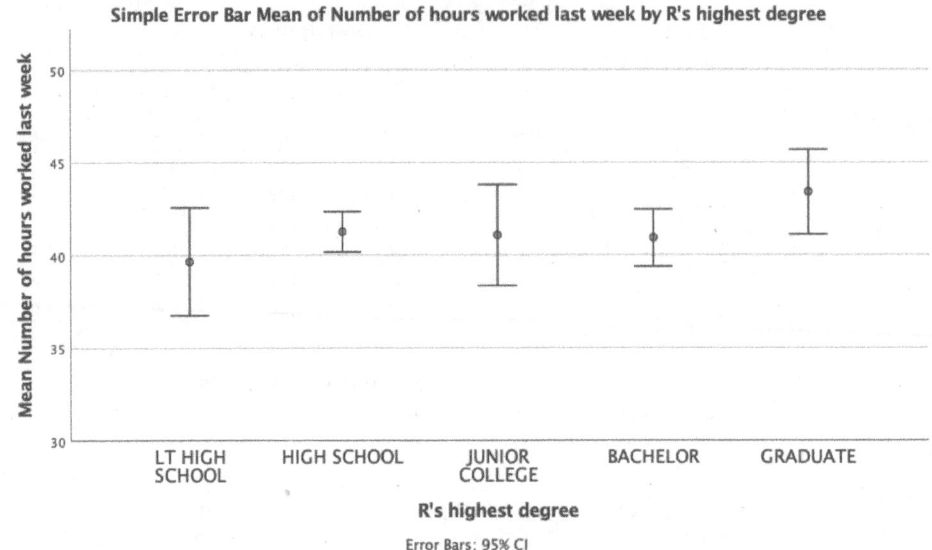

10 All degree groups overlap except the high school and graduate degree groups, which is consistent with the findings in the Multiple Comparisons table. This result suggests that significant differences exist between the high school degree group, which has significantly more time to relax than the graduate degree group.

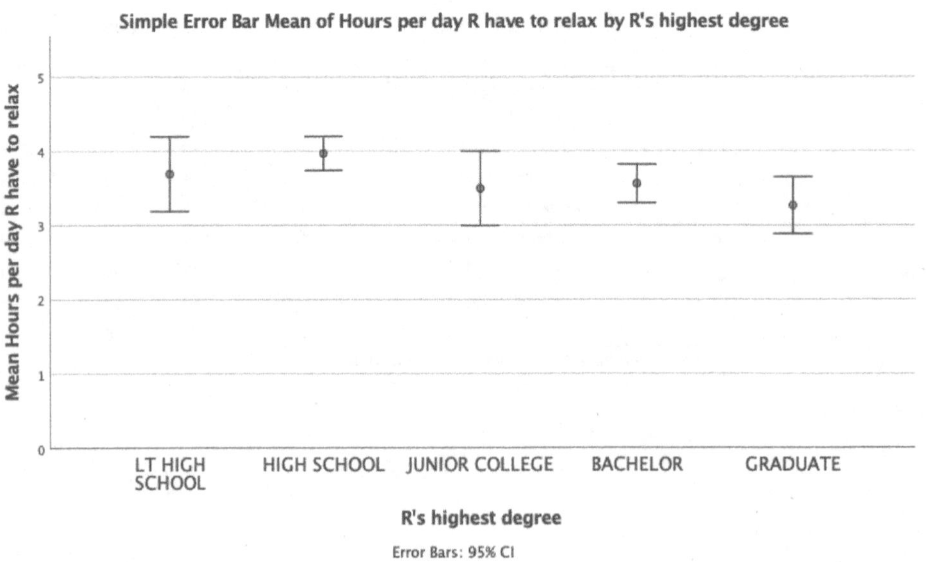

11 All degree groups overlap except the high school and the bachelor and graduate degree groups, which is consistent with the findings in the Multiple Comparisons table. This result suggests that significant differences exist between the high school degree group, which is significantly more conservative than the other groups, and the bachelor and graduate degree groups.

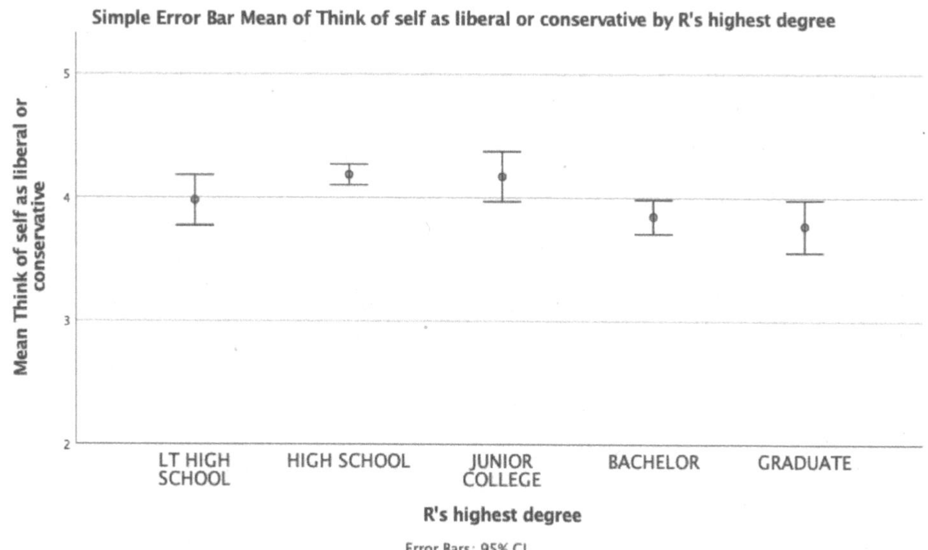

12 The less than high school, high school, and junior college degree groups overlap and the bachelor and graduate degree groups overlap. However, no overlap exists between the less than high school, high school, and junior college degree groups and the bachelor and graduate degree groups, so these groups differ from each other.

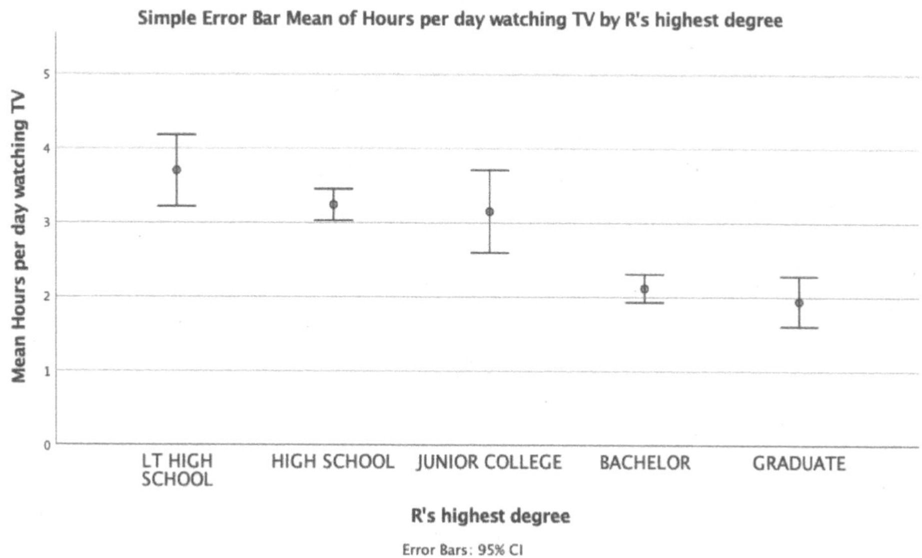

IN THIS CHAPTER

» Viewing relationships with scatterplots

» Assessing relationships with the Pearson correlation

Chapter 10
Testing Relationships with Correlation

A correlation studies the relationship between two continuous variables to determine whether one variable increases or decreases in relation to another variable. For example, you might study the relationship between the number of items purchased and the total amount spent, and find that as the number of items purchased increases, the amount spent increases. The variables are *correlated* with each other because changes in one variable affect the other.

The Pearson correlation coefficient measures the extent — the strength and direction — of the linear (straight-line) relationship between two continuous variables. For example, you might want to know whether higher SAT scores are associated with higher first-year college GPAs; or whether eating more often at fast-food restaurants is related to more frequent shopping at convenience stores; or whether lower levels of depression are associated with higher self-esteem scores.

In this chapter, you use a scatterplot to display the relationship between two continuous variables. You then use correlation to quantify this relationship.

Viewing Relationships

Graphs organize data and help you detect patterns. *Scatterplots,* which have a horizontal dimension (the x-axis) and a vertical dimension (the y-axis), visually present the relationship between two continuous variables. Correlation procedures (which we discuss later in this chapter) and linear regression procedures (described in Chapter 11) are appropriate only when a linear relationship exists between the variables. Therefore, before running either of these procedures, you must create a scatterplot, which will show the type of pattern between variables. (For more on creating scatterplots, see Chapter 13.)

TIP

In a scatterplot, note the following:

- Whether there is a linear relationship
- The direction of the relationship, if there is one
- Any outliers

The following steps show you how to construct a simple scatterplot:

1. **Download and open GSS2018.sav file.**

   The file contains data from the General Social Survey (GSS), a nationally representative survey of adults in the United States that collects data on contemporary opinions, attitudes, and behaviors. To download the file, go to the book's companion website at `www.dummies.com/go/spssstatisticsworkbookfd`.

   To open the file, choose File⇨Open⇨Data and then select the GSS2018.sav file.

2. **Choose Graphs⇨Chart Builder.**
3. **In the Choose From list, select Scatter/Dot.**
4. **Select the first scatterplot diagram (with the Simple Scatter tooltip), and drag it to the panel at the top.**
5. **In the Variables list, select HEIGHT and drag it to the rectangle labeled X-Axis in the diagram.**
6. **In the Variables list, select WEIGHT and drag it to the rectangle labeled Y-Axis in the diagram.**
7. **In the Element Properties tab, select Total under the Linear Fit Lines section.**
8. **Click OK.**

   The chart in Figure 10-1 appears.

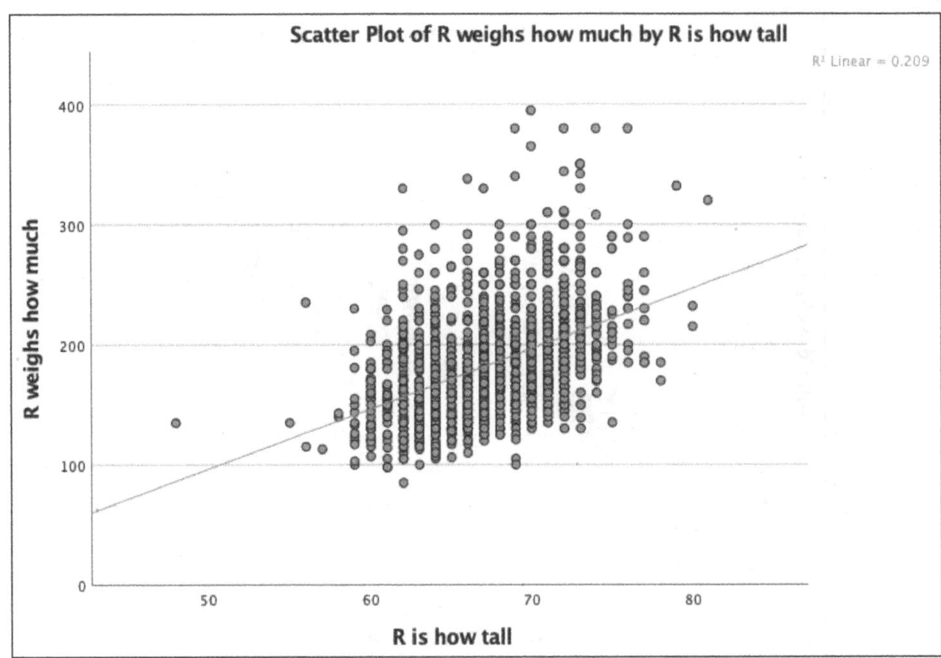

FIGURE 10-1: The scatterplot of weight and height.

The scatterplot of height and weight shows the relationship between these two variables. Note that, for the most part, shorter people are associated with lower weight and taller people are associated with higher weight. This type of relationship is a *positive linear relationship:* as you increase one variable, you increase the other variable, so low numbers for one variable relate to low numbers for another variable and high numbers for one variable are associated with high numbers for another variable.

You can also have a *negative linear relationship,* which means that as you increase one variable, you decrease the other variable, so low numbers on one variable go with high numbers on the other variable.

WARNING

You can use the Bivariate procedure, which we demonstrate in the following example, whenever you have a positive or negative linear relationship. Don't use the Bivariate procedure when you have a nonlinear relationship because the results would be misleading.

See the following for an example of making a scatterplot.

EXAMPLE

Q. Using the Cars.sav file, create a scatterplot depicting the relationship between engine size (x-axis) and weight (y-axis). Describe the relationship between the two variables.

A. A very strong positive linear relationship between engine size and vehicle weight exists, with smaller engines associated with lower weight and larger engines associated with heavier cars.

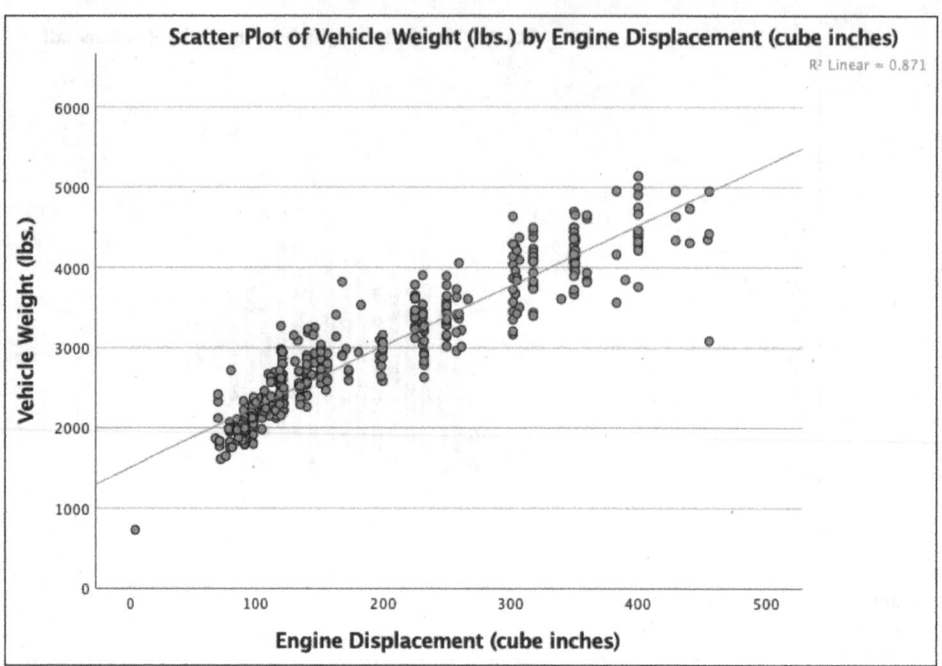

1. Using the Cars.sav file, create a scatterplot showing the relationship between horsepower (x-axis) and acceleration (y-axis). Describe the relationship between the two variables.

2. Using the Cars.sav file, create a scatterplot showing the relationship between mpg (x-axis) and acceleration (y-axis). Describe the relationship between the two variables.

3 Using the Cars.sav file, create a scatterplot showing the relationship between horsepower (x-axis) and mpg (y-axis). Describe the relationship between the two variables.

Using the GSS2018.sav file, create a scatterplot showing the relationship between hrs1 (x-axis) and wwwhr (y-axis). Describe the relationship between the two variables.

Running the Bivariate Procedure

Whereas a scatterplot shows the relationship between two continuous variables, the *Pearson correlation coefficient* quantifies the strength and direction of that relationship. The Pearson correlation coefficient measures the extent of the linear (straight-line) relationship between two variables, using values between –1 and +1. The larger the absolute value, the stronger the correlation. A correlation of 0 means no straight-line relationship exists. Note that SPSS refers to running correlations as the *Bivariate procedure.*

When examining the distributions of two continuous variables, you want to know whether an observed relationship is likely to exist in the target population or is caused by random sampling variation.

A statistical test can determine whether a relationship between two or more variables is statistically significant — that is, whether the correlation differs from 0 (indicating no linear association) in the population. You can set up two hypotheses to test:

- **Null hypothesis:** The variables are not linearly related to each other. That is, the variables are independent.
- **Alternative hypothesis:** The variables are linearly related to each other. That is, the variables are associated.

To perform a correlation, follow these steps:

1. **Choose File ⇨ Open ⇨ Data and load the GSS2018.sav data file.**

   You can download the file at `www.dummies.com/go/spssstatisticsworkbookfd`.

2. **Choose Analyze ⇨ Correlate ⇨ Bivariate.**

   The Bivariate Correlations dialog box appears. In this example, you will study the relationship between height and weight.

3. **Select the HEIGHT and WEIGHT variables and place them in the Variables box.**

4. **Click OK.**

   SPSS calculates the correlations between the variables, as shown in Figure 10-2.

Correlations

		R is how tall	R weighs how much
R is how tall	Pearson Correlation	1	.457**
	Sig. (2-tailed)		.000
	N	1402	1374
R weighs how much	Pearson Correlation	.457**	1
	Sig. (2-tailed)	.000	
	N	1374	1380

**. Correlation is significant at the 0.01 level (2-tailed).

FIGURE 10-2: The Correlations table.

Note that the table is symmetric, so the same information is represented above and below the major diagonal (the diagonal that runs from left to right). Also, the correlations in the major diagonal are 1 because these are the correlations of each variable with itself.

TIP

If you want to remove the redundancy in the table, along with the correlation of a variable with itself in the major diagonal, select the Show Only the Lower Triangle option in the Bivariate Correlations dialog box.

The Correlations table provides three pieces of information:

- **Pearson correlation:** The Pearson correlation ranges from +1 to –1. The farther away the value is from 0, the stronger the relationship.
- **Two-tailed significance level:** All correlations with a significance level less than 0.05 will have an asterisk next to the coefficient, indicating that the correlation is statistically significant. In other words, the correlation is not 0.
- **N:** N is the sample size.

In the example, you have a moderate positive correlation (0.457) that is statistically significant between height and weight. Note that the probability of the null hypothesis being true for this relationship is extremely small (less than 0.01). Therefore, you can reject the null hypothesis and conclude that a positive linear relationship exists between these variables.

REMEMBER

Every statistical test has assumptions. The better you meet these assumptions, the more you can trust the results of the test. The Pearson correlation coefficient has the following assumptions:

- The variables are continuous.
- The variables are linearly related (linearity).
- The variables are normally distributed (normality).
- There are no influential outliers.
- Similar variation exists throughout the regression line (homoscedasticity).

See the following for an example of running a correlation.

EXAMPLE

Q. Using the Cars.sav file, run a correlation depicting the relationship between engine size and weight. Describe the relationship between the two variables.

A. A very strong positive linear relationship of .933 exists between engine size and vehicle weight. The relationship is statistically significant; smaller engines are associated with lower weight and larger engines are associated with heavier cars.

Correlations

		Engine Displacement (cube inches)	Vehicle Weight (lbs.)
Engine Displacement (cube inches)	Pearson Correlation	1	.933**
	Sig. (2-tailed)		.000
	N	406	406
Vehicle Weight (lbs.)	Pearson Correlation	.933**	1
	Sig. (2-tailed)	.000	
	N	406	406

**. Correlation is significant at the 0.01 level (2-tailed).

5 Using the Cars.sav file, run a correlation showing the relationship between horsepower and acceleration. Describe the relationship between the two variables.

6 Using the Cars.sav file, run a correlation showing the relationship between mpg and acceleration. Describe the relationship between the two variables.

7 Using the Cars.sav file, run a correlation showing the relationship between horsepower and mpg. Describe the relationship between the two variables.

8 Using the Cars.sav file, run a correlation showing the relationship between horsepower, engine size, weight, acceleration, year, number of cylinders, and mpg. Which relationships are statistically significant? Which variables have the strongest relationship? Which variables have the weakest relationship?

9 Using the GSS2018.sav file, run a correlation showing the relationship between hrs1 and wwwhr. Describe the relationship between the two variables.

Answers to Problems in Testing Relationships with Correlation

1 A very negative linear relationship exists between horsepower and time to accelerate. Cars with less horsepower take longer to accelerate than cars with more horsepower.

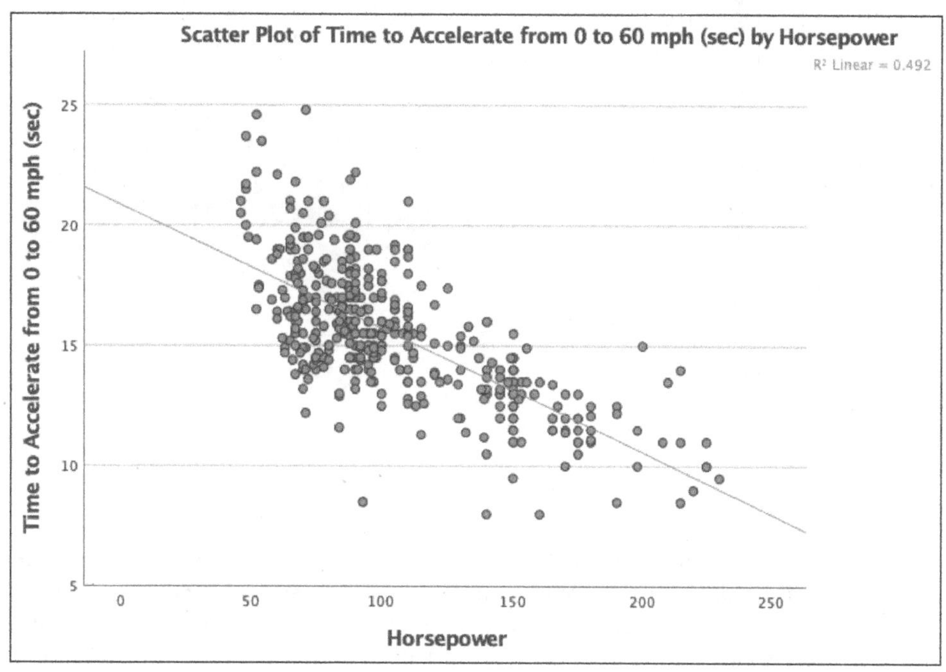

2. A moderate to strong positive linear relationship exists between miles per gallon and acceleration. In other words, cars that average fewer miles per gallon take less time to speed up to 60 mph, and cars that average more miles per gallon take longer to reach an appropriate speed.

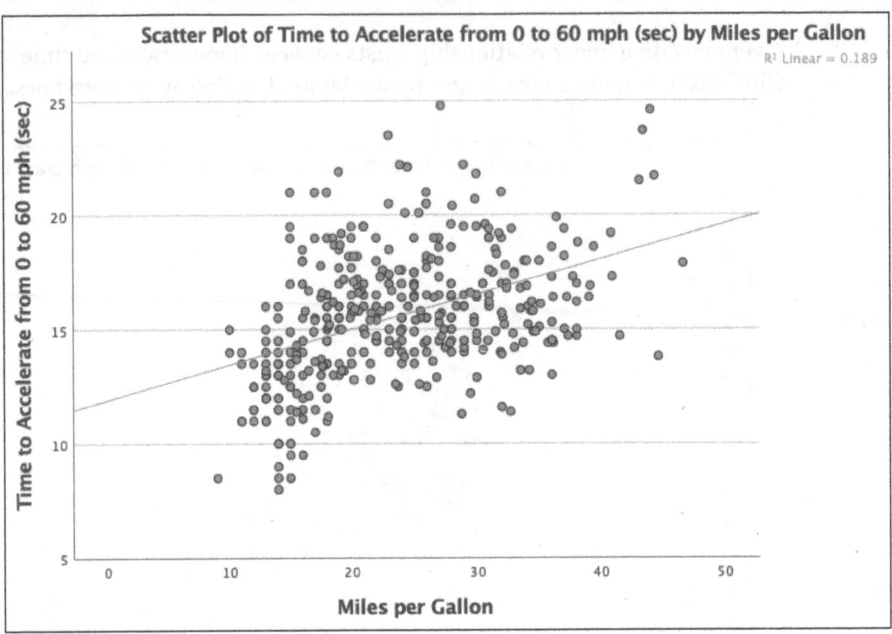

3. A very negative relationship exists between horsepower and miles per gallon. Cars with less horsepower average more miles per gallon, and cars with more horsepower average fewer miles per gallon. However, the relationship is not entirely linear because there is a bit of curvature at both ends of the distribution. Because of the curvature, using traditional correlation or linear regression might misrepresent the true nature of this association.

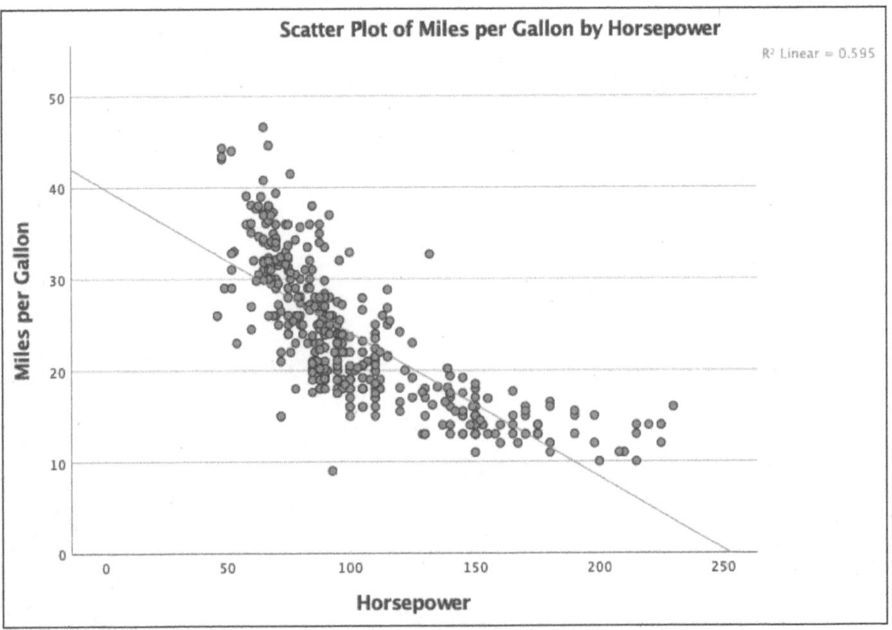

4. The scatterplot shows no linear relationship between the variables because there is no pattern in the data. In other words, knowing how many hours someone worked in the last week provides no information as to how many hours that person was online.

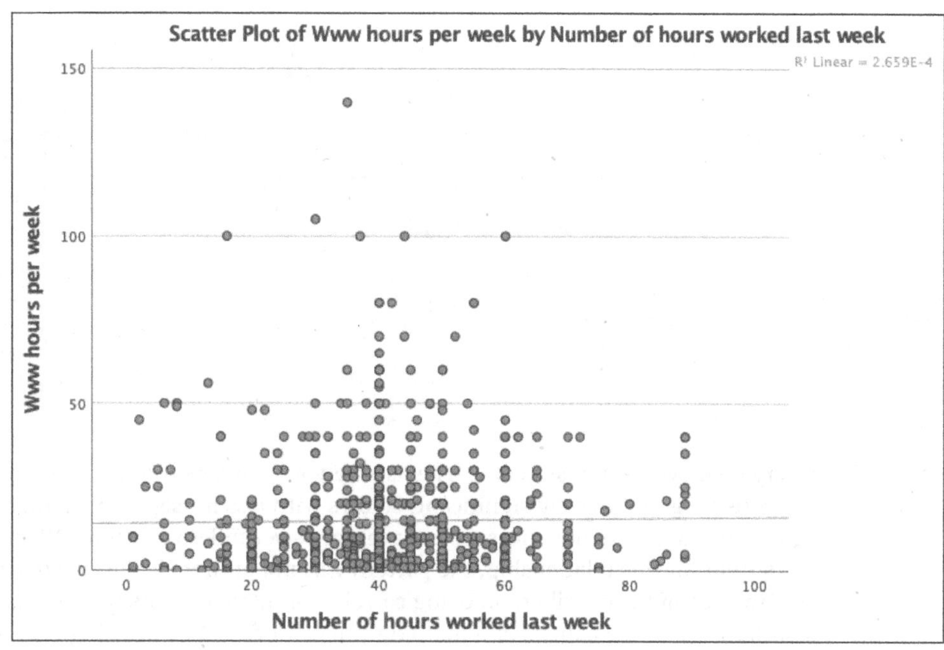

5. A very negative linear relationship of –.701 exists between horsepower and time to accelerate. The relationship is statistically significant, so cars with less horsepower take longer to accelerate than cars with more horsepower.

Correlations

		Time to Accelerate from 0 to 60 mph (sec)	Horsepower
Time to Accelerate from 0 to 60 mph (sec)	Pearson Correlation	1	-.701**
	Sig. (2-tailed)		.000
	N	406	400
Horsepower	Pearson Correlation	-.701**	1
	Sig. (2-tailed)	.000	
	N	400	400

**. Correlation is significant at the 0.01 level (2-tailed).

6 A moderate positive linear relationship of .434 exists between miles per gallon and acceleration. The relationship is statistically significant, so cars that average fewer miles per gallon take less time to speed up to 60 mph, and cars that have more miles per gallon take longer to reach an appropriate speed.

Correlations

		Miles per Gallon	Time to Accelerate from 0 to 60 mph (sec)
Miles per Gallon	Pearson Correlation	1	.434**
	Sig. (2-tailed)		.000
	N	398	398
Time to Accelerate from 0 to 60 mph (sec)	Pearson Correlation	.434**	1
	Sig. (2-tailed)	.000	
	N	398	406

**. Correlation is significant at the 0.01 level (2-tailed).

7 A very negative relationship of –.771 exists between horsepower and miles per gallon. The relationship is statistically significant, so cars with less horsepower average more miles per gallon and cars with more horsepower average fewer miles per gallon. When you ran the scatterplot for this relationship, the pattern was not entirely linear due to a bit of curvature at both ends of the distribution. Using correlation misrepresents the true nature of this association, and the pattern is something you can't detect by focusing only on the correlation coefficient itself; you need a scatterplot to truly appreciate this relationship.

Correlations

		Miles per Gallon	Horsepower
Miles per Gallon	Pearson Correlation	1	-.771**
	Sig. (2-tailed)		.000
	N	398	392
Horsepower	Pearson Correlation	-.771**	1
	Sig. (2-tailed)	.000	
	N	392	400

**. Correlation is significant at the 0.01 level (2-tailed).

8 All relationships are statistically significant. The strongest relationship is between engine size and the number of cylinders, with a correlation of .952. The weakest relationship, although still statistically significant, is between acceleration and the year, with a correlation of .308.

Correlations

		Miles per Gallon	Engine Displacemen t (cube inches)	Horsepower	Vehicle Weight (lbs.)	Time to Accelerate from 0 to 60 mph (sec)	Model Year (modulo 100)	Number of Cylinders
Miles per Gallon	Pearson Correlation	1	-.789**	-.771**	-.807**	.434**	.576**	-.774**
	Sig. (2-tailed)		.000	.000	.000	.000	.000	.000
	N	398	398	392	398	398	397	397
Engine Displacement (cube inches)	Pearson Correlation	-.789**	1	.897**	.933**	-.545**	-.379**	.952**
	Sig. (2-tailed)	.000		.000	.000	.000	.000	.000
	N	398	406	400	406	406	405	405
Horsepower	Pearson Correlation	-.771**	.897**	1	.859**	-.701**	-.419**	.844**
	Sig. (2-tailed)	.000	.000		.000	.000	.000	.000
	N	392	400	400	400	400	399	399
Vehicle Weight (lbs.)	Pearson Correlation	-.807**	.933**	.859**	1	-.415**	-.310**	.895**
	Sig. (2-tailed)	.000	.000	.000		.000	.000	.000
	N	398	406	400	406	406	405	405
Time to Accelerate from 0 to 60 mph (sec)	Pearson Correlation	.434**	-.545**	-.701**	-.415**	1	.308**	-.528**
	Sig. (2-tailed)	.000	.000	.000	.000		.000	.000
	N	398	406	400	406	406	405	405
Model Year (modulo 100)	Pearson Correlation	.576**	-.379**	-.419**	-.310**	.308**	1	-.357**
	Sig. (2-tailed)	.000	.000	.000	.000	.000		.000
	N	397	405	399	405	405	405	405
Number of Cylinders	Pearson Correlation	-.774**	.952**	.844**	.895**	-.528**	-.357**	1
	Sig. (2-tailed)	.000	.000	.000	.000	.000	.000	
	N	397	405	399	405	405	405	405

**. Correlation is significant at the 0.01 level (2-tailed).

9 No linear relationship exists between the variables because there is no pattern in the data; the correlation is not statistically significant with a value of .016. In other words, knowing how many hours someone worked in the last week provides no information as to how many hours that person was online.

Correlations

		Number of hours worked last week	Www hours per week
Number of hours worked last week	Pearson Correlation	1	.016
	Sig. (2-tailed)		.631
	N	1381	868
Www hours per week	Pearson Correlation	.016	1
	Sig. (2-tailed)	.631	
	N	868	1362

IN THIS CHAPTER

» Running a simple linear regression

» Practicing with multiple linear regression

» Interpreting regression coefficients

Chapter 11 Making Predictions Using Linear Regression

In this chapter, you practice producing linear regression output. And just as importantly, you practice interpreting the results as well.

Performing Simple Linear Regression

In Chapter 10, you learn that correlation helps establish the strength of the relationship between two variables. After you establish a relationship, you can use one variable to predict the other. To use the language of regression, you predict the dependent variables with one or more independent variables. We use the phrase *simple linear regression* when there is only one independent variable.

In this example, you use engine liters to predict miles per gallon:

1. **Choose File ⇨ Open ⇨ Data and load the Cars_Regression.sav file.**

   To download the file, go to the book's companion website at `www.dummies.com/go/spssstatisticsworkbookfd`.

2. **Choose Analyze ⇨ Regression ⇨ Linear.**

   The Regression dialog appears.

3. **Select the mpg variable and place it in the Dependent box.**
4. **Select the engine_liters variable and place it in the Block 1 of 1 box.**
5. **Click OK.**

   The tables shown in Figure 11-1 appear.

Regression

Variables Entered/Removed[a]

Model	Variables Entered	Variables Removed	Method
1	engine_liters[b]	.	Enter

a. Dependent Variable: Miles per Gallon
b. All requested variables entered.

Model Summary

Model	R	R Square	Adjusted R Square	Std. Error of the Estimate
1	.789[a]	.622	.621	4.812

a. Predictors: (Constant), engine_liters

ANOVA[a]

Model		Sum of Squares	df	Mean Square	F	Sig.
1	Regression	15082.701	1	15082.701	651.345	<.001[b]
	Residual	9169.874	396	23.156		
	Total	24252.575	397			

a. Dependent Variable: Miles per Gallon
b. Predictors: (Constant), engine_liters

Coefficients[a]

Model		Unstandardized Coefficients		Standardized Coefficients	t	Sig.
		B	Std. Error	Beta		
1	(Constant)	34.873	.506		68.889	<.001
	engine_liters	-3.596	.141	-.789	-25.521	<.001

a. Dependent Variable: Miles per Gallon

FIGURE 11-1: The simple linear regression results.

We focus on only the most important results, starting with the Model Summary table. R is 0.789 and is a correlation, as we discuss in Chapter 10. However, when doing regression, we more often talk about the square of this number, R2. In our example, R2 is 0.622. A common way to phrase this result is that engine_liters explains 62.2% of the variance in our dependent mpg.

Next, let's discuss the ANOVA table in the figure. Its role here is a bit different than in Chapter 9 because in regression you are not comparing groups. For our purposes in this chapter, we simply check that Sig. is < 0.05 to see if our model is significant. It is.

The next table, Coefficients, contains information about each of the independent variables. Remember that when doing a simple linear regression, you have only one independent variable, in this case engine_liters. Its Sig. is below 0.05, so it is significant. So, now let's discuss the coefficient itself. In the Unstandardized B column, the engine_liters variable's unstandardized coefficient is −3.596, which means that the MPG drops 3.596 for each additional liter of engine size.

To predict possible dependent variable values, you use a modified version of the formula of the equation of a line: y = mx + b. Conventionally, in regression, the components of this formula are referred to as y = b1 (the coefficient for the independent variable) * x (the independent variable) + b0 (the constant).

See the following for an example of running simple linear regression.

Q. Using the Cars_Regression.sav file, run a simple linear regression predicting horsepower with engine_liters. What would be the predicted horsepower for a car with a 6-liter engine? (Rounding is okay if you want to work without a calculator.)

Regression

Variables Entered/Removed[a]

Model	Variables Entered	Variables Removed	Method
1	engine_liters[b]	.	Enter

a. Dependent Variable: Horsepower
b. All requested variables entered.

Model Summary

Model	R	R Square	Adjusted R Square	Std. Error of the Estimate
1	.897[a]	.805	.804	17.044

a. Predictors: (Constant), engine_liters

ANOVA[a]

Model		Sum of Squares	df	Mean Square	F	Sig.
1	Regression	476479.798	1	476479.798	1640.249	<.001[b]
	Residual	115615.980	398	290.492		
	Total	592095.778	399			

a. Dependent Variable: Horsepower
b. Predictors: (Constant), engine_liters

Coefficients[a]

Model		Unstandardized Coefficients B	Std. Error	Standardized Coefficients Beta	t	Sig.
1	(Constant)	41.002	1.792		22.884	<.001
	engine_liters	19.966	.493	.897	40.500	<.001

a. Dependent Variable: Horsepower

A. The predicted horsepower is 41 + (6 * 20) = 161.

In this workbook, you've had a lot of practice generating output. It's also important to practice retrieving information from SPSS output so that you draw conclusions from it.

Regression

Variables Entered/Removed[a]

Model	Variables Entered	Variables Removed	Method
1	engine_liters[b]	.	Enter

a. Dependent Variable: Time to Accelerate from 0 to 60 mph (sec)

b. All requested variables entered.

Model Summary

Model	R	R Square	Adjusted R Square	Std. Error of the Estimate
1	.545[a]	.297	.295	2.369

a. Predictors: (Constant), engine_liters

ANOVA[a]

Model		Sum of Squares	df	Mean Square	F	Sig.
1	Regression	955.981	1	955.981	170.365	<.001[b]
	Residual	2266.989	404	5.611		
	Total	3222.970	405			

a. Dependent Variable: Time to Accelerate from 0 to 60 mph (sec)

b. Predictors: (Constant), engine_liters

Coefficients[a]

Model		Unstandardized Coefficients B	Unstandardized Coefficients Std. Error	Standardized Coefficients Beta	t	Sig.
1	(Constant)	18.329	.247		74.240	<.001
	engine_liters	-.891	.068	-.545	-13.052	<.001

a. Dependent Variable: Time to Accelerate from 0 to 60 mph (sec)

1 **Using the Cars_Regression.sav file, run a simple linear regression keeping engine_ liters as the independent variable and acceleration as the dependent variable. What are R and R2? Is the overall model significant according to the ANOVA table?**

Is your independent variable significant?

 Approximately how much faster does a car accelerate (in seconds) for each additional liter of engine size?

 What do you notice about the significance of the overall model, according to the ANOVA table, and the significance of the engine liters variable? Are there any similarities between both results?

Performing Multiple Linear Regression

Multiple linear regression is an extension of simple linear regression. The only difference is that you have more than one independent variable. The steps in the dialog are nearly the same, but interpretation is a bit more complicated.

First, as you add or drop independent variables, the values for all coefficients change, sometimes dramatically. You can't assume that the coefficient you observed when there was only one independent variable will remain constant when you add another. Second, the formula now has to take into account more than one coefficient:

$$y = b1(x1) + b2(x2) + \ldots + b0$$

In this section, you rehearse using multiple linear regression by adding a second independent variable to your examples. Follow these steps to perform multiple linear regression:

1. **Choose File ⇨ Open ⇨ Data and load the Cars_Regression.sav file.**

   To download the file, go to the book's companion website at `www.dummies.com/go/spssstatisticsworkbookfd`.

2. **Choose Analyze ⇨ Regression ⇨ Linear.**

   The Regression dialog appears.

3. **Select the mpg variable and place it in the Dependent box.**
4. **Select the engine_liters and American_car variables, and place them in the Block 1 of 1 box.**

5. **Click OK.**

The results are shown in Figure 11-2.

Regression

Variables Entered/Removed[a]

Model	Variables Entered	Variables Removed	Method
1	American car, engine_liters[b]	.	Enter

a. Dependent Variable: Miles per Gallon
b. All requested variables entered.

Model Summary

Model	R	R Square	Adjusted R Square	Std. Error of the Estimate
1	.806[a]	.650	.648	4.624

a. Predictors: (Constant), American car, engine_liters

ANOVA[a]

Model		Sum of Squares	df	Mean Square	F	Sig.
1	Regression	15617.907	2	7808.953	365.257	<.001[b]
	Residual	8423.465	394	21.379		
	Total	24041.372	396			

a. Dependent Variable: Miles per Gallon
b. Predictors: (Constant), American car, engine_liters

Coefficients[a]

Model		Unstandardized Coefficients B	Unstandardized Coefficients Std. Error	Standardized Coefficients Beta	t	Sig.
1	(Constant)	35.205	.490		71.891	<.001
	engine_liters	-3.437	.179	-.754	-19.194	<.001
	American car	-1.236	.631	-.077	-1.959	.051

a. Dependent Variable: Miles per Gallon

FIGURE 11-2: Multiple regression results with MPG and two independent variables.

Let's discuss the key results. The ANOVA table shows a significance of < 0.001, so your model is significant. The R2 of 0.65 indicates that 65% of the variance in the dependent variable has been explained by the independent variables. One of the independent variables, engine_liters, is significant, but the other independent variable, american_car, is not.

TECHNICAL STUFF

When you have a p-value that is just above 0.05. it's customary to acknowledge that fact in the prose of your report. Although 0.051 is not significant at 95% confidence, it's close enough that you would want to alert other researchers (or your professor or boss), who might have a hypothesis about it and might want to investigate further. Additionally, it is customary to run the regression again with non-significant variables removed.

If you want to predict future values of mpg by using the two independent variables, the formula would be

$$\text{Predicted MPG} = -3.347(\text{engine\_liters}) - 1.236(\text{american\_car}) + 35.205$$

Note that you subtract –1.236 from the predicted MPG when the car is American but make no adjustment for other cars. European and Japanese cars have a value of 0 for american_car.

Practice using a different dependent variable.

EXAMPLE

Q. Using the Car_Regression.sav file, rerun the multiple regression using horsepower as the dependent variable and engine_liters and american_car as the independent variables. Summarize the key findings.

Regression

Variables Entered/Removed[a]

Model	Variables Entered	Variables Removed	Method
1	American car, engine_liters[b]	.	Enter

a. Dependent Variable: Horsepower
b. All requested variables entered.

Model Summary

Model	R	R Square	Adjusted R Square	Std. Error of the Estimate
1	.909[a]	.827	.826	16.089

a. Predictors: (Constant), American car, engine_liters

ANOVA[a]

Model		Sum of Squares	df	Mean Square	F	Sig.
1	Regression	489447.422	2	244723.711	945.395	<.001[b]
	Residual	102507.996	396	258.859		
	Total	591955.419	398			

a. Dependent Variable: Horsepower
b. Predictors: (Constant), American car, engine_liters

Coefficients[a]

Model		Unstandardized Coefficients B	Unstandardized Coefficients Std. Error	Standardized Coefficients Beta	t	Sig.
1	(Constant)	40.868	1.702		24.015	<.001
	engine_liters	22.714	.622	1.016	36.531	<.001
	American car	-14.099	2.213	-.177	-6.372	<.001

a. Dependent Variable: Horsepower

A. The model is statistically significant and has an R2 of 0.827. Both independent variables are significant.

Regression

Variables Entered/Removed[a]

Model	Variables Entered	Variables Removed	Method
1	American car, engine_liters[b]	.	Enter

a. Dependent Variable: Time to Accelerate from 0 to 60 mph (sec)

b. All requested variables entered.

Model Summary

Model	R	R Square	Adjusted R Square	Std. Error of the Estimate
1	.577[a]	.333	.330	2.294

a. Predictors: (Constant), American car, engine_liters

ANOVA[a]

Model		Sum of Squares	df	Mean Square	F	Sig.
1	Regression	1058.156	2	529.078	100.526	<.001[b]
	Residual	2115.762	402	5.263		
	Total	3173.918	404			

a. Dependent Variable: Time to Accelerate from 0 to 60 mph (sec)

b. Predictors: (Constant), American car, engine_liters

Coefficients[a]

Model		Unstandardized Coefficients B	Std. Error	Standardized Coefficients Beta	t	Sig.
1	(Constant)	18.398	.241		76.459	<.001
	engine_liters	-1.102	.088	-.676	-12.545	<.001
	American car	1.005	.312	.174	3.225	.001

a. Dependent Variable: Time to Accelerate from 0 to 60 mph (sec)

5 Using the Car_Regression.sav file, run a multiple linear regression with acceleration as the dependent variable and engine_liters and american_car as the independent variables. What are R and R2? Is the overall model significant according to the ANOVA table?

6 Are both independent variables significant? Which one appears to be more important?

7 Write the formula to predict acceleration. What is the prediction for an American car with a 5-liter engine?

8 Write the formula to predict acceleration. What is the prediction for a 5-liter car that was not made in America? What does that tell us about European or Japanese cars?

Answers to Problems in Regression

1. The output shows an R of 0.545 and an R2 of 0.297. Therefore, engine size explains nearly 30% of the variance in acceleration. The ANOVA table shows that you have a significant model. You can conclude that it is a good model, and that engine size has a significant and substantial capability to predict acceleration, which would also be consistent with what you would anticipate about these two variables.

2. Yes, your independent variable is significant. The output shows the significance of acceleration as an independent variable as < 0.01.

3. The coefficient is 0.897, so the acceleration is reduced by approximately one second (9/10s of a second) for each additional liter of engine size.

4. It's noteworthy that the p-value for both is the same, < 0.001. This is occurring because the two statistical tests are determining the significance of the same thing. When there is only one independent variable, the tests of the overall model and of the independent variable are equivalent.

TECHNICAL STUFF

You can double-click an output table, and then double-click again to see more decimal places. Doing so will reveal the value in scientific notion. If you were to do this for both the ANOVA table and the independent variable, you would find that they are the same to more than 30 decimal places.

Coefficients[a]

Model		Unstandardized Coefficients		Standardized Coefficients		
		B	Std. Error	Beta	t	Sig.
1	(Constant)	18.329	.247		74.240	.000
	engine_liters	-.891	.068	-.545	-13.0	9.8376E-33

a. Dependent Variable: Time to Accelerate from 0 to 60 mph (sec)

5. R is 0.577 and R2 is 0.333. Therefore, more than 33% of the variance in acceleration is explained by the combination of engine size and whether or not the car is American.

TECHNICAL STUFF

When you're using multiple linear regression, R is no longer quite the same concept as the correlation in Chapter 10. It is a referred to as "multiple R." R2 becomes the more important number to share and is the most widely reported with multiple regression results.

6. They are both significant, with values < 0.05. Two clues tell you that engine_liters is the more important of the two. First, the significance value is slightly lower. The output reports0.001 for american_car but < 0.001 for engine_liters. Second, the coefficients are nearly the same magnitude, but the coefficient for engine_liters is multiplied times engine size. The coefficient for american_car is multiplied by only 0 or 1, so the product of the two numbers can't become large.

REMEMBER

It's always safer to compare the standardized coefficients to determine which of two independent variables is more important. If you refer to the unstandardized coefficient, you have to carefully take into account the metric of the variable. If two variables are not measured in the same way, you can get in trouble. For example, a car's acceleration is measured in seconds, but engine size is measured in liters.

7. The predicted acceleration is (–1.102 * 5) + 1.005 + 18.398.

8. The predicted acceleration is (–1.102 * 5) + 18.398.

   Note that you simply omit the adjustment for american_car because the coefficient is multiplied times 0. This calculation does not (and cannot) help you differentiate between Japanese and European cars. It can tell you only how American cars differ from all other cars in the dataset.

TECHNICAL STUFF

To get a more complete picture of car origin, you would have to create another variable that's 1 when a car is European and 0 when it's not. Then you could add both car variables. You would have a coefficient that compared American cars to Japanese cars, and a coefficient that compared European to Japanese. Japanese cars would be the reference group and would not get a coefficient assigned to them.

4

Producing and Editing Output

IN THIS PART . . .

Display your data visually.

Create simple and complex graphs.

Edit your graphs and tables.

IN THIS CHAPTER

» Creating bar charts

» Building line charts

» Looking at pie charts

» Making sense of histograms

Chapter 12

Building Graphs with Bars, Lines, and Wedges

SPSS can display your data in a variety of charts. Additionally, each type of chart can have multiple appearances. For example, a bar chart can have a two- or three-dimensional appearance, represent data in different colors, and contain simple lines or error bars. The choice of layouts is almost endless.

TIP

In the world of SPSS, the terms *chart* and *graph* mean the same thing and are used interchangeably.

Chart Builder is the most common way to create graphs. It begins by presenting various chart types. After you select a chart, you specify which variables you will use.

TIP

When you use Chart Builder, you can drag and drop into your graph any variables you want to see. If the variable doesn't make sense in a particular location, the drop will fail. Chart Builder in SPSS will tell you what will and won't work.

Building Bar Graphs

A *bar graph* is a comparison of relative magnitudes. Simple bar graphs are the most common way of charting statistics. Follow these steps to generate a simple bar graph:

1. **Choose File ⇨ Open ⇨ Data and load the GSS2018.sav file.**

   You can download the file from the book's companion website at `www.dummies.com/go/spssstatisticsworkbookfd`. This file contains data from the General Social Survey (GSS), a nationally representative survey of adults in the United States that collects data on contemporary opinions, attitudes, and behaviors.

2. **Choose Graphs ⇨ Chart Builder.**

   A warning appears, informing you that before you use this dialog, the measurement level should be set properly for each variable in your chart. (We've set the correct measurement level, so you can proceed.)

3. **In the Choose From list, select Bar.**
4. **Select the first graph image (with the Simple Bar tooltip) and drag it to the panel at the top of the window.**
5. **In the Variables list, select CLASS and drag it to the X-Axis rectangle.**

REMEMBER

   The graphic display in the Chart Builder preview window *never* represents your actual data, even after you insert variable names. This preview window simply displays a diagram that demonstrates the composition and appearance of the graph that will be produced.

6. **Click OK.**

   The simple bar graph in Figure 12-1 appears. Most people in the sample are either in the working- or middle-class categories.

See the following for an example of creating a bar chart.

EXAMPLE

Q. Using the GSS2018.sav file, create a bar chart showing the distribution of the CONBUS (confidence in major companies) variable. In this example, display percentages instead of counts.

A. More than 60% of respondents chose the Only Some response option. To display percentages instead of counts, you must go to the Elements tab and select Percentage in the Statistic section.

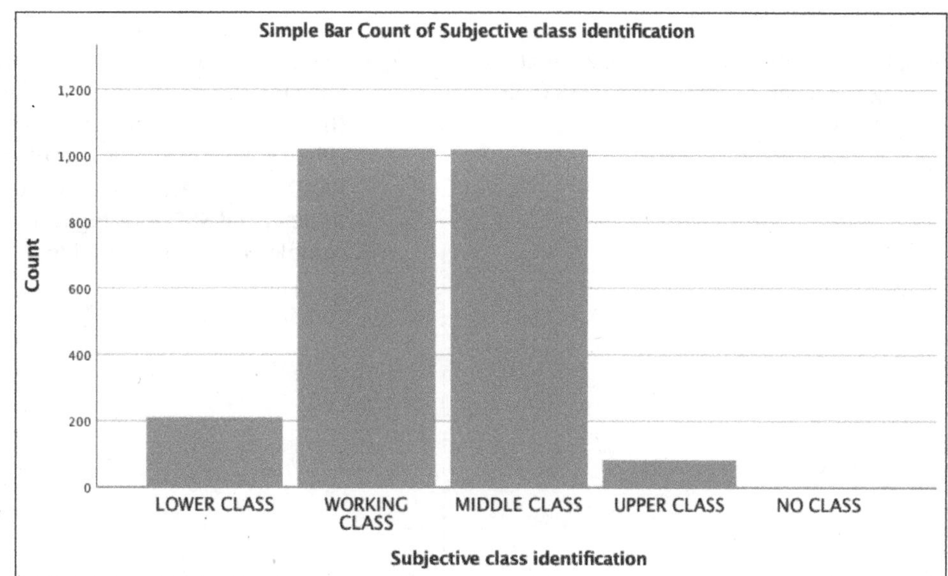

FIGURE 12-1: Bar graph of the Class variable.

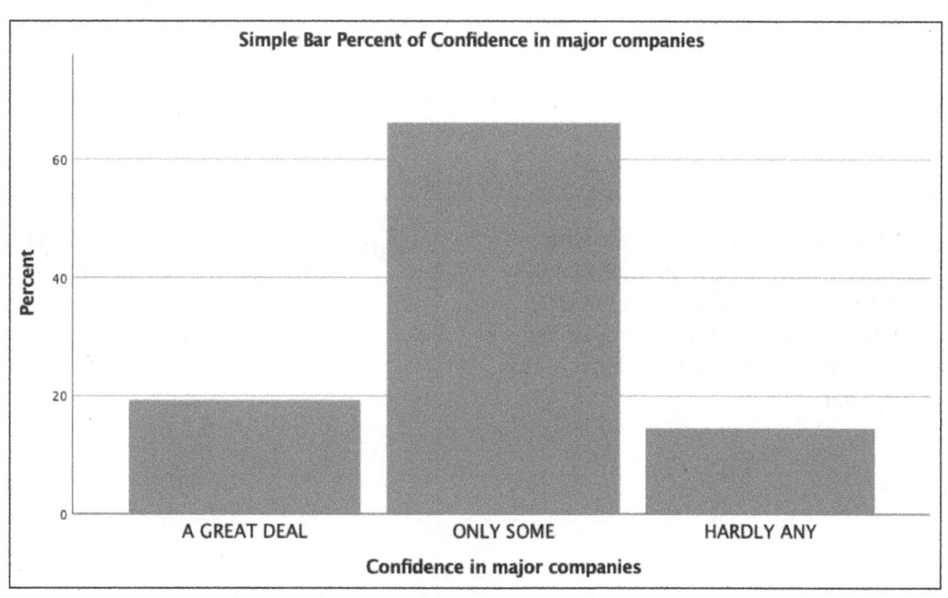

1 Using the GSS2018.sav file, create a bar chart showing the mean age for each CLASS (subjective class identification).

2 Using the GSS2018.sav file, create a bar chart showing the mean education for the EDUC (highest year of school completed), MAEDUC (highest year of school completed, mother), PAEDUC (highest year of school completed, father), and SPEDUC (highest year of school completed, spouse) variables.

3 Using the GSS2018.sav file, create a bar chart showing the distribution of CLASS (subjective class identification) but only for those people who have a value of Yes for the BORN (was respondent born in this country) variable.

4 Using the GSS2018.sav file, create a bar chart showing the distribution of DEGREE (respondent's highest degree) but change the title to *Highest Degree.*

Creating Line Charts

Line charts are popular because they are easy to read and work well as a visual summary of categorical values. They're also useful for displaying timelines because they demonstrate up and down trends so well. The following steps generate a basic line graph:

1. **Choose File ⇨ Open ⇨ Data and load the GSS2018.sav file.**

   You can download the file from the book's companion website at `www.dummies.com/go/spssstatisticsworkbookfd`.

2. **Choose Graphs ⇨ Chart Builder.**
3. **In the Choose From list, select Line to specify the general type of graph to be constructed.**
4. **Select the first diagram (with the Simple Line tooltip) and drag it to the panel at the top.**
5. **In the Variables list, select CLASS and drag it to the X-Axis rectangle.**
6. **Click OK.**

   The simple line graph in Figure 12-2 appears. Most people in the sample are in either the working- or middle-class categories, and the graph is very similar to the bar chart we created earlier.

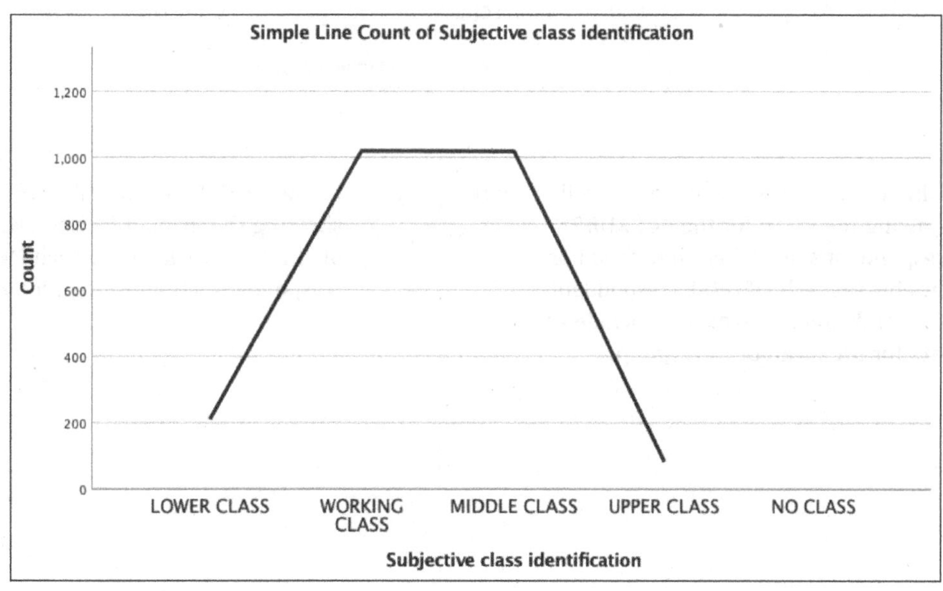

FIGURE 12-2: Line graph of the Class variable.

See the following for an example of creating a line chart.

EXAMPLE

Q. Using the GSS2018.sav file, create a line chart showing the mean for the AGEKDBRN (respondent's age when first child born) variable for each DEGREE (respondent's highest degree).

A. As education increases, the average age when respondents had their first child increases. To create this graph, you'll need to place the DEGREE variable on the x-axis and the AGEKDBRN variable on the y-axis.

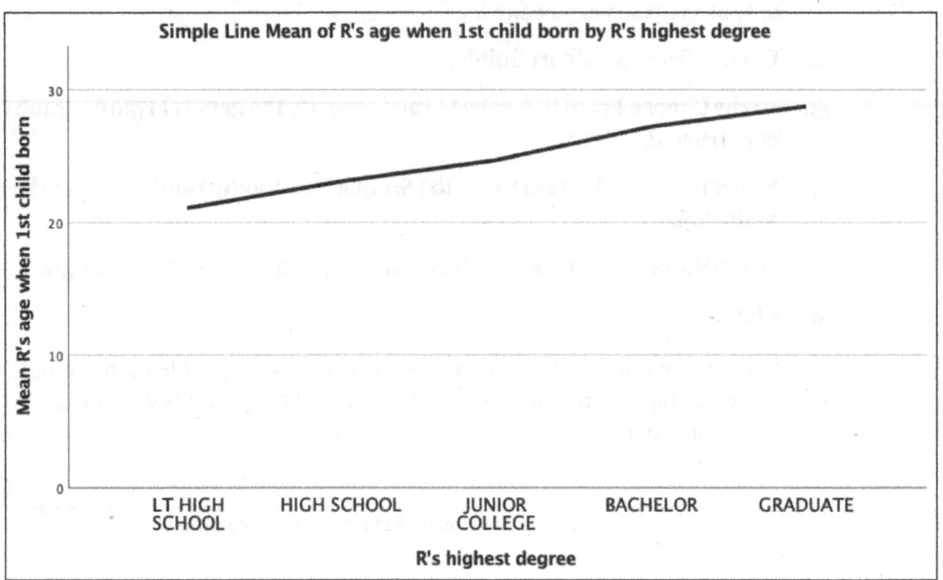

5 Using the GSS2018.sav file, create a line chart showing the mean for the AGEKDBRN (respondent's age when first child born) variable for each DEGREE (respondent's highest degree) variable but include error bars for the mean of each group.

6 Using the GSS2018.sav file, create a line chart showing the median for the CHILDS (number of children) variable for each DEGREE (respondent's highest degree) variable.

7 Using the GSS2018.sav file, create a line chart showing the mean for the CHILDS (number of children) and CHLDIDEL (ideal number of children) variables.

8 Using the GSS2018.sav file, create a line chart showing the mean age for each category of the BORN (was respondent born in this country) variable but add a footnote *Data from 2018.*

Making Pie Charts

Pie charts show how something (the whole) is divided into pieces — whether two, ten, or any other number. Each slice in the pie chart represents its percentage or count of the whole. In the following steps, you construct a pie chart:

1. **Choose File ⇨ Open ⇨ Data and load the GSS2018.sav file.**

   You can download the file from the book's companion website at `www.dummies.com/go/spssstatisticsworkbookfd`.

2. **Choose Graphs ⇨ Chart Builder.**
3. **In the Choose From list, select Pie/Polar.**
4. **Drag the pie diagram to the panel at the top of the window.**
5. **In the Variables list, drag DEGREE to the Slice By rectangle at the bottom of the panel.**
6. **Click OK.**

   The pie chart shown in Figure 12-3 appears. The high school degree group makes up about 50% of the sample.

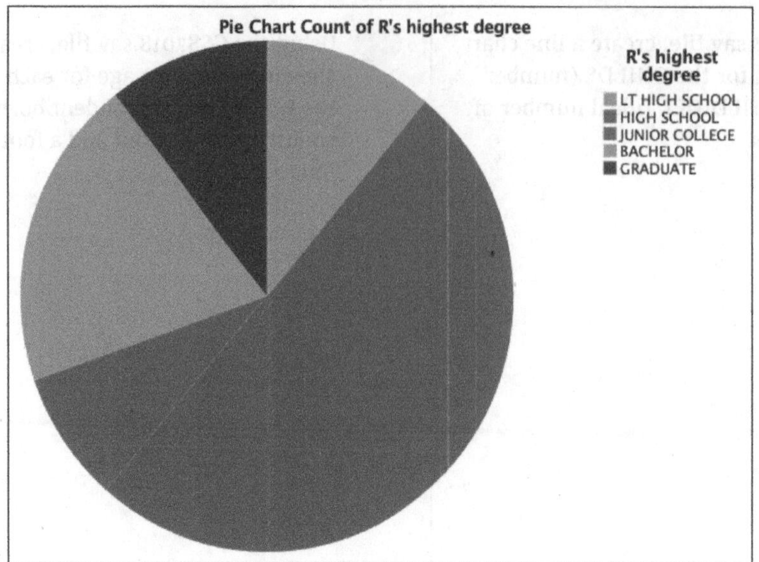

FIGURE 12-3: Pie chart of the Degree variable.

See the following for an example of creating a pie chart.

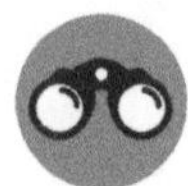

EXAMPLE

Q. Using the GSS2018.sav file, create a pie chart showing the distribution of the MARITAL (marital status) variable. Display percentages instead of the count.

A. The married group accounts for almost half the sample. To display percentages instead of counts, go to the Elements tab and select Percentage in the Statistic section.

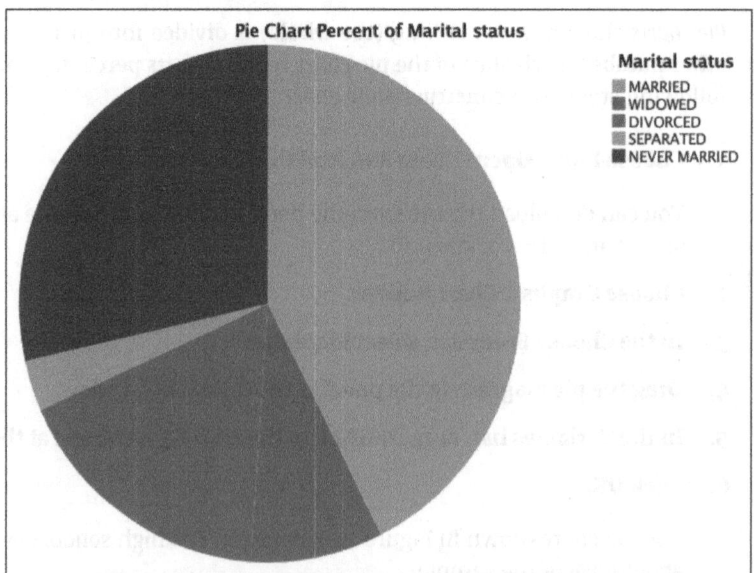

9 Using the GSS2018.sav file, create a pie chart showing the distribution of the DEGREE (respondent's highest degree) variable broken out by where respondents were born (BORN). Make sure to place the BORN variable in separate columns.

10 Using the GSS2018.sav file, create a pie chart showing the distribution of the DEGREE (respondent's highest degree) variable broken out by where respondents were born (BORN). Make sure to place the BORN variable in separate rows in the Groups/Point ID tab.

11 Using the GSS2018.sav file, create a pie chart showing the distribution of the DEGREE (respondent's highest degree) variable broken out by where respondents were born (BORN) and CLASS (subjective class identification). Make sure to place the BORN variable in separate rows and the CLASS variable in separate columns.

12 Using the GSS2018.sav file, create a pie chart showing the distribution of the HAPPY (general happiness) variable but add the subtitle *Before Intervention.*

Constructing Histograms

A *histogram* represents the number of items that appear within a range of values. You can use a histogram to look at a graphic representation of the frequency distribution of a continuous variable's values. In the following steps, you construct a histogram:

1. **Choose File ⇨ Open ⇨ Data and load the GSS2018.sav file.**

   You can download the file from the book's companion website at `www.dummies.com/go/spssstatisticsworkbookfd`.

2. **Choose Graphs ⇨ Chart Builder.**
3. **In the Choose From list, select Histogram.**
4. **Drag the first graph diagram (with the Simple Histogram tooltip) to the panel at the top of the window.**
5. **In the Variables list, select the AGE variable and drag to the X-Axis rectangle in the panel.**
6. **Click OK.**

   The histogram shown in Figure 12-4 appears. (Although the graph looks like a bar chart, it's a histogram.) The average age is almost 49 years.

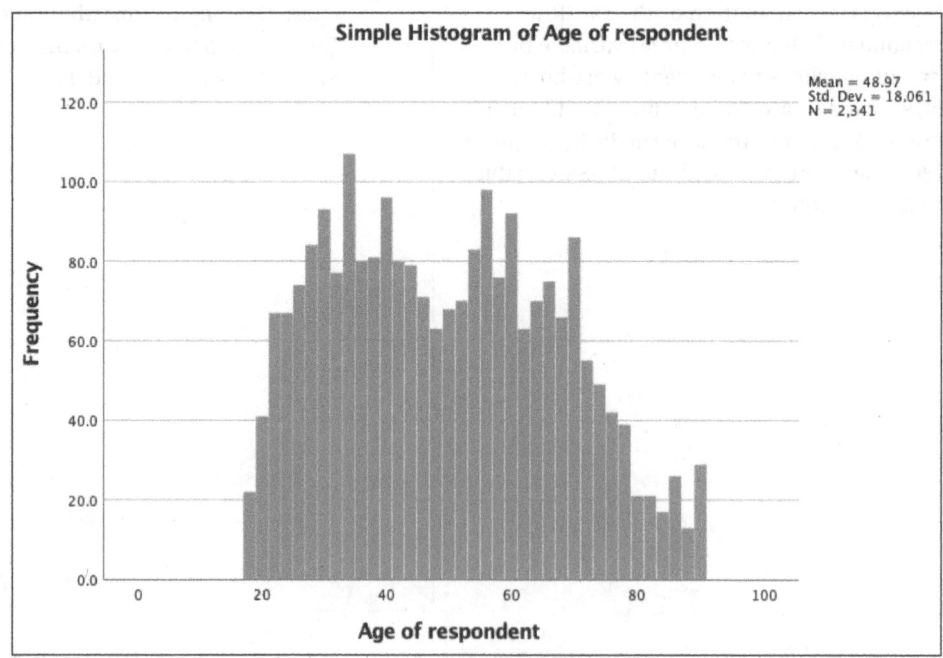

FIGURE 12-4: A histogram displaying distribution of age.

See the following for an example of creating a histogram.

EXAMPLE

Q. Using the GSS2018.sav file, create a histogram showing the distribution of the AGEKDBRN (respondent's age when first child born) variable. Display a normal curve.

A. The majority of respondents had their first child when they were in their late teens through mid-twenties. To display a normal curve, go to the Elements tab and select Display Normal Curve.

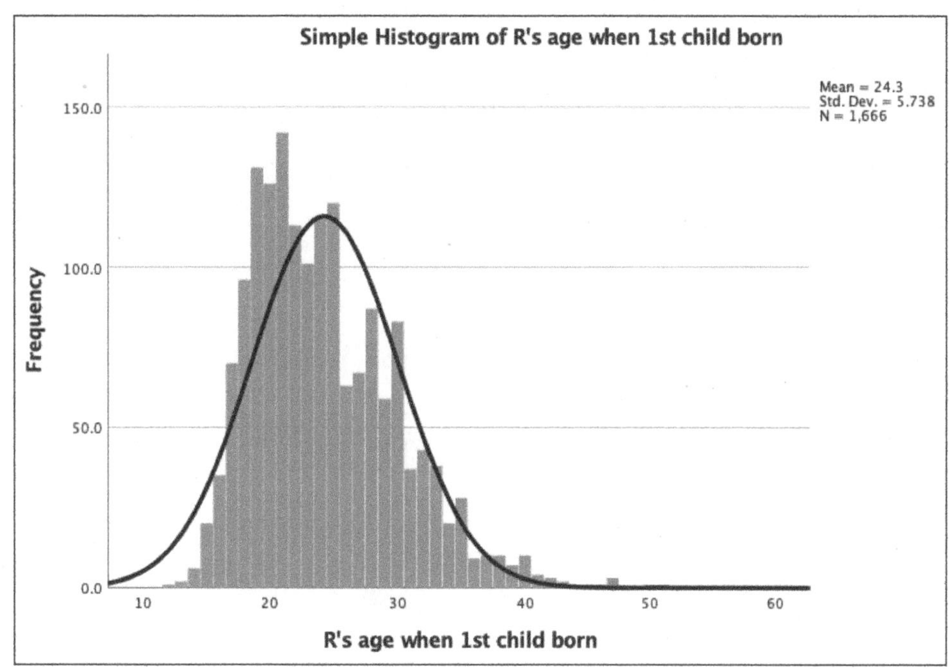

13 Using the GSS2018.sav file, create a histogram showing the distribution of the EDUC (highest year of school completed) variable but display percentages instead of counts.

14 Using the GSS2018.sav file, create a histogram showing the distribution of the CHILDS (number of children) variable but only for those people who have a value of Yes for the BORN (was respondent born in this country) variable.

15 Using the GSS2018.sav file, create a histogram showing the mean for the CHILDS (number of children) and CHLDIDEL (ideal number of children) variables.

16 Using the GSS2018.sav file, create a histogram showing the mean for the CHILDS (number of children) and CHLDIDEL (ideal number of children) variables but include error bars variables.

Answers to Problems in Building Graphs with Bars, Lines, and Wedges

1. The average age is highest for the upper-class group, while the working-class group is the youngest. To create this bar chart, place the CLASS variable on the x-axis and the AGE variable on the y-axis.

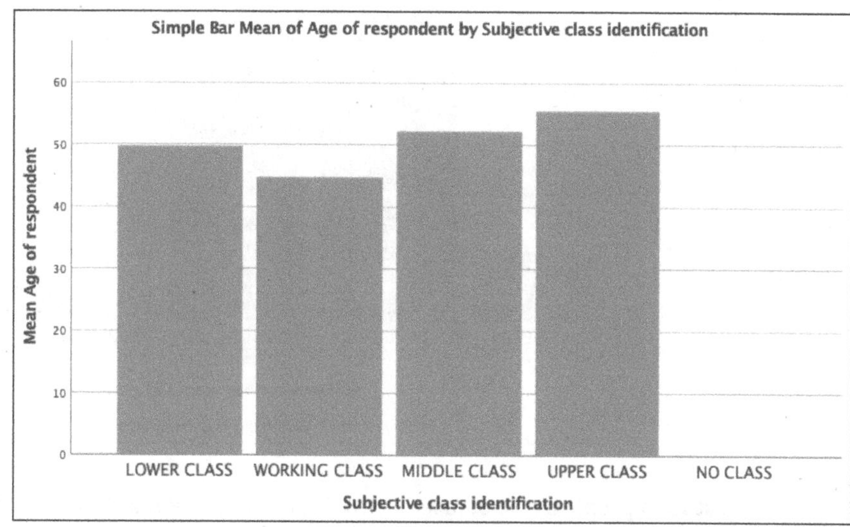

2. The average education for mothers and fathers is similar. In addition, the education for the respondents and their spouses are similar. To create this bar chart, you need to place all the education variables on the y-axis.

Be careful how you drop variables on the y-axis. When adding variables, drop them on the little box containing the plus sign. If you drop additional variables on top of the variable already on the y-axis, the original variable will be replaced.

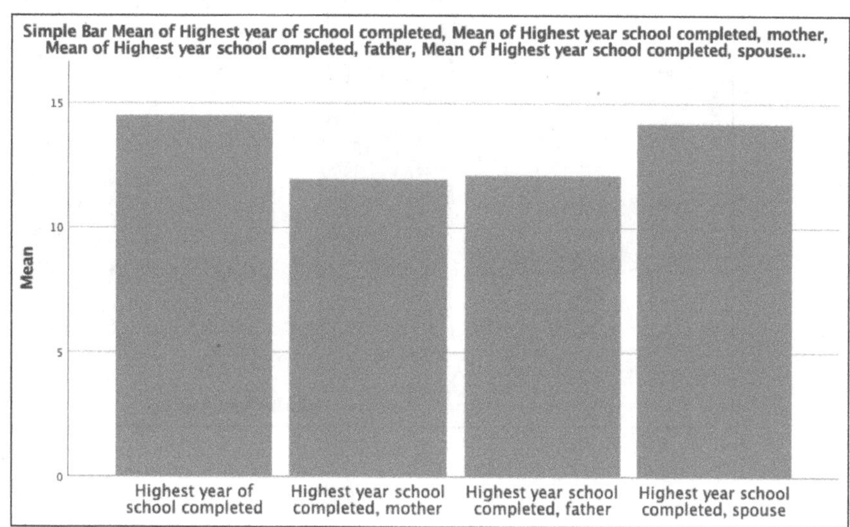

3. The graph shows that most people are in either the working- or middle-class group. To create this graph, place the CLASS variable on the x-axis and place the BORN variable in the Filter box. Then, in the Element Properties section, click the No value to select it, and then click the red X to exclude the No value from the graph.

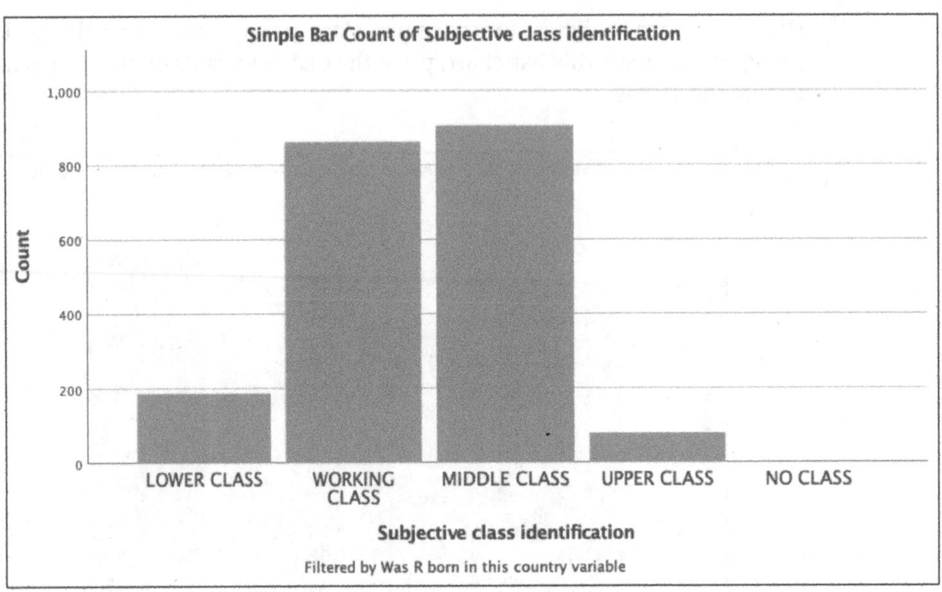

4. Almost 1200 people in the dataset have a high school degree. To create this graph, place the DEGREE variable on the x-axis. Then click the Titles/Footnote tab and select the Title 1 check box. Then, in the Element Properties section, click the Custom text box and type the title *Highest Degree.*

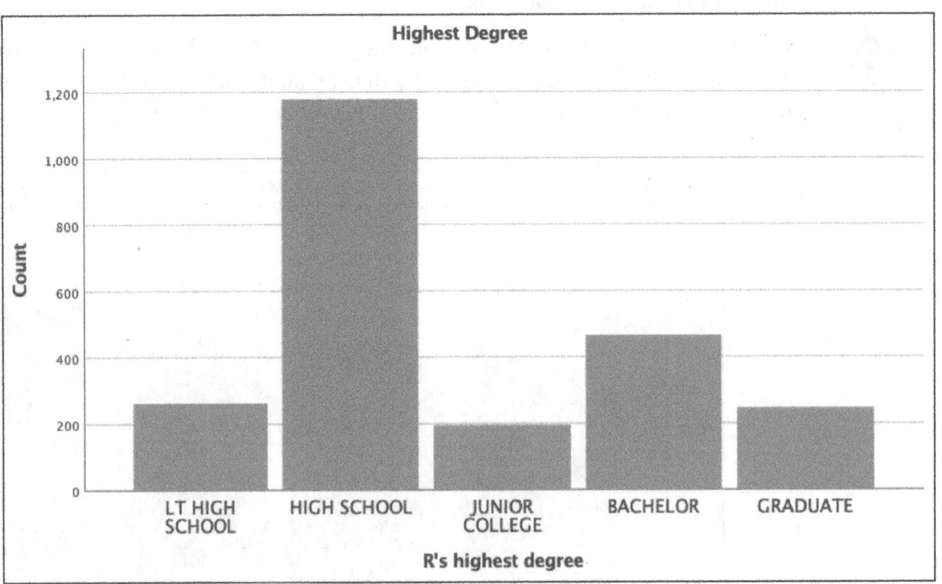

5 The graph shows that the average age of having a first child increases as education increases. You can also see that the high school group has less variability compared to the other education categories. To create this line graph, place the DEGREE variable on the x-axis and place the AGEKDBRN variable on the y-axis. Then, in the Element Properties section, click the Display Error Bars option.

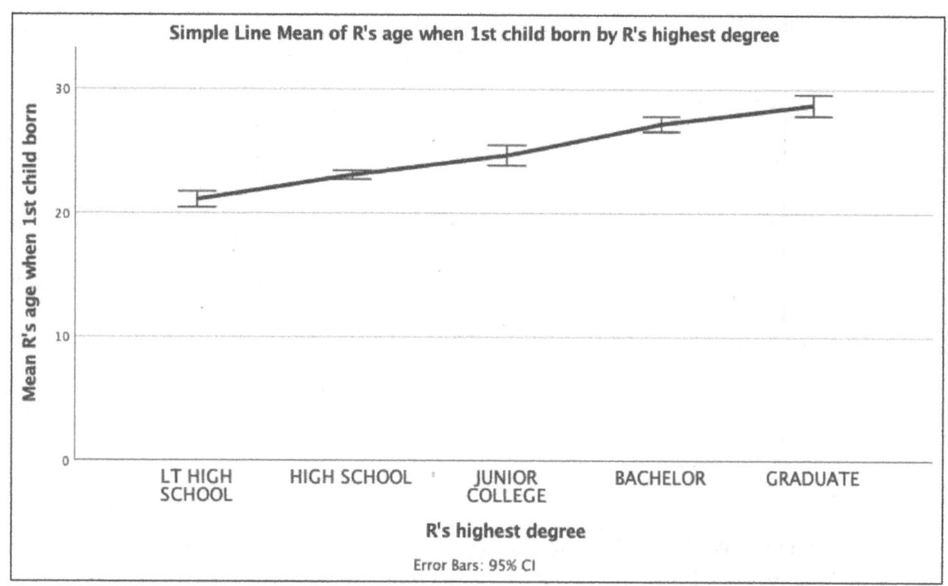

6 The median number of children for every degree group is 2, except for the less than high school group, which has a median of 3. To create this line graph, place the DEGREE variable on the x-axis and place the CHILDS variable on the y-axis. Then, in the Element Properties section, change the Statistic from Mean to Median.

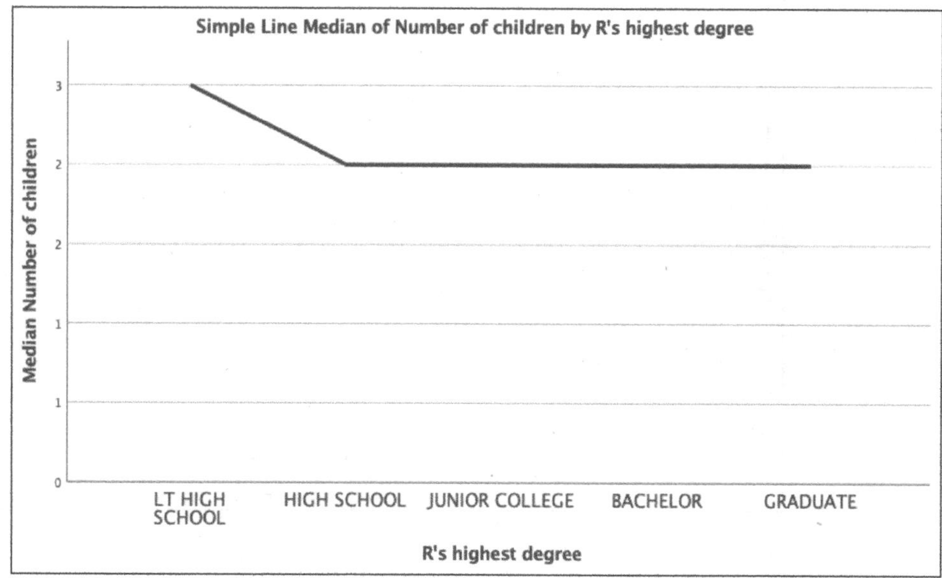

7. The graph shows that the ideal number of children is much higher than the actual number of children. To create this line chart, place all the variables on the y-axis.

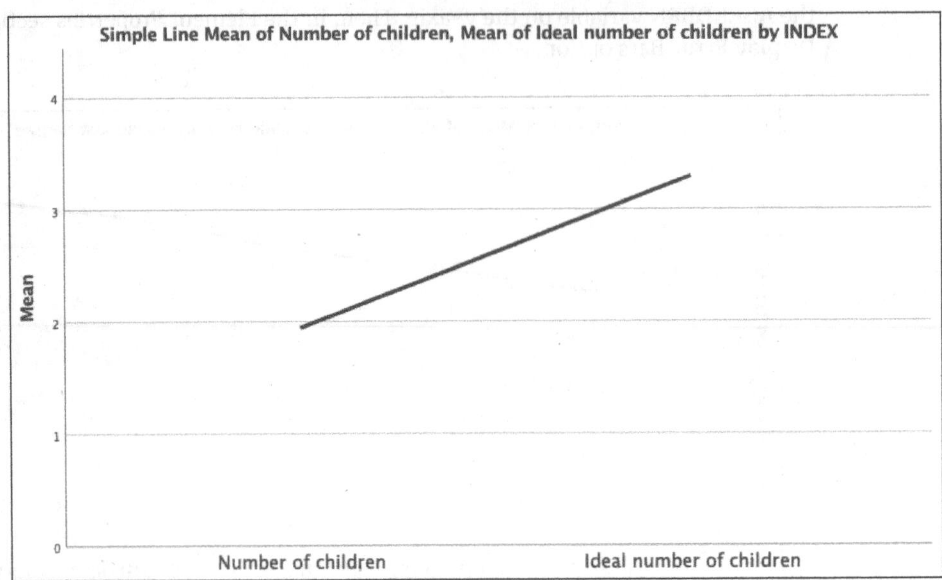

8. The graph shows that the average age in the sample is similar regardless of where respondents were born. To create this line chart, place the BORN variable on the x-axis and place the AGE variable on the y-axis. Click the Titles/Footnote tab and select the Footnote 1 check box. Then, in the Element Properties section, click the Custom text box and type the footnote *Data from 2018*.

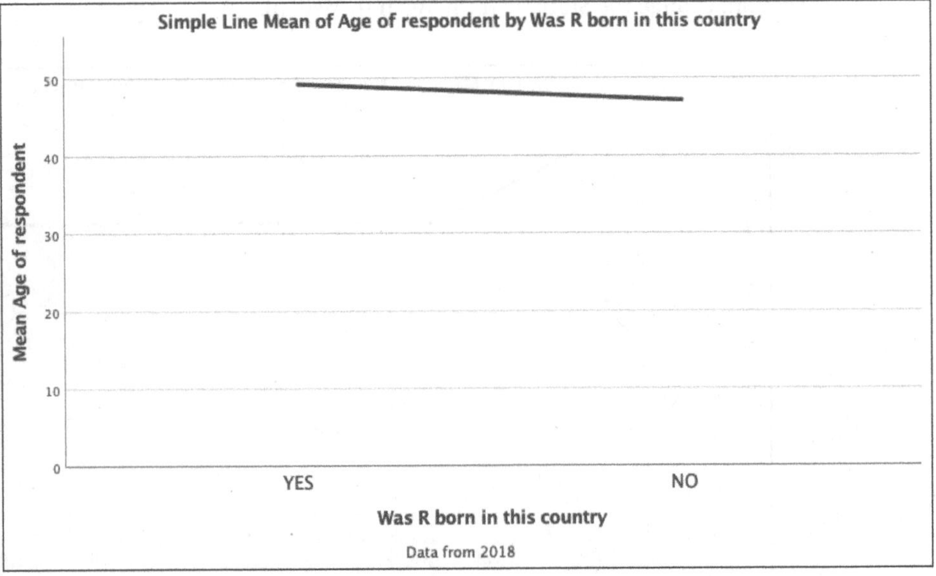

9 When comparing the two graphs, more people with a high school degree were born in this country. Additionally, more people with less than a high school degree were not born in this country. To create this pie chart, place the DEGREE variable in the Slice By box. Then click the Groups/Point ID tab, select the Columns panel variable check box, and place the BORN variable in the Panel box.

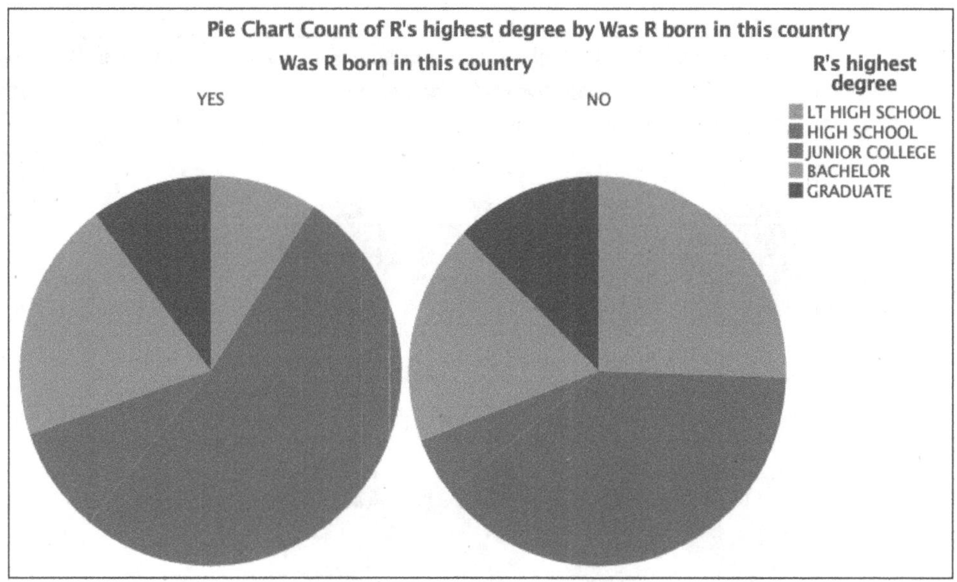

10 When comparing the two graphs, more people with a high school degree were born in this country. Additionally, more people with less than a high school degree were not born in this country. To create this pie chart, place the DEGREE variable in the Slice By box. Then click the Groups/Point ID tab, select the Rows panel variable check box, and place the BORN variable in the Panel box.

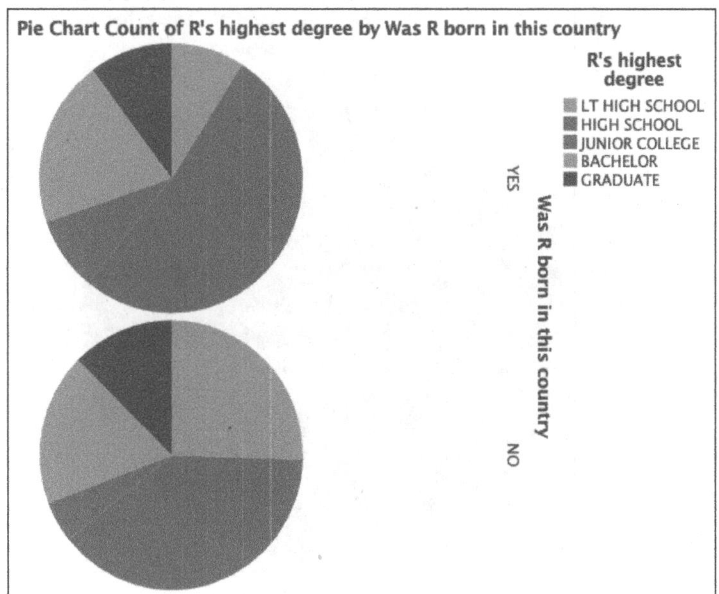

11 There is certainly a variety of differences for each of the combinations, but the combination that stands out the most is for people in the upper class who were not born in this country, because this group is equally comprised of graduate and high school degrees. To create this pie chart, place the DEGREE variable in the Slice By box. Then click the Groups/Point ID tab, select the Rows panel variable check box, and place the BORN variable in the Panel box. Next, select the Columns panel variable check box, and place the CLASS variable in the Panel box.

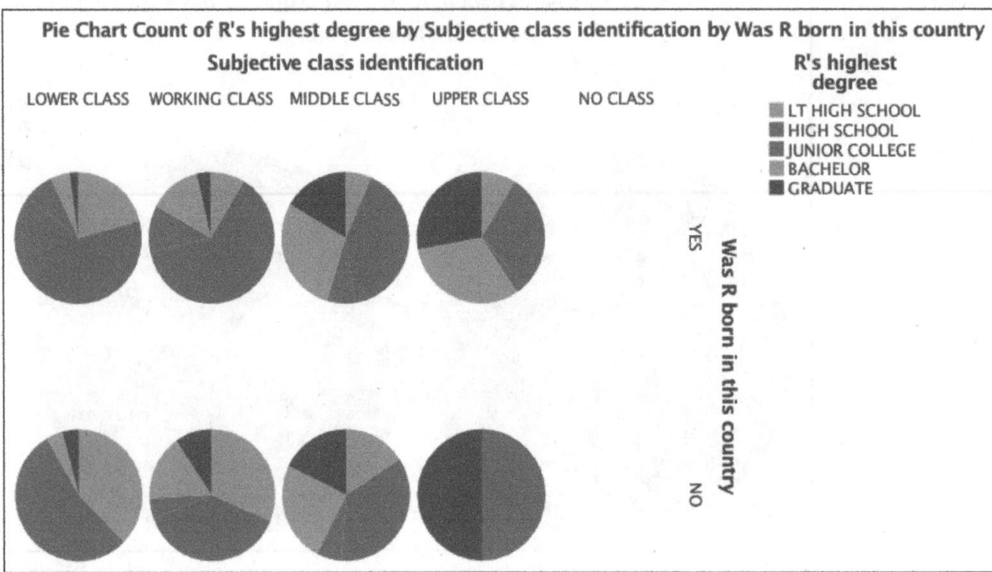

12 The pretty happy group makes up about 50% of the sample. To create this pie chart, place the HAPPY variable in the Slice By box. Click the Titles/Footnote tab, and select the Subtitle check box. Then, in the Element Properties section, click the Custom text box and type the subtitle *Before Intervention.*

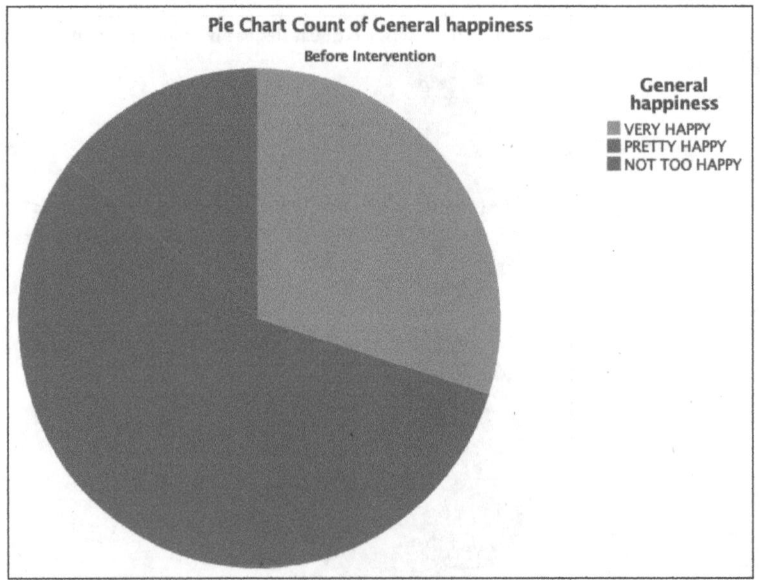

13 The histogram shows that the most common response is 12 years of education; almost 30% of the sample has this value. To create this histogram, place the EDUC variable on the x-axis. Then, in the Element Properties section, change the Statistic from Histogram to Histogram Percent.

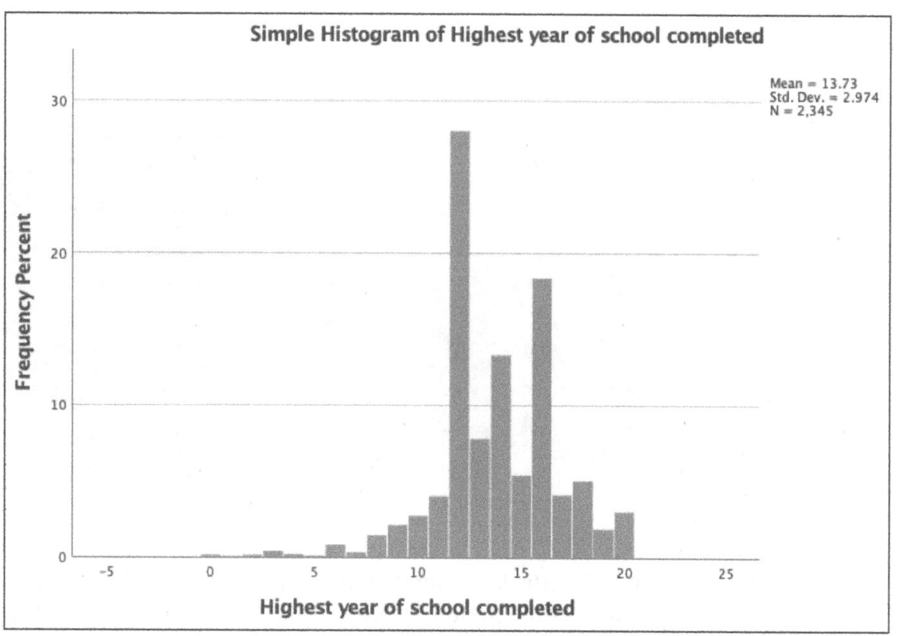

14 The histogram shows that most people have either 0 or 2 children. To create this histogram, place the CHILDS variable on the x-axis, and place the BORN variable in the Filter box. Then, in the Element Properties section, click the No value to select it, and then click the red X to exclude the No value from the graph.

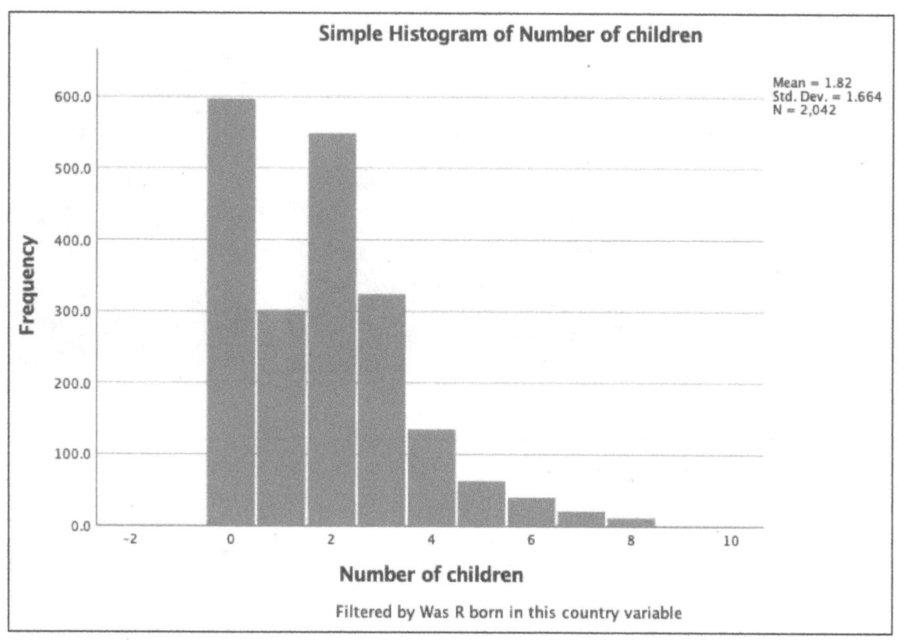

15 The graph shows that the ideal number of children is much higher than the actual number of children. To create this histogram, place all the variables on the y-axis.

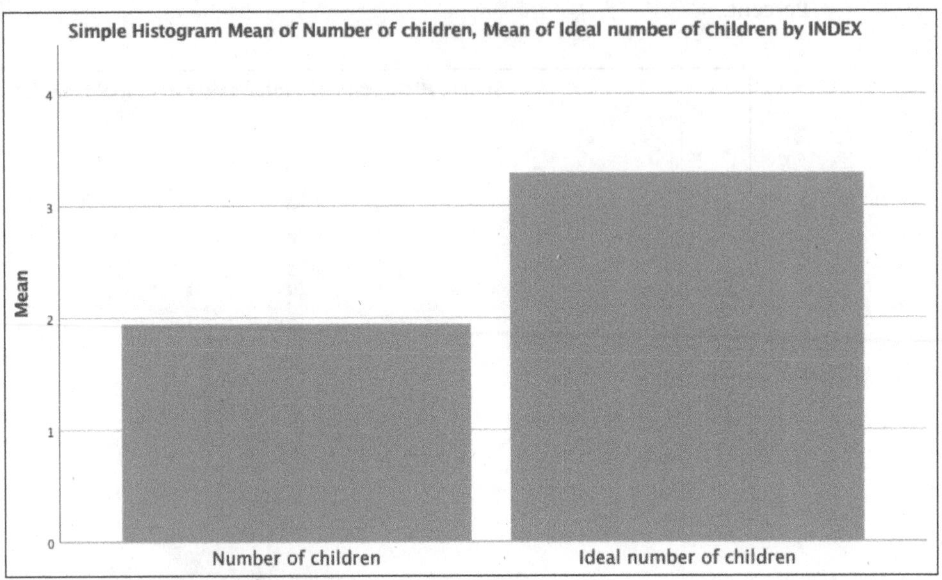

16 The graph shows that the ideal number of children is much higher than the actual number of children. To create this histogram, place all variables on the y-axis. Then, in the Element Properties section, click the Display Error Bars option.

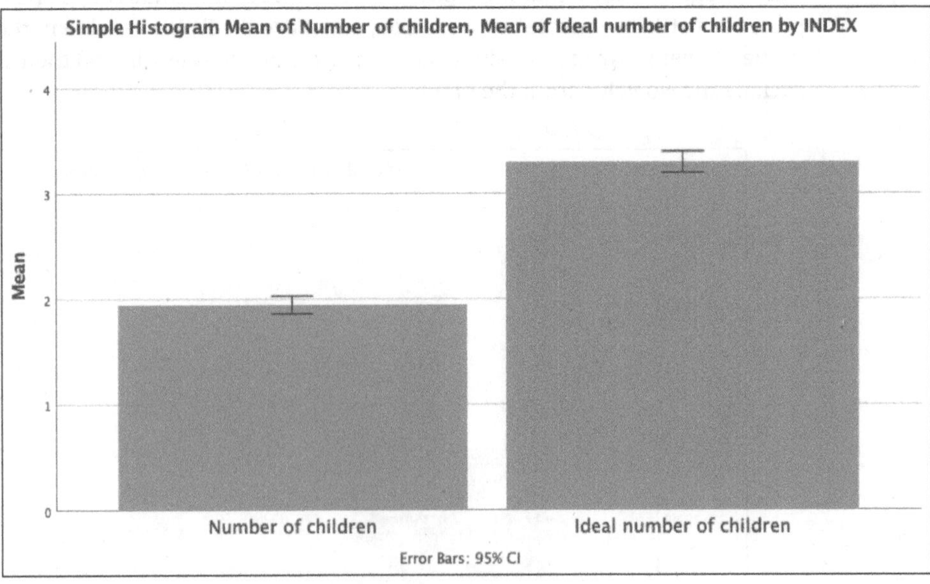

IN THIS CHAPTER

» Working with clustered charts

» Building multiple line charts

» Understanding error bar charts

» Making sense of scatterplots

» Creating box and whiskers plots

Chapter 13
Building Slightly More Complex Graphs

A wide variety of chart types are available in SPSS. In this chapter, you discover how to create some of these different graph types. Although we don't cover every combination of the chart features, you can use the steps here to produce some useful graphs.

You may be familiar with some of the simple graphs, but you'll also see examples of charts that may be less familiar and aren't as simple.

Building Clustered Graphs

As shown in Chapter 7, a *clustered bar chart* is the most effective graph for displaying the results of a cross tabulation. In this chapter, we show a few different types of clustered charts.

Follow these steps to create a clustered bar chart:

1. **Choose File ⇨ Open ⇨ Data and load the GSS2018.sav file.**

   You can download the file from the book's companion website at `www.dummies.com/go/spssstatisticsworkbookfd`. This file contains data from the General Social Survey (GSS), a nationally representative survey of adults in the United States that collects data on contemporary opinions, attitudes, and behaviors.

2. **Choose Graphs ⇨ Chart Builder.**
3. **In the Choose From list, select Bar.**
4. **Select the second graph image (with the Clustered Bar tooltip), and drag it to the panel at the top of the window.**
5. **Select the DEGREE variable, and place it in the Cluster on X: Set Color box.**
6. **Select the OWNGUN variable, and place it in the X-Axis box.**
7. **Click OK.**

   The graph in Figure 13-1 appears. As you can see, many more people have a high school degree than any other education category.

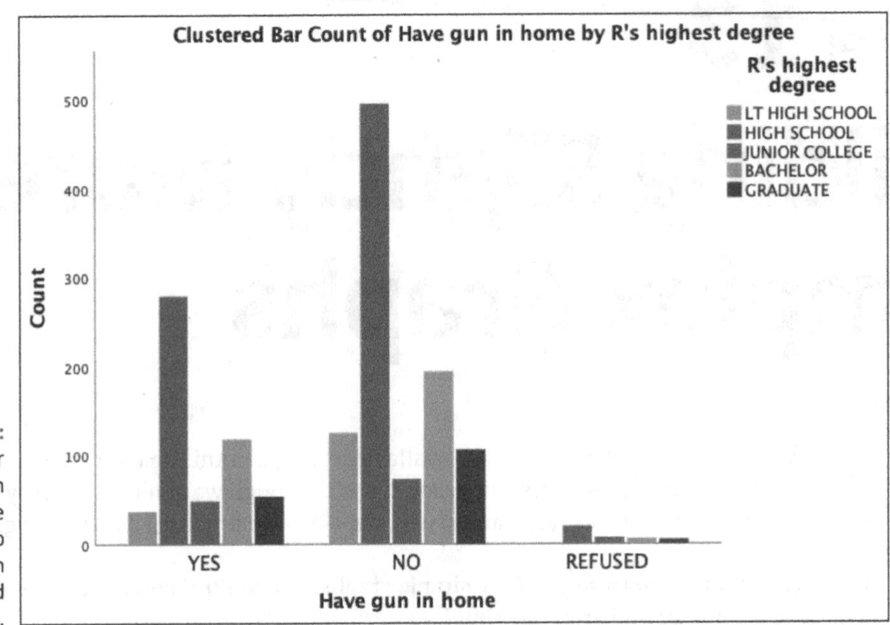

FIGURE 13-1: A cluster bar graph displaying the relationship between degree and owning a gun.

See the following for an example of creating a clustered bar chart.

Q. Using the GSS2018.sav file, create a clustered bar chart showing the relationship between DEGREE (Cluster Set Color box) and OWNGUN (x-axis), as in the preceding example. In this example, however, display percentages instead of counts for the DEGREE variable. Describe the results.

A. Those with less than a high school degree are the least likely people to own a gun. The relationship is much clearer when percentages are used instead of counts. To create this graph, do the following:

a. Place the DEGREE variable on the Cluster Set Color box.

b. Place the OWNGUN variable on the x-axis.

c. In the Element Properties section, change Statistic to Percentage.

d. Click the Set Parameters button and select Total for Each Legend Variable Category (same fill color) as the percentage that you'll use.

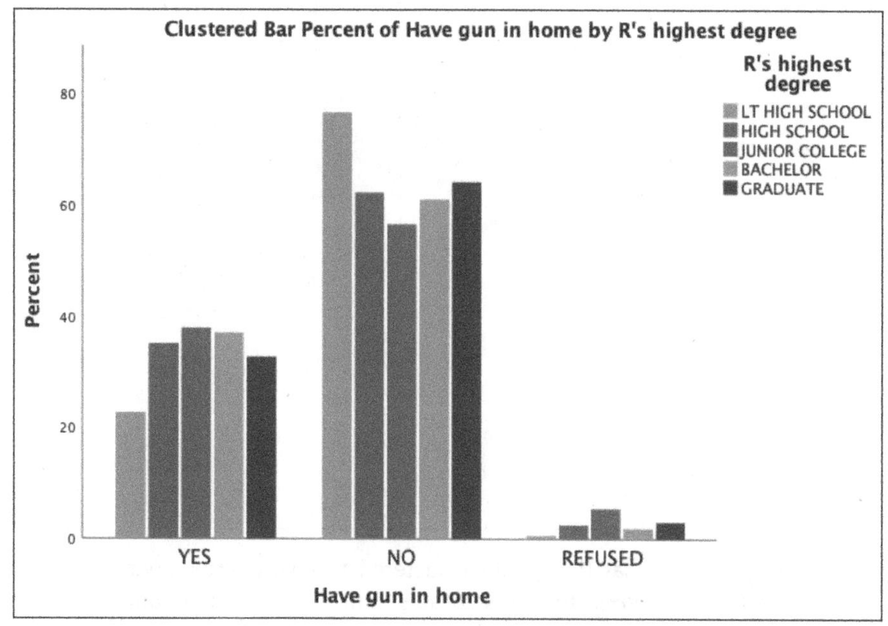

---

1 Using the GSS2018.sav file, create a clustered error bar chart showing the relationship between OWNGUN (Cluster Set Color box), CLASS (x-axis), and EDUC (y-axis). Describe the results.

2 Using the GSS2018.sav file, recreate the previous clustered bar chart but this time change the placement of the variables: CLASS (Cluster Set Color box), OWNGUN (x-axis), and EDUC (y-axis). Describe the results.

3 Using the GSS2018.sav file, create a clustered 3-D bar chart showing the relationship between DEGREE (Cluster Set Color box), OWNGUN (x-axis), and CLASS (z-axis). Describe the results.

4 Using the GSS2018.sav file, create a clustered 3-D bar chart showing the relationship between DEGREE (Cluster Set Color box), OWNGUN (x-axis), and CLASS (z-axis) but this time request percentages based off the legend variable instead of counts. Describe the results.

Creating Multiple Line Charts

In Chapter 12, you create a simple line chart. In this chapter, you extend your graphing capabilities by adding more than one line on a chart. The only caveat is that the variables must contain a similar range of values so they can be properly represented by the same axis. For example, if one variable ranges from 0 to 100 pounds and another variable ranges from 1 to 2 pounds, the values of the second variable will show up as a straight line, regardless of how much the values fluctuate.

The following steps generate a basic multiline graph:

1. **Choose File ⇨ Open ⇨ Data and load the GSS2018.sav file.**

   You can download the file from the book's companion website at `www.dummies.com/go/spssstatisticsworkbookfd`.

2. **Choose Graphs ⇨ Chart Builder.**
3. **In the Choose From list, select Line to specify the general type of graph to be constructed.**
4. **To specify that this graph should contain multiple lines, select the second diagram (with the Multiple Line tooltip) and drag it to the panel at the top.**
5. **In the Variables list, select CLASS and drag it to the X-Axis rectangle in the panel.**
6. **In the Variables list, select SEX and drag it to the Set Color rectangle in the panel at the top.**
7. **Click OK.**

   The chart shown in Figure 13-2 appears. The number of males and females in each class group is very similar, except more females than males are in the working class group.

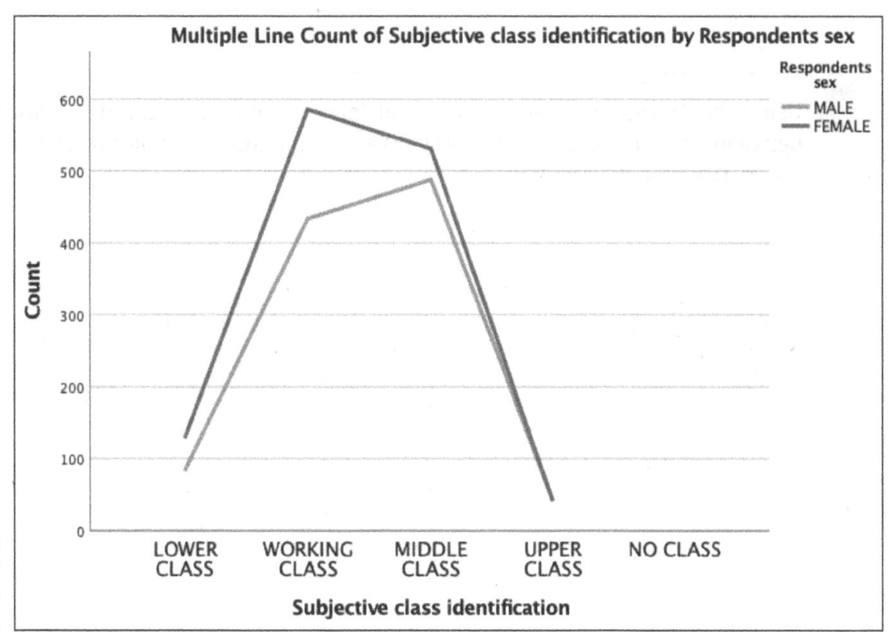

FIGURE 13-2: Distribution of the CLASS by SEX variables.

See the following for an example of creating a multiple line chart.

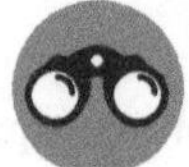

EXAMPLE

Q. Using the GSS2018.sav file, create a multiple line chart showing the frequency for the CHILDS (number of children) variable for each DEGREE (respondent's highest degree).

A. The frequency is much higher for the high school group than for any other education category, and the most common number of children for each group is either zero or two. To create this graph, place the DEGREE variable on the Set Color box and the CHILDS variable on the x-axis.

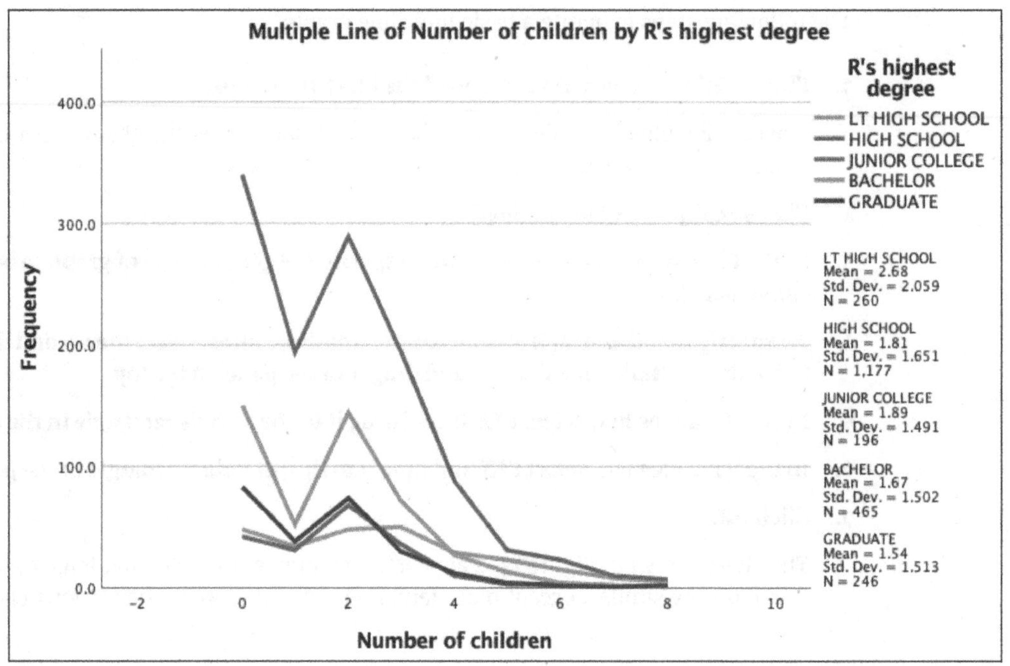

5 Using the GSS2018.sav file, create a multiple line chart showing the relationship between HEIGHT (x-axis), WEIGHT (y-axis), and SEX (Set Color box). Display the mean. Describe the results.

6 Using the GSS2018.sav file, create a multiple line chart showing the relationship between HEIGHT (x-axis), WEIGHT (y-axis), and SEX (Set Color box). Display the mean and add error bars. Describe the results.

7 Using the GSS2018.sav file, create a multiple line chart showing the relationship between DEGREE (respondent's highest degree) on the x-axis and the CHILDS (number of children) and CHLDIDEL (ideal number of children) variables on the y-axis. Describe the results.

WARNING

Be careful how you drop variables on the y-axis. You want to drop additional variables on the little box containing the plus sign. If you drop the new variable on top of the one that's already there, the original variable could be replaced.

8 Using the GSS2018.sav file, create a multiple line chart showing the relationship between DEGREE (respondent's highest degree) on the x-axis and the CHILDS (number of children) and CHLDIDEL (ideal number of children) variables on the y-axis. Add error bars and describe the results.

WARNING

Make sure you drop additional variables on the little box containing the plus sign, not on top of an existing variable.

Using Error Bar Charts to Compare Means

In Chapters 8 and 9, you use error charts to compare group means. In this chapter, you extend your use of error bar charts by including more than one continuous variable, creating a clustered error bar chart, and by modifying how the results are displayed. Follow these steps to create an error bar chart:

1. **Choose File ➪ Open ➪ Data and load the GSS2018.sav data file.**

   Download the file at `www.dummies.com/go/spssstatisticsworkbookfd`.

2. **Choose Graphs ➪ Chart Builder.**
3. **In the Choose From list, select Bar.**
4. **Select the seventh graph image (with the Simple Error Bar tooltip) and drag it to the panel at the top of the window.**
5. **Select the AGEKDBRN variable (age when respondents had their first child), and place it in the Y-axis box.**
6. **Select the SEX variable, and place it in the X-axis box.**
7. **Click OK.**

   The graph in Figure 13-3 appears.

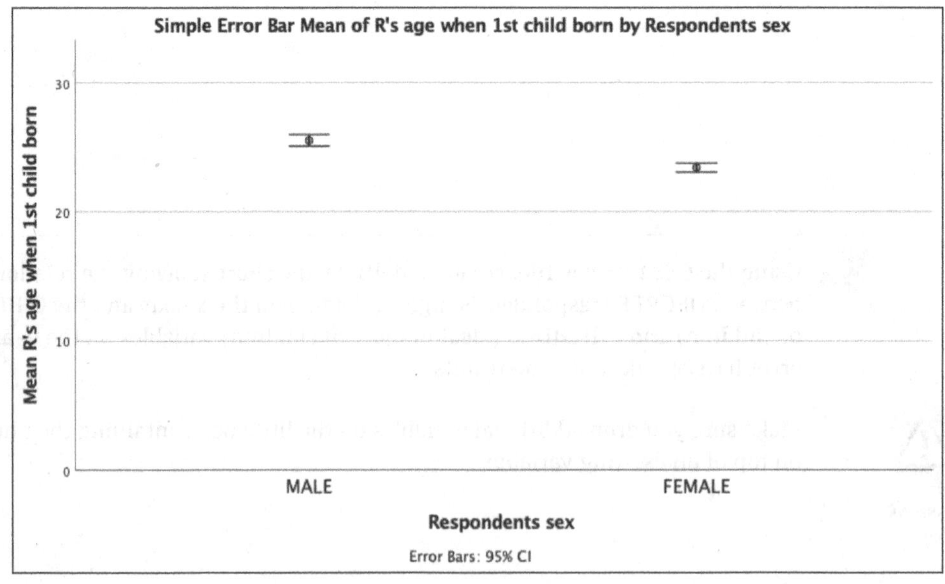

FIGURE 13-3: Error bar graph displaying the age when respondents had their first child, broken down by sex.

This chart represents the mean age when participants had their first child for each sex, along with 95% confidence intervals. The confidence intervals for the genders do not overlap, which indicates that the groups are significantly different, with women having their first child at an earlier age than men.

See the following for an example of making an error bar chart.

Q. Using the GSS2018.sav file, create a simple error bar chart depicting the relationship between the AGEKDBRN (age when respondents had their first child) variable and the DEGREE (respondent's highest degree) variable. Describe your findings.

A. The mean age of when participants had their first child for each degree along with 95% confidence intervals is represented in this chart. The confidence intervals for the degree groups do not overlap, which indicates that all the groups are significantly different from each other.

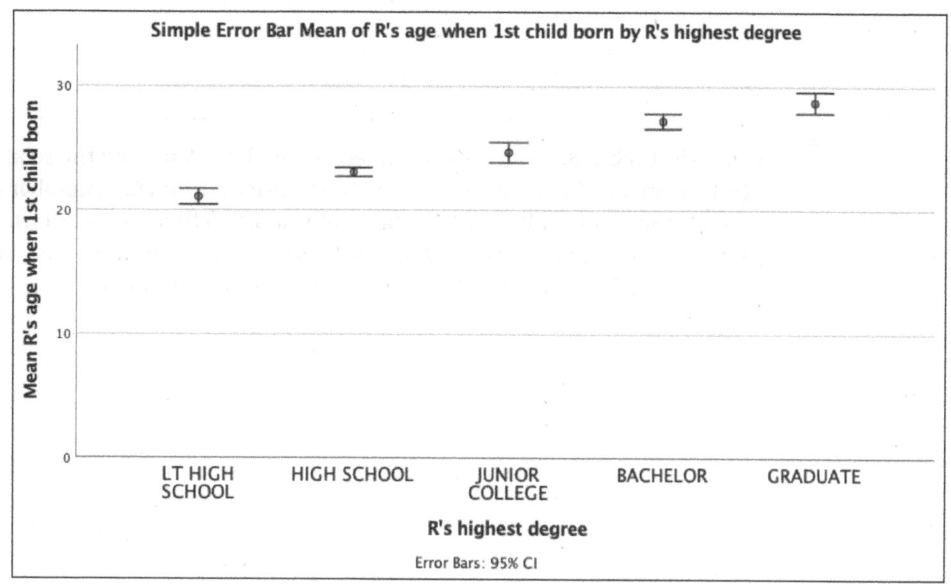

9 Using the GSS2018.sav file, create a clustered error bar chart depicting the relationship between AGEKDBRN (age when respondents had their first child) and the DEGREE (respondent's highest degree) and SEX variables. Place AGEKDBRN on the y-axis, DEGREE on the x-axis, and SEX in the Cluster on X: Set Color box. Describe your findings.

 Using the GSS2018.sav file, create an error bar chart depicting the relationship between CHILDS (number of children) and CHLDIDEL (ideal number of children) and DEGREE (respondent's highest degree). Describe your findings.

TIP

When creating an error bar chart to depict the findings of multiple continuous variables, add both variables to the y-axis.

---

11 Using the GSS2018.sav file, create an error bar chart depicting the relationship between EDUC (number of years of respondent's education), MAEDUC (number of years of mother's education), PAEDUC (number of years of father's education), SPEDUC (number of years of spouse's education), and CLASS (subjective class identification). Place the CLASS variable on the x-axis. Describe your findings.

---

12 Using the GSS2018.sav file, create an error bar chart depicting the relationship between EDUC (number of years of respondent's education), MAEDUC (number of years of mother's education), PAEDUC (number of years of father's education), SPEDUC (number of years of spouse's education), and CLASS (subjective class identification). Place the CLASS variable on the x-axis. Exclude the No Class group (because it has no data), and change the order of the education variables so that the mother's and father's education are next to each other, and the respondent's and spouse's education are next to each other. Describe your findings.

Viewing Relationships between Continuous Variables

In Chapter 10, you use scatterplots to visually present the relationship between two continuous variables. In this chapter, you extend your use of scatterplots by including more than two continuous variables, by using colored scatterplots to incorporate a categorical variable, and by creating a *scatterplot matrix*, which is a group of scatterplots combined into a single graphic image. Follow these steps to create a scatterplot:

1. **Choose File ⇨ Open ⇨ Data and load the GSS2018.sav data file.**

   Download the file at `www.dummies.com/go/spssstatisticsworkbookfd`.

2. **Choose Graphs ⇨ Chart Builder.**
3. **In the Choose From list, select Scatter/Dot.**
4. **Select the first scatterplot diagram (with the Simple Scatter tooltip), and drag it to the panel at the top.**
5. **In the Variables list, select HEIGHT and drag it to the rectangle labeled X-Axis in the diagram.**

REMEMBER

   In a scatterplot, both the x-axis and y-axis variables are scale. Look for the ruler icon to identify scale variables.

6. **In the Variables list, select WEIGHT and drag it to the rectangle labeled Y-Axis in the diagram.**
7. **Click OK.**

   The chart in Figure 13-4 appears.

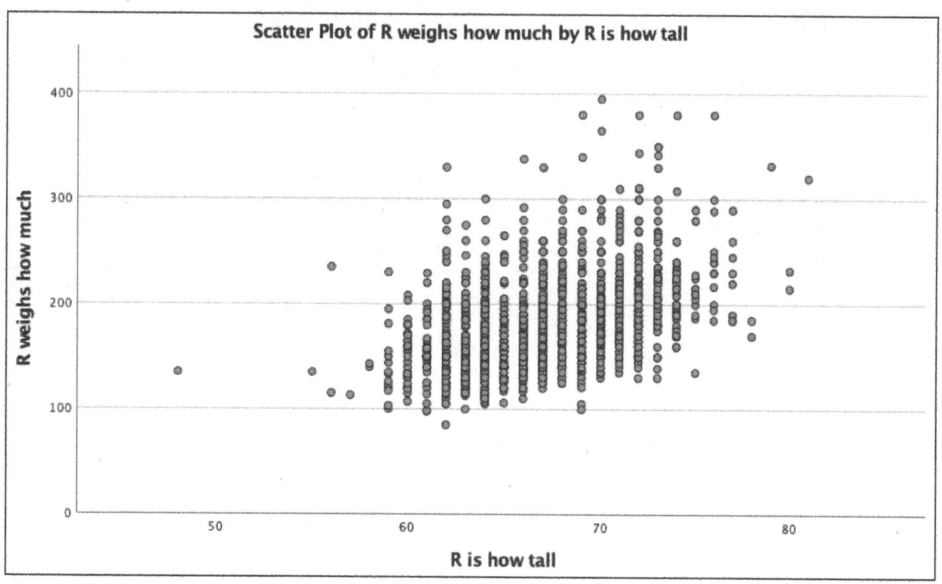

FIGURE 13-4: The scatterplot of weight and height.

The scatterplot of height and weight shows the relationship between these two variables. For the most part, shorter people are associated with lower weight and taller people are associated with higher weight. (See Chapter 10 for additional information pertaining to scatterplots.)

See the following for an example of making a scatterplot.

EXAMPLE

Q. Using the GSS2018.sav file, create a simple scatterplot depicting the relationship between HEIGHT (x-axis) and WEIGHT (y-axis), but add a regression line.

A. As you saw previously, the scatterplot of height and weight shows that shorter people are associated with lower weight and taller people are associated with higher weight. However, adding the regression makes it easier to see the pattern. To create this graph, do the following:

a. Place the HEIGHT variable on the x-axis.

b. Place the WEIGHT variable on the y-axis.

c. In the Element Properties section, click the Total check box in the Linear Fit Lines section.

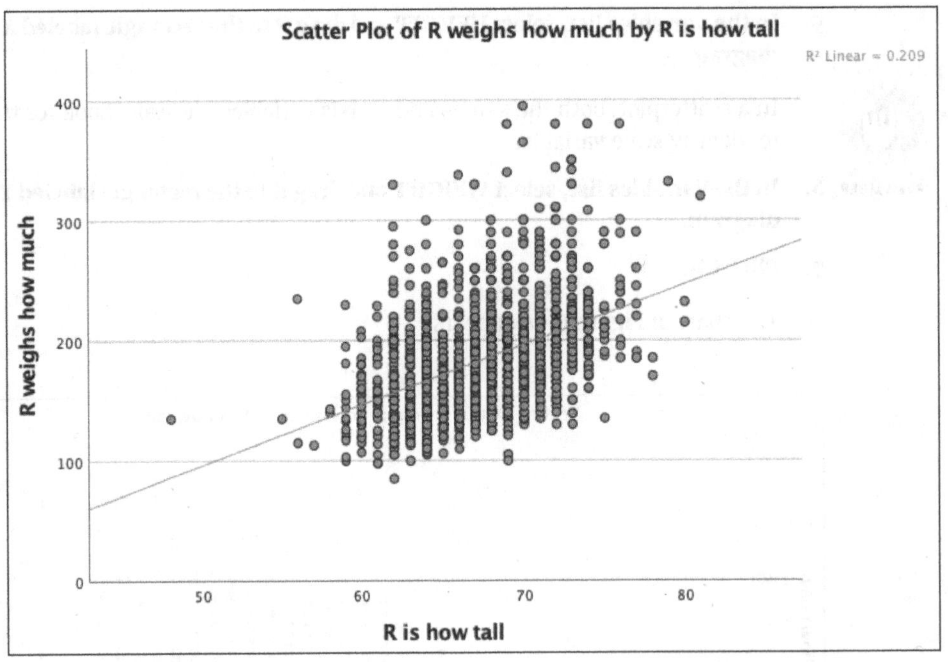

13. Using the GSS2018.sav file, create a simple scatterplot depicting the relationship between HEIGHT (x-axis) and WEIGHT (y-axis). For the third variable, include the SEX variable, using the Set Color option. Also request a regression line, as well as a regression line for each subgroup. Describe your findings.

14. Using the GSS2018.sav file, create a simple scatterplot depicting the relationship between HEIGHT (x-axis) and WEIGHT (y-axis). For the third variable, include the SEX variable, using the Set Size option. Also request a regression line. Describe your findings.

15. Using the GSS2018.sav file, create a simple 3-D scatterplot depicting the relationship between HEIGHT (x-axis), WEIGHT (y-axis), and SEX (z-axis). Describe your findings.

16. Using the GSS2018.sav file, create a scatterplot matrix showing the relationship between EDUC (number of years of respondent's education), MAEDUC (number of years of mother's education), PAEDUC (number of years of father's education), and SPEDUC (number of years of spouse's education). Describe your findings.

Identifying Outliers with Boxplots

Boxplots, also referred to as box and whisker plots, are particularly useful for obtaining an overall feel for a distribution. In addition, the boxplot graphically identifies outliers. Follow these steps to create a boxplot:

1. **Choose File ⇨ Open ⇨ Data and load the GSS2018.sav data file.**

   Download the file at `www.dummies.com/go/spssstatisticsworkbookfd`.

2. **Choose Graphs ⇨ Chart Builder.**
3. **In the Choose From list, select Boxplot.**
4. **Select the third graph image (with the 1-D Boxplot tooltip) and drag it to the panel at the top of the window.**
5. **Select the TVHOURS variable (hours per day watching tv), and place it in the X-Axis box.**
6. **Click OK.**

   The graph in Figure 13-5 appears.

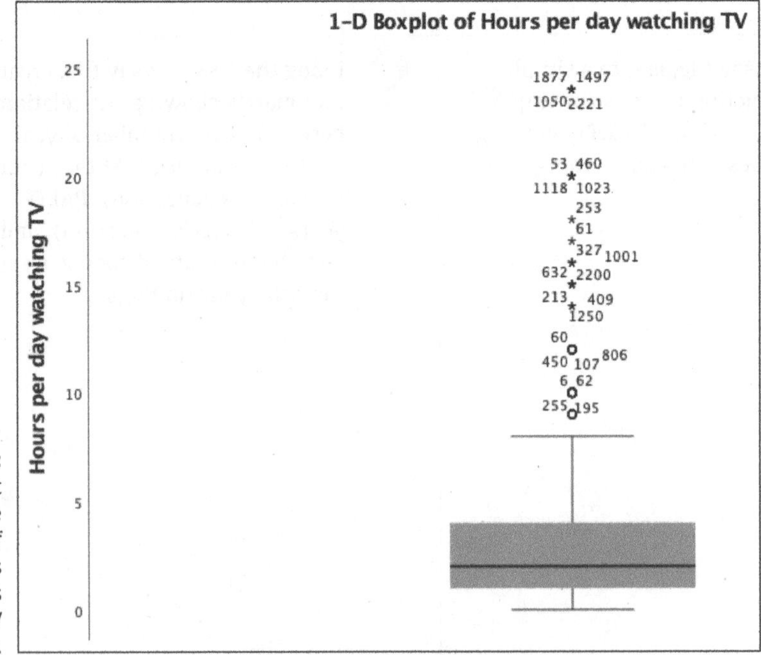

FIGURE 13-5: Boxplot displaying the number of hours respondents watch TV per day.

The vertical axis represents the scale for the continuous variable. The solid line inside the box represents the median, or 50th percentile. The top and bottom borders (referred to as *hinges*) of the box correspond to the 75th and 25th percentile values, respectively, and thus define the interquartile range (IQR). In other words, the middle 50% of data values fall within the box. The length of the box indicates the amount of spread within the data.

The *whiskers* (vertical lines extending from the top and bottom of the box) are the last data values that lie within 1.5 box lengths (or IQRs) of the respective hinges (borders of box). Data points more than 1.5 box lengths from a hinge are considered outliers. These points are marked with a circle. Points more than 3 box lengths (IQR) from a hinge are considered extreme value points and are marked with an asterisk. Note that the whiskers are not fully extended because they stop at the minimum and maximum values.

This plot has many outliers. The number that appears next to an outlier is the case sequence number, which aids in data checking.

If the distribution were symmetrical, the median would be centered within the box. In the plot in Figure 13-5, the median is toward the bottom of the box, indicating a negatively skewed distribution.

See the following for an example of making a boxplot.

EXAMPLE

Q. Using the GSS2018.sav file, create a simple boxplot depicting the distribution of the TVHOURS (hours per day watching TV) variable within RACE. Describe your findings.

A. All three racial categories have outliers, but the Other group has the least number of outliers and the distribution seems a little more symmetrical. To create this simple boxplot, place the RACE variable on the x-axis and the TVHOURS variable on the y-axis.

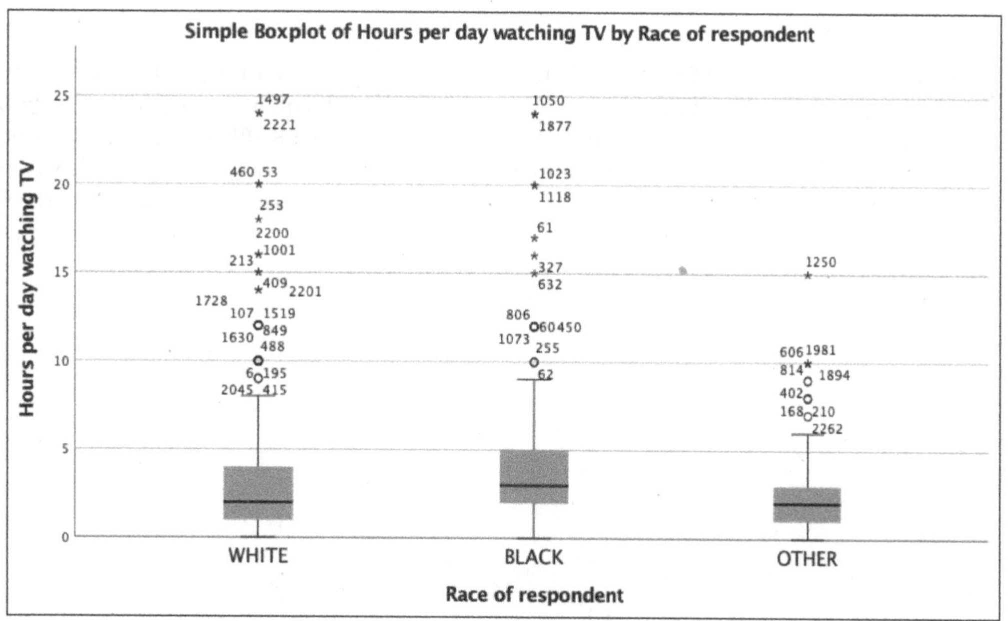

17 Using the GSS2018.sav file, create a 1-D boxplot depicting the distribution of the AGEKDBRN (age when respondents had their first child) variable. Describe your findings.

18 Using the GSS2018.sav file, create a simple boxplot depicting the distribution of the AGEKDBRN (age when respondents had their first child) variable within SEX. Describe your findings.

19 Using the GSS2018.sav file, create a simple boxplot depicting the distribution of the AGEKDBRN (age when respondents had their first child) variable within DEGREE. Describe your findings.

20 Using the GSS2018.sav file, create a clustered boxplot depicting the distribution of the AGEKDBRN (age when respondents had their first child) variable within DEGREE on the x-axis and SEX on the Cluster on X: Set Color box. Describe your findings.

Answers to Problems in Building Slightly More Complex Graphs

1. No significant differences exist in the number of years of education within each Class because there is an overlap for the gun groups. To create this clustered error bar chart, do the following:

   a. Place the OWNGUN variable in the Cluster Set Color box.

   b. Place the CLASS variable on the x-axis.

   c. Place the EDUC variable on the y-axis.

TIP

For clustered error bar charts, make sure the variables on the y-axis are set to scale.

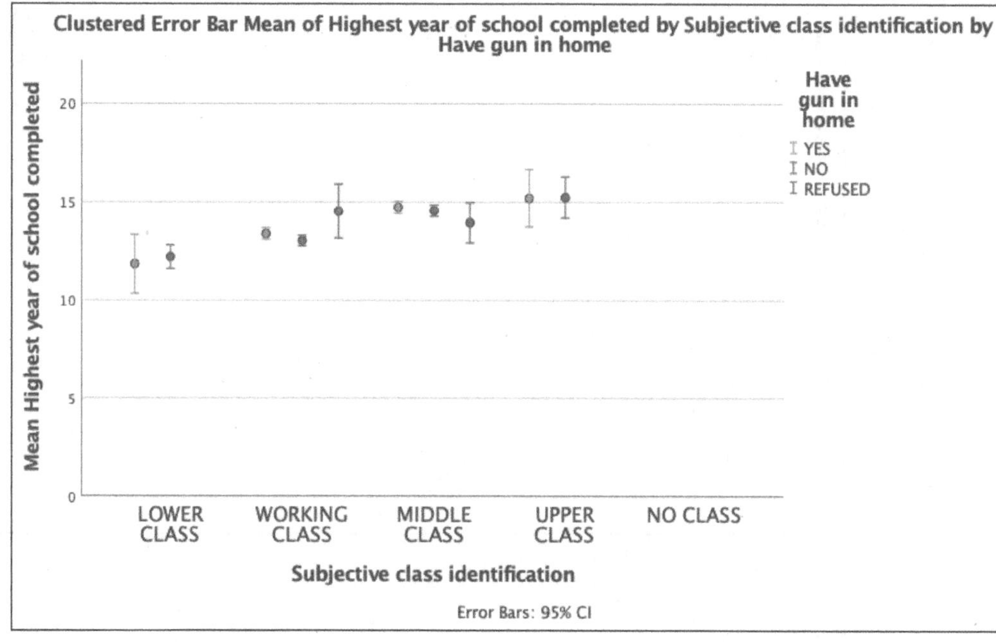

2 Contrary to the prior example, now you are investigating whether significant differences exist in the number of years of education within each gun group when comparing classes. Regardless of whether someone owns or does not own a gun, the middle- and upper-class groups have significantly more education than the lower- and working-class groups. To create this clustered error bar chart, do the following:

a. Place the CLASS variable in the Cluster Set Color box.

b. Place the OWNGUN variable on the x-axis.

c. Place the EDUC variable on the y-axis.

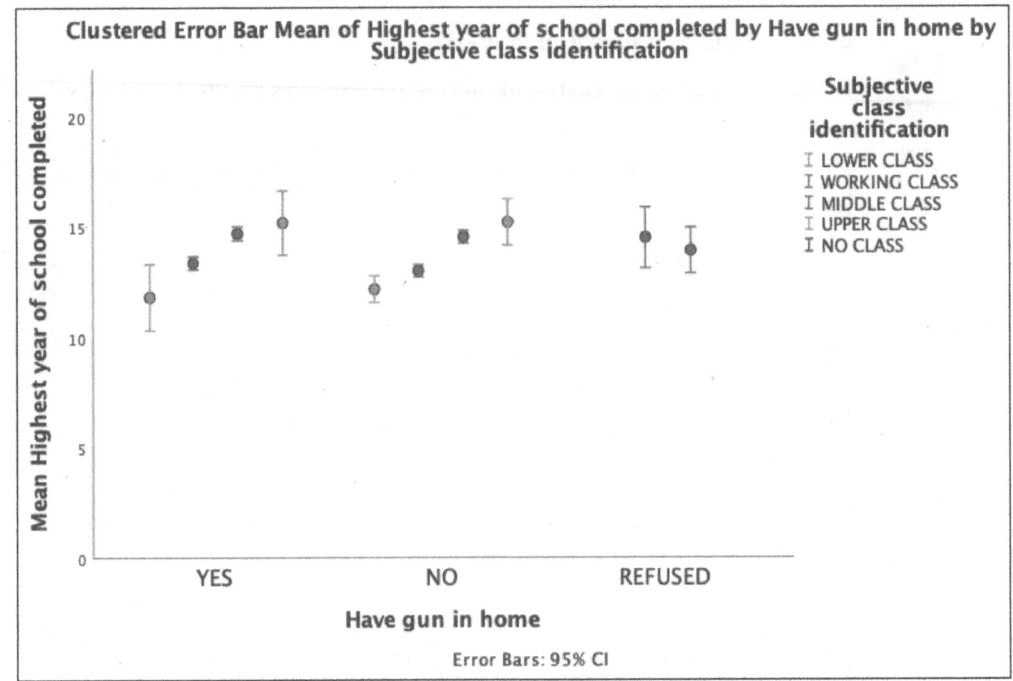

3 When showing the frequency or count, the results from the high school group stand out because that group is the most numerous. To create this clustered 3-D bar chart, do the following:

a. Place the DEGREE variable in the Cluster Set Color box.

b. Add the OWNGUN variable to the x-axis.

c. Add the CLASS variable to the z-axis.

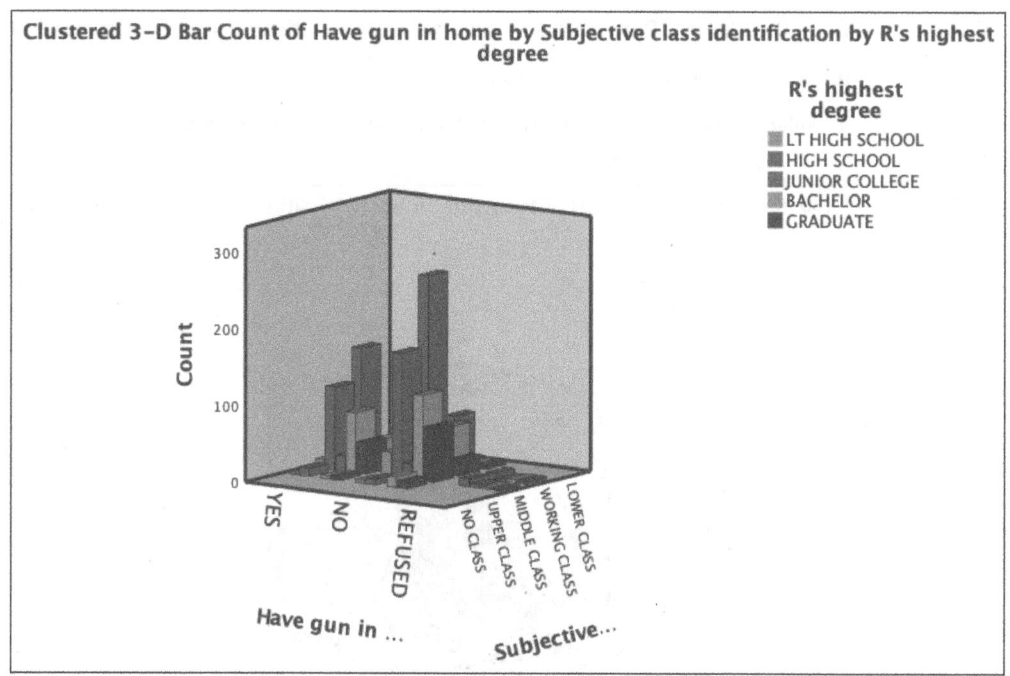

4 When showing percentages, the results are different from the previous example and you can start to appreciate where groups differ from each other. For instance, a large percentage of people have a junior college degree, are in the middle-class category, and do not own guns. To create this clustered 3-D bar chart, do the following:

a. Place the DEGREE variable in the Cluster Set Color box.

b. Add the OWNGUN variable to the x-axis.

c. Add the CLASS variable to the z-axis.

d. In the Element Properties section, change the Statistic from Count to Percentage.

e. Click the Set Parameters button.

f. Select Total for Each Legend Variable Category (same fill color).

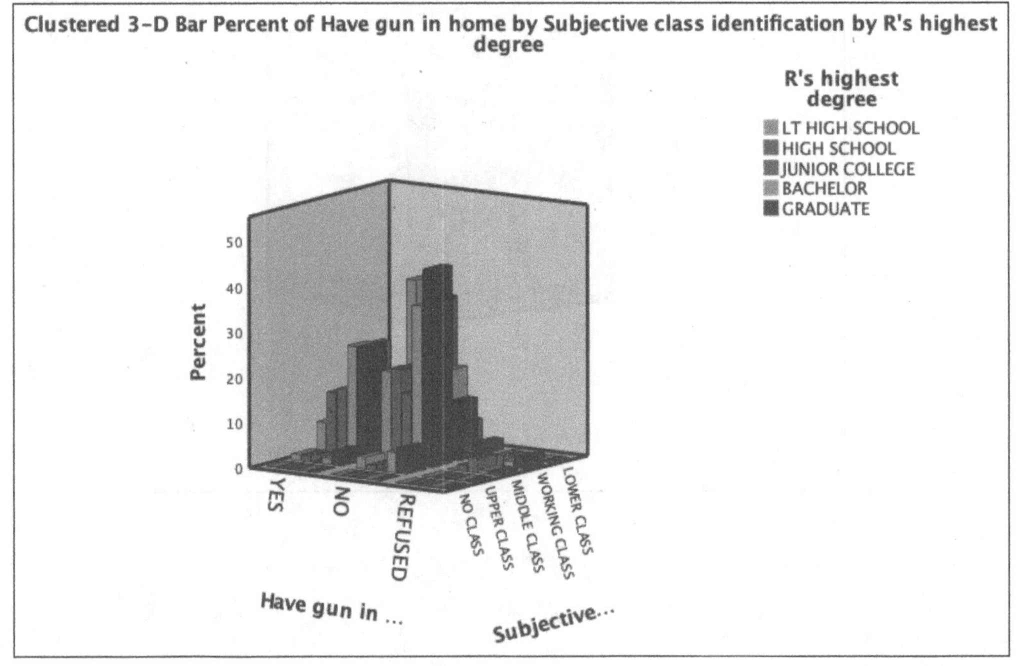

5 As height increases, so does weight. The data for women is lower on the distribution than that for men. (In general, women are shorter and weigh less than men, although the genders show considerable overlap.) To create this multiple line graph, do the following:

a. Place the HEIGHT variable on the x-axis.

b. Place the WEIGHT variable on the y-axis.

c. Add the SEX variable to the Set Color rectangle.

d. In the Element Properties section, change the Statistic from Count to Mean.

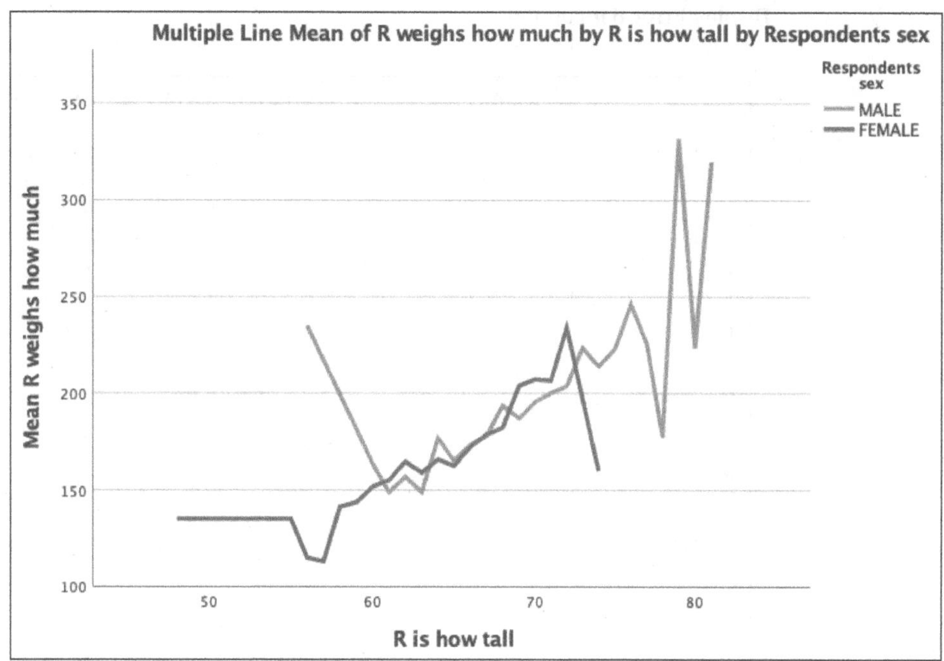

6 As in the previous example, the graph shows that as height increases, so does weight. The data for women is lower on the distribution than that for men. (In general, women are shorter and weigh less than men, although the genders have considerable overlap.) To create this multiple line graph, do the following:

a. Place the HEIGHT variable on the x-axis.

b. Place the WEIGHT variable on the y-axis.

c. Add the SEX variable to the Set Color rectangle.

d. In the Element Properties section, change the Statistic from Count to Mean and click the Display Error Bars option.

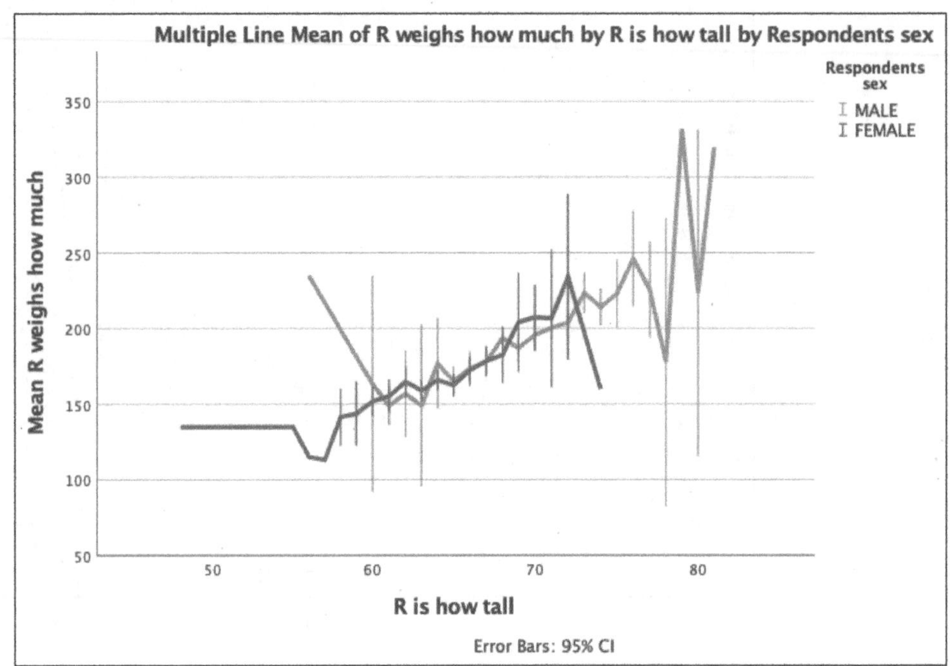

7. The ideal number of children is much higher than the actual number of children for all degree groups. To create this multiple line chart, do the following:

   a. Place the DEGREE variable on the x-axis.

   b. Place the CHILDS and CHLDIDEL variables on the y-axis.

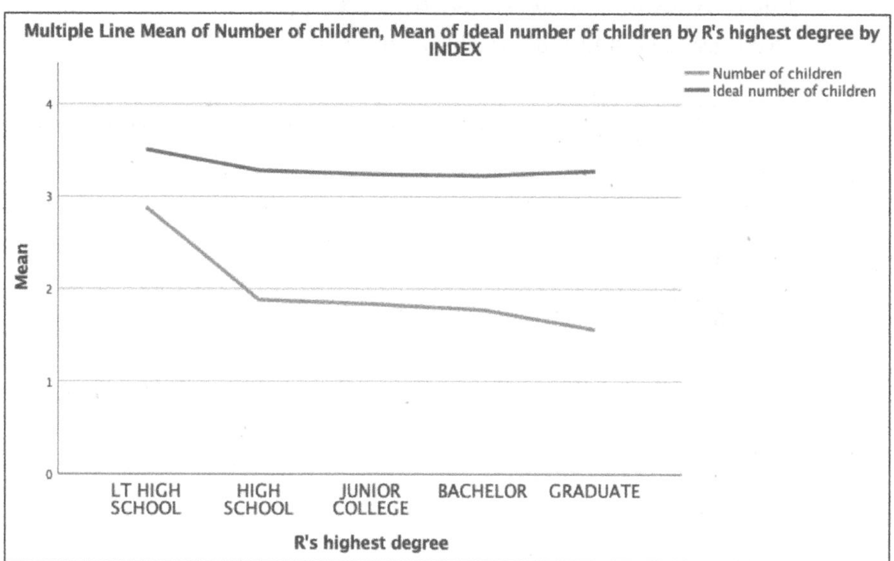

8. The ideal number of children is much higher than the actual number of children for all degree groups and there is no overlap between actual and ideal number of children. To create this multiple line chart, do the following:

   a. Place the DEGREE variable on the x-axis.

   b. Place the CHILDS and CHLDIDEL variables on the y-axis.

   c. In the Element Properties section, click the Display Error Bars option.

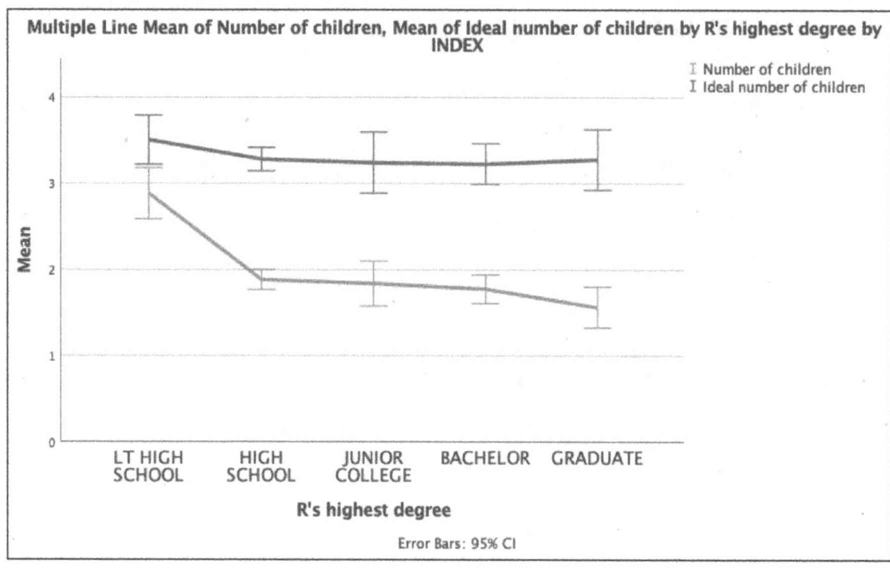

9 No overlap exists between the genders with a high school education with regard to how old they were when they had their first child. Women were significantly younger than men, and this is the only significant difference among these comparisons. To create this clustered error bar chart, do the following:

a. Place the SEX variable in the Cluster Set Color box.

b. Place the DEGREE variable on the x-axis.

c. Place the AGEKDBRN variable on the y-axis.

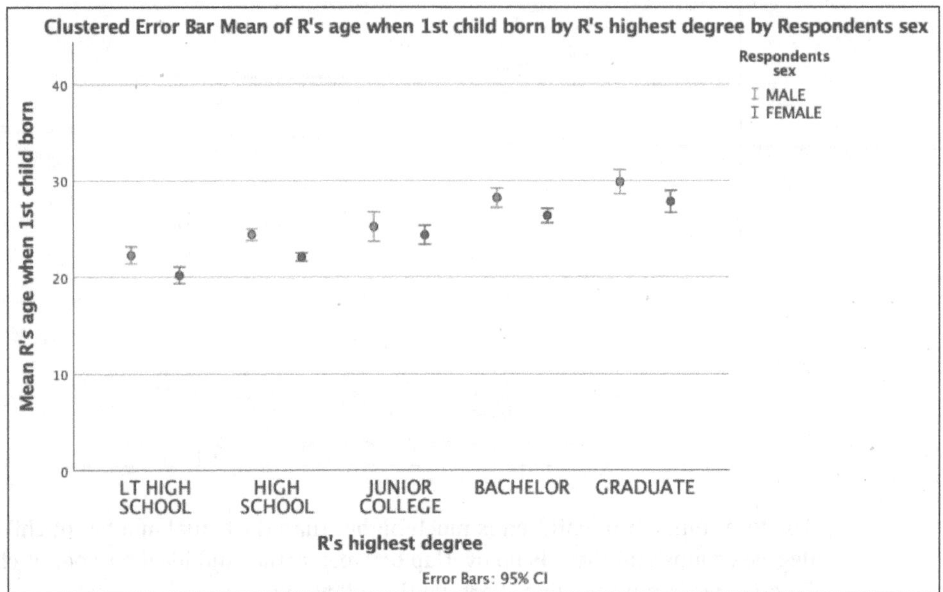

10 The ideal number of children is much higher than the actual number of children for all degree groups, and no overlap exists between the actual and ideal number of children. To create this clustered error bar chart, do the following:

a. Place the CHILDS and CHLDIDEL variables on the y-axis.

b. Add the DEGREE variable to the x-axis.

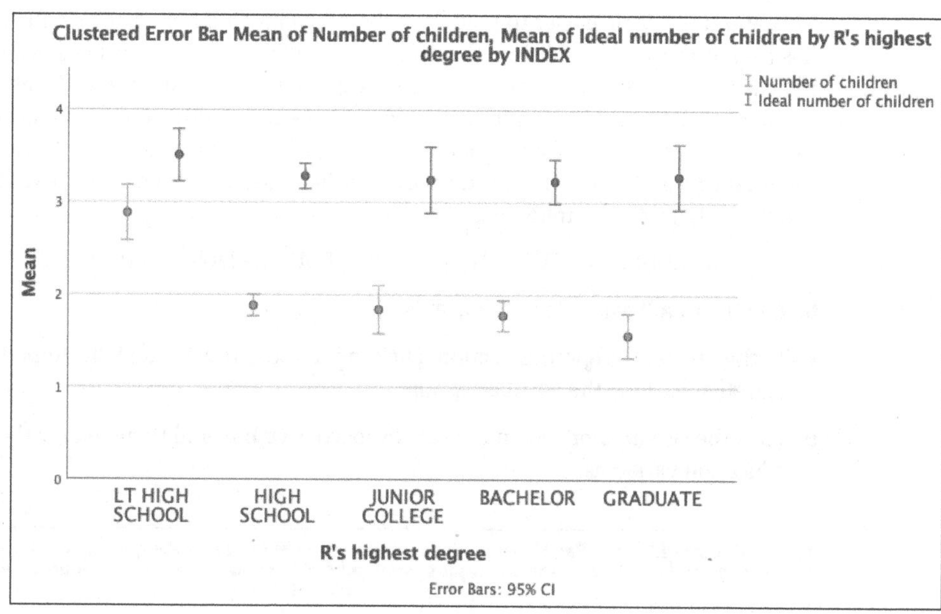

11 No differences in education exist for the lower-class group. However, the respondent's and spouse's education is significantly higher than the mother's and father's education for all other class groupings. To create this clustered error bar chart, do the following:

a. Place the EDUC, MAEDUC, PAEDUC, and SPEDUC variables on the y-axis.

b. Add the CLASS variable to the x-axis.

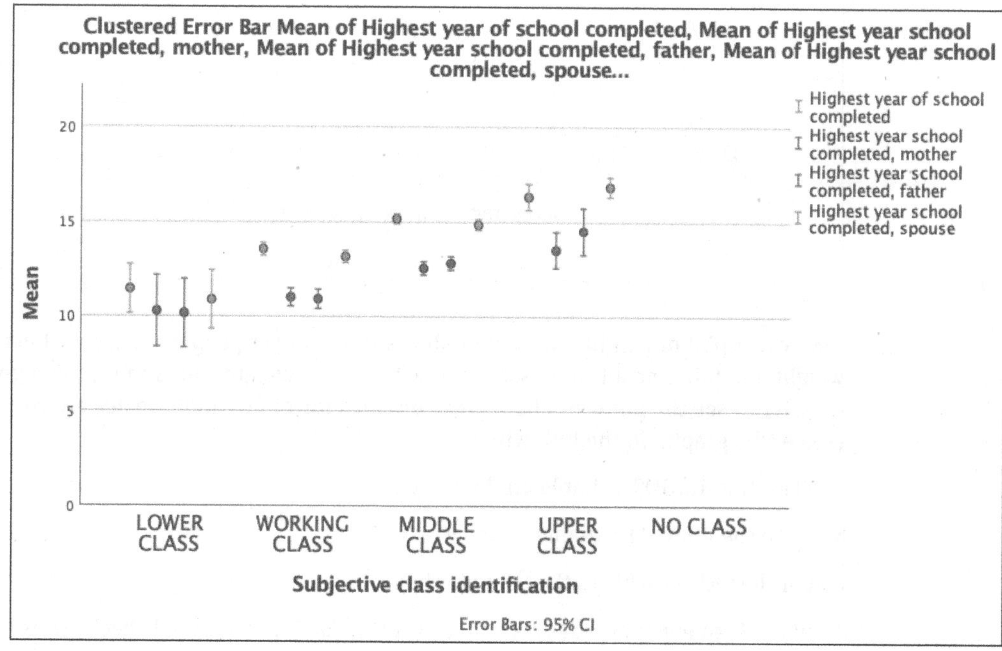

12 This graph is a little easier to read than the preceding example because the No Class group has been removed. In addition, the order of the education variables has been changed so that it's easier to see the similarities between respondent's and spouse's education as well as mother's and father's education. As before, there are no differences in education for the lower-class group, but the respondent's and spouse's education is significantly higher than the mother's and father's education for all other class groupings. To create this clustered error bar chart, do the following:

a. Place the EDUC, MAEDUC, PAEDUC, and SPEDUC variables on the y-axis.

b. Add the CLASS variable to the x-axis.

c. In the Element Properties section, click the X-Axis1 row in the Edit Properties Of box and then exclude the No Class group.

d. Click the GroupColor row in the Edit Properties Of box and then change the order of the education variables.

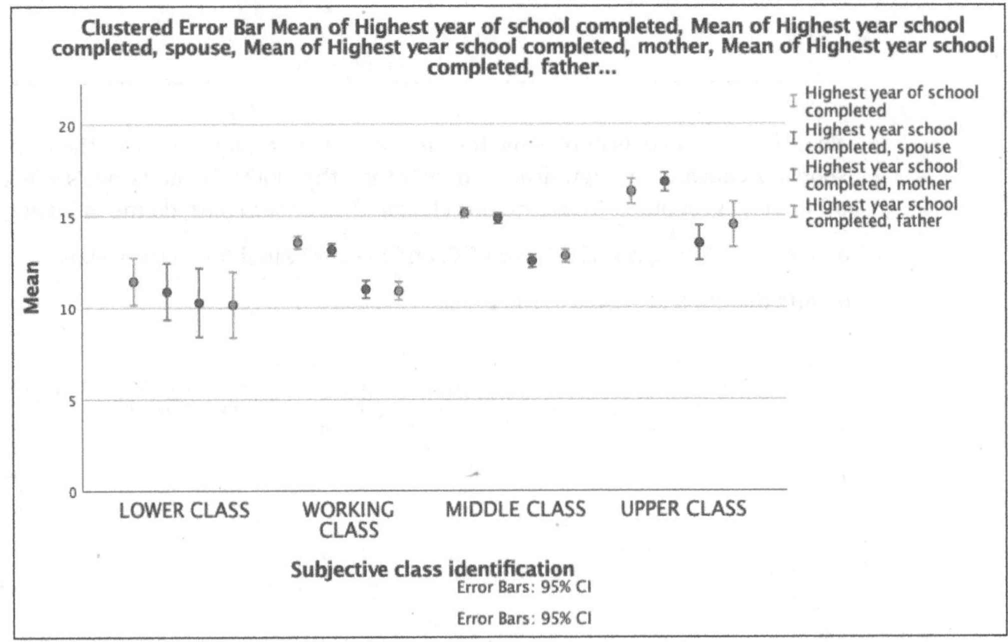

13 The scatterplot of height and weight shows that shorter people are associated with lower weight and taller people are associated with higher weight, but adding the regression makes it easier to see the pattern. The regression lines for males and females are very similar. To create this graph, do the following:

a. Place the HEIGHT variable on the x-axis.

b. Place the WEIGHT variable on the y-axis.

c. Put the SEX variable in the Set Color box.

d. In the Element Properties section, select the Total and Subtotal check boxes in the Linear Fit Lines section.

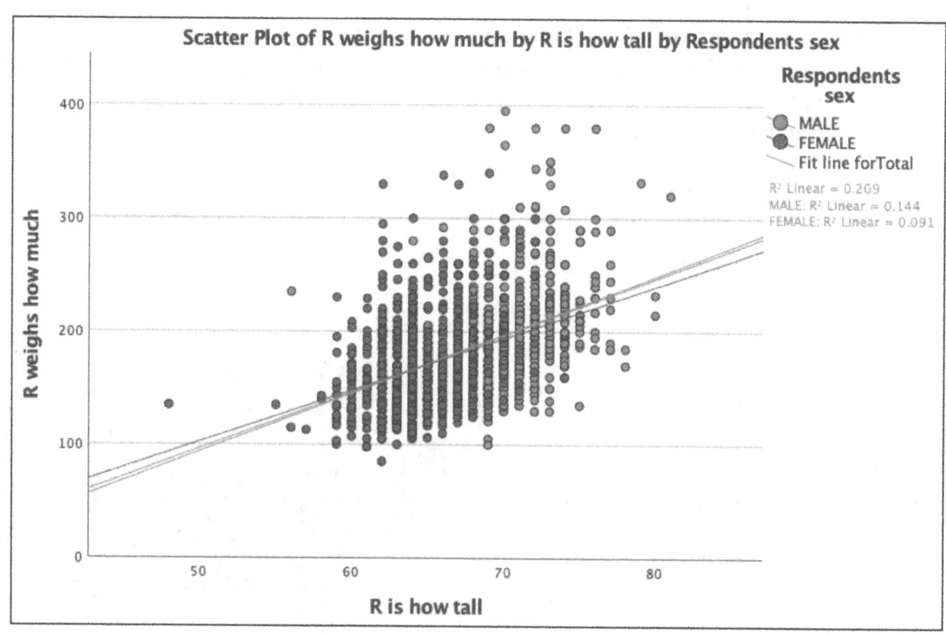

14 The scatterplot of height and weight shows that shorter people are associated with lower weight and taller people are associated with higher weight, but adding the regression makes it easier to see the pattern. The size of the bubbles adds an interesting element in place of the different colors for each group. To create this graph, do the following:

a. Place the HEIGHT variable on the x-axis.

b. Place the WEIGHT variable on the y-axis.

c. Put the SEX variable in the Set Size box.

d. In the Element Properties section, select the Total check box in the Linear Fit Lines section.

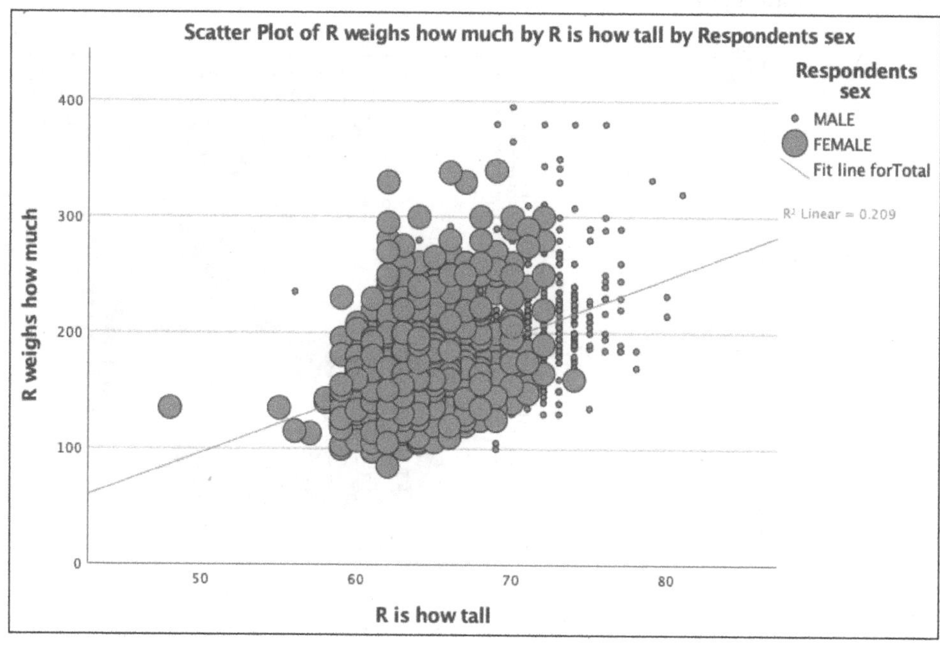

15 The scatterplot of height and weight shows that shorter people are associated with lower weight and taller people are associated with higher weight, but adding the regression makes it easier to see the pattern. Adding a third dimension allows you to tease apart the data for males and females. To create this 3-D scatterplot graph, place the HEIGHT variable on the x-axis.

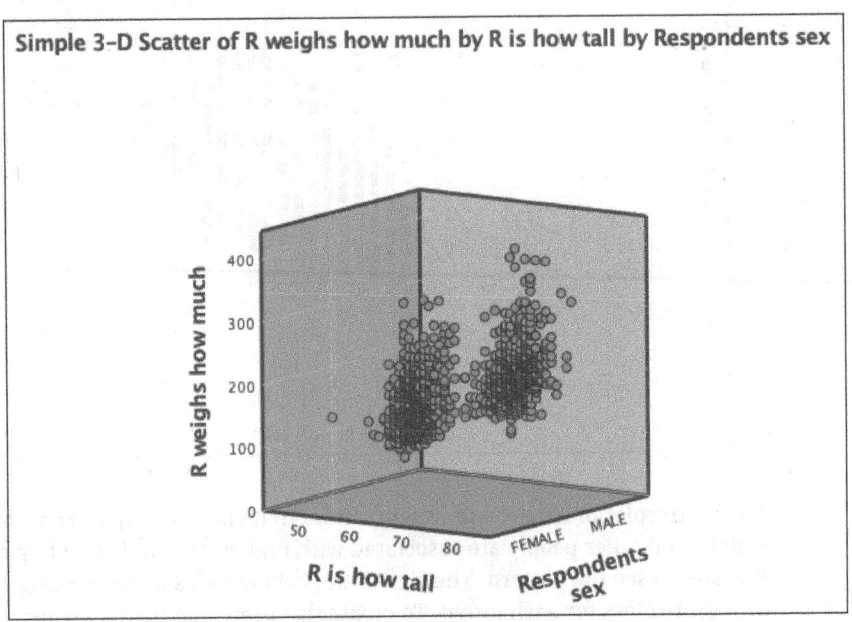

16 Each variable is plotted against each of the others. Note that the scatterplots along the diagonal from the upper left to the lower right are blank because it's useless to plot a variable against itself. The graph shows that all variables have positive linear relationships, with the strongest relationship being between mother's and father's education. To create this scatterplot matrix, add all the educations variables (EDUC, MAEDUC, PAEDUC, and SPEDUC) to the Scattermatrix box.

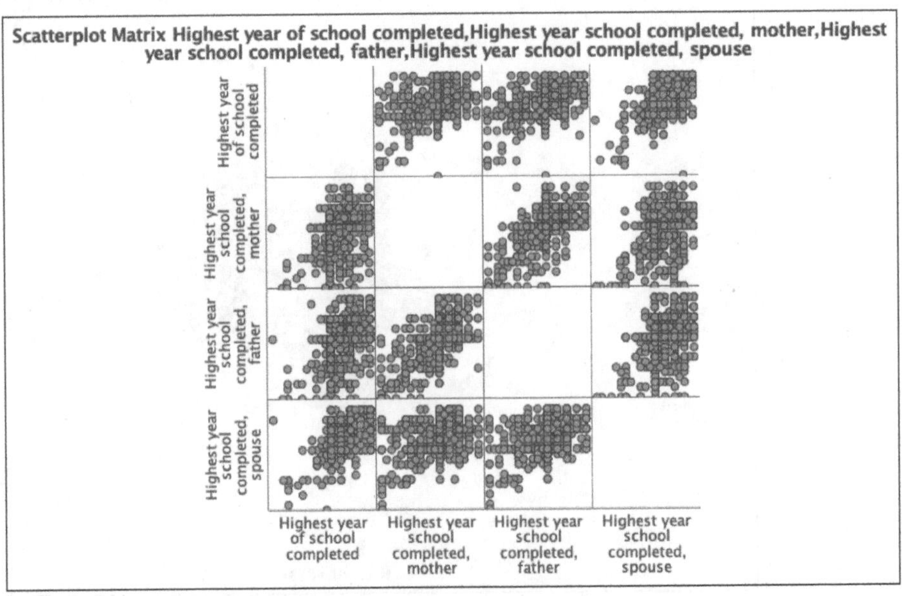

17 All the outliers for the AGEKDBRN variable are at the upper end of the distribution, starting with values above 40. To create this 1-D boxplot, place the AGEKDBRN variable on the x-axis.

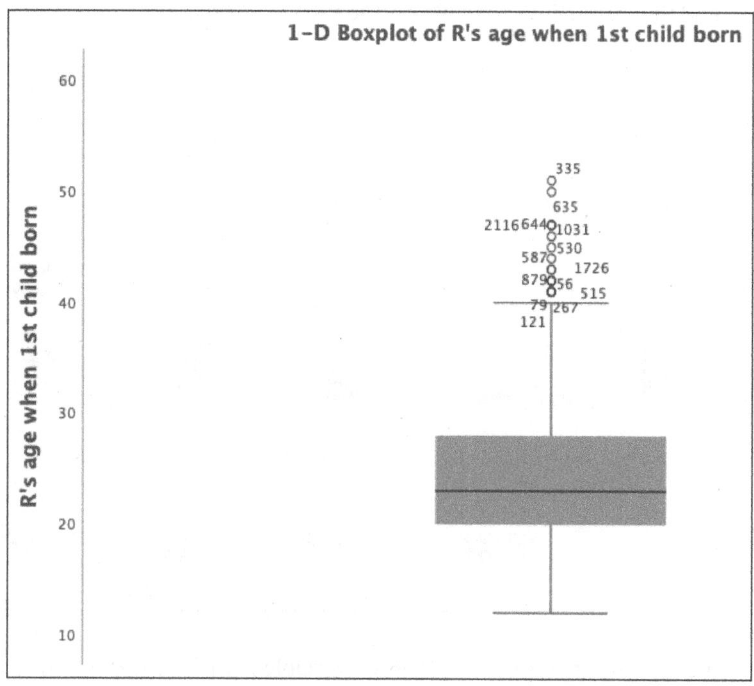

18 All the outliers for the AGEKDBRN variable are at the upper end of the distribution, starting with values above 41 for men and 39 for women. The median age when men have their first child is a little higher than it is for women. To create this simple boxplot, place the SEX variable on the x-axis and the AGEKDBRN variable on the y-axis.

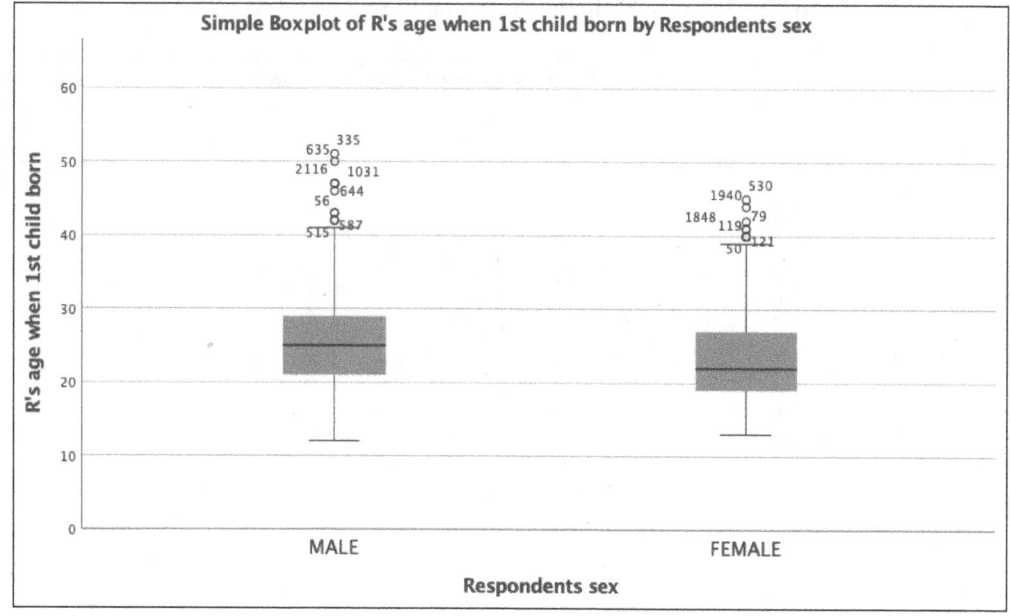

19 All the outliers for the AGEKDBRN variable are in the upper end of the distribution, except for one person who has a graduate degree. The median age for having a first child increases as education increases. To create this simple boxplot, place the DEGREE variable on the x-axis and the AGEKDBRN variable on the y-axis.

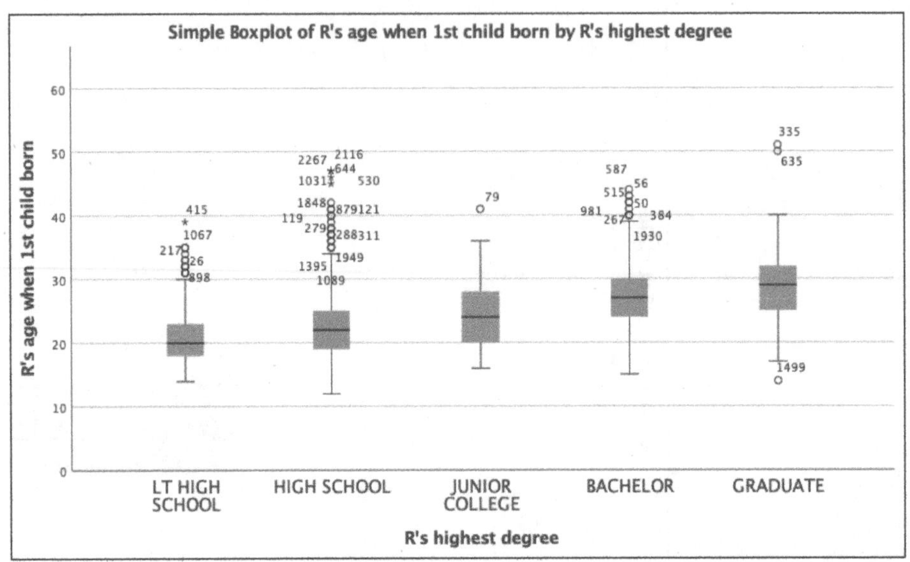

20 All the outliers for the AGEKDBRN variable are in the upper end of the distribution, except for a few people with a bachelor's degree. The median age for having a first child increases as education increases. Within each degree group, the median age when men have their first child is a little higher than it is for women. To create this clustered boxplot, do the following:

a. Place the DEGREE variable on the x-axis.

b. Place the AGEKDBRN variable on the y-axis.

c. Add the SEX variable to the Cluster on X: Set Color box.

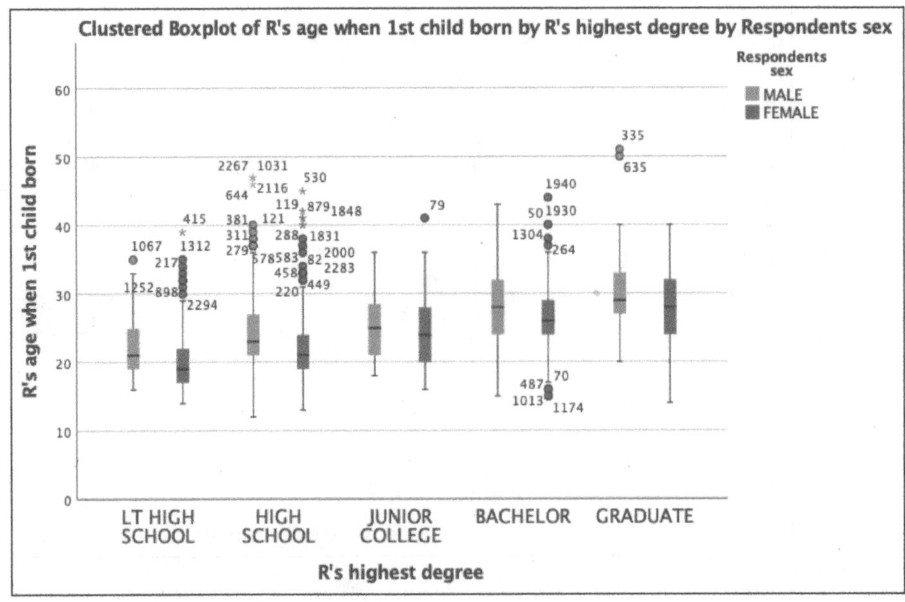

IN THIS CHAPTER

» Editing charts

» Editing tables

Chapter 14
Editing Output

Almost all SPSS output comes in the form of charts or tables. Throughout this book, you create a variety of charts and tables using an array of procedures. In this chapter, you learn how to edit charts and tables so you can customize them to have the look and information you want.

Editing Graphs

In Chapters 12 and 13, you create a variety of graphs. In this section, you see how to edit those graphs to create a better visual representation of the information in the chart. Although this section doesn't cover every type of edit you can make to a chart, you can use the steps here as a baseline to explore additional editing options.

Chart Editor provides an easy-to-use environment for customizing charts and exploring data. You can quickly select and edit parts of a chart by using toolbar menus, context menus, and toolbar options, and you can enter text directly in a chart.

To use Chart Editor, you select the element you want to modify and then use the Properties dialog to change the properties for that element. The tabs you see in the Properties dialog change based on your current selection.

TIP

Chart Editor works in a drill-down fashion, so you might need to click a specific element multiple times to arrive to the element level you want to modify.

In Chapter 13, you create a simple error bar chart by following these steps:

1. **Choose File ⇨ Open ⇨ Data and load the GSS2018.sav data file.**

   Download the file at `www.dummies.com/go/spssstatisticsworkbookfd`.

2. **Choose Graphs⇨Chart Builder.**
3. **In the Choose From list, select Bar.**
4. **Select the seventh graph image (the one with the Simple Error Bar tooltip) and drag it to the panel at the top of the window.**
5. **Select the AGEKDBRN variable (age when respondents had their first child), and place it in the Y-Axis box.**
6. **Select the SEX variable, and place it in the X-Axis box.**
7. **Click OK.**

   The Viewer window appears. You'll be making changes in the window in the next set of steps.

This chart represents the mean age when participants had their first child for each sex along with 95% confidence intervals. However, there's a lot empty space between the bottom of the graph and where the information begins along the y-axis. In the following steps, you edit the y-axis so that the range of the axis begins at 20:

1. **In the SPSS Statistics Viewer window, navigate to the graph and double-click the chart.**
2. **Click any value on the y-axis.**

   The Properties dialog appears.
3. **Click the Scale tab.**
4. **In the Range section, deselect the Auto Check box next to Minimum, and change the number in the Custom box to 20.**
5. **Click Apply.**

   The chart in Figure 14-1 appears. The y-axis has changed to eliminate unnecessary space in the graph.

Editing graphs like this allows you to create clearer graphs and improve how the information is visualized.

See the following for an example of creating a bar chart.

Q. Using the GSS2018.sav file, create a bar chart showing the distribution of the CLASS (subjective class identification) variable. Middle class is the main group of interest, so display this group using a different color.

A. Do the following:

   a. Use Chart Builder to create a simple bar chart, and place the CLASS variable on the x-axis.

   b. Double-click the graph and click any one of the bars, so that all the bars are selected.

   c. Click the bar for Middle Class to select only this group.

   d. Click the Fill & Border tab in the Properties box, select the color you'd like for the Middle Class group, and then click Apply.

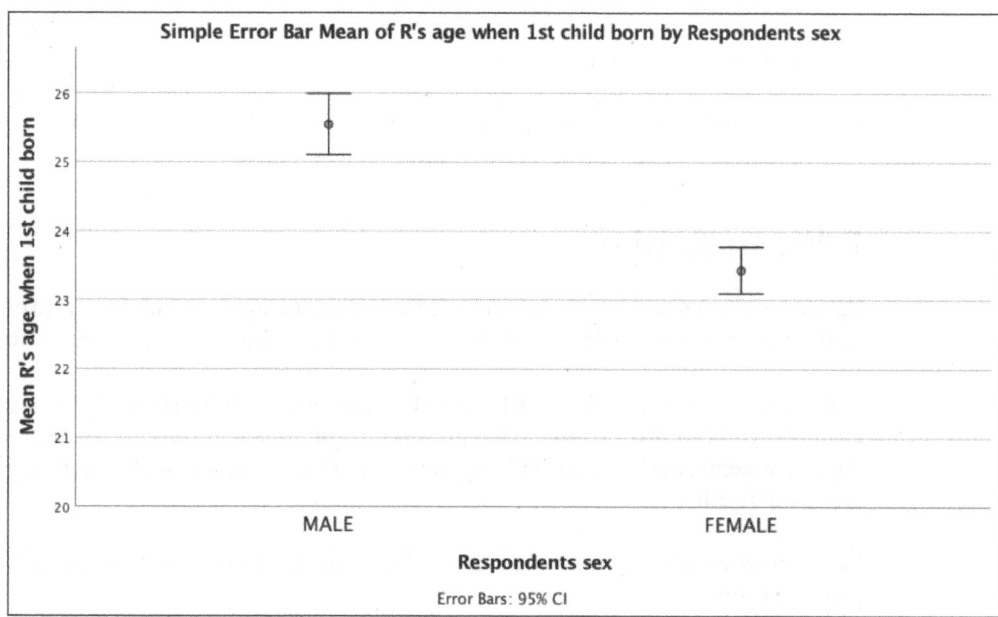

FIGURE 14-1: The graph with the edited y-axis.

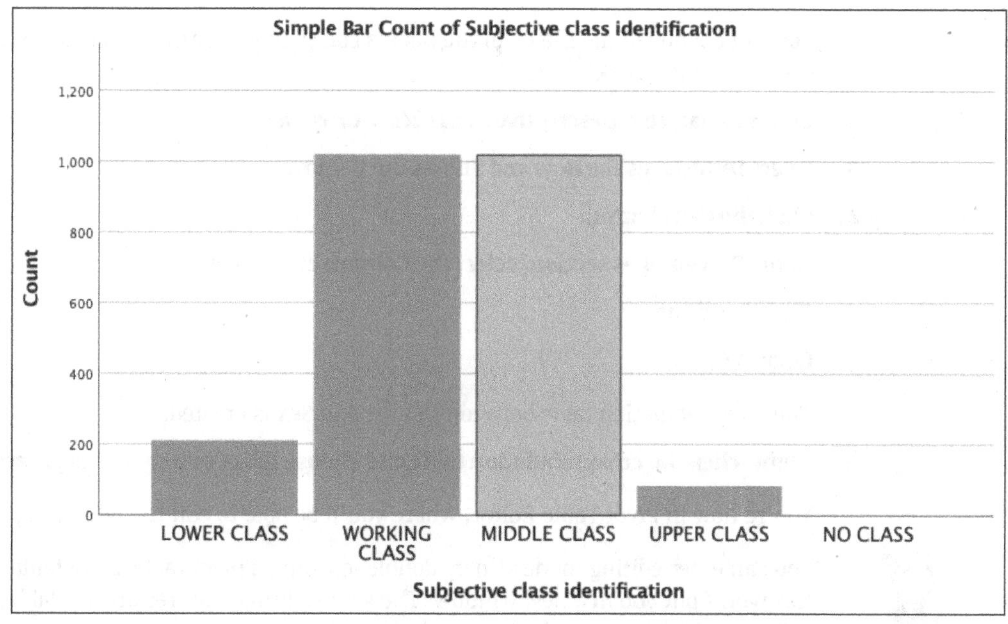

1. Using the GSS2018.sav file, recreate the simple bar chart in the previous example but add the data values to each bar.

2. Using the GSS2018.sav file, recreate the simple bar chart in the previous example but remove the comma from the data values and place the data values at the top of each bar.

3. Using the GSS2018.sav file, recreate the simple bar chart in the previous example but remove the label on the x-axis.

4. Using the GSS2018.sav file, recreate a template of the changes you made in the previous example and apply it to a new graph.

Editing Tables

Tables are the most common output in SPSS. In this section, we explain how to edit them. Many more options are available than those in this short section, but we get you off to a good start.

The menus that we describe here have many additional options, so be sure to explore them. For example, you can pivot a table (change rows to columns and columns to rows, or move individual elements, such as a variable or set of statistics. Changing the layout of the table does not affect the results.

In this next example, you create a cross tabulation table and then you edit the table by using pivoting trays:

1. **Choose File ⇨ Open ⇨ Data and load the GSS2018.sav file.**

   You can download the file from the book's companion website at `www.dummies.com/go/spssstatisticsworkbookfd`.

2. **Choose Analyze ⇨ Descriptive Statistics ⇨ Crosstabs.**
3. **Select DEGREE as the Row and SEX as the Column.**
4. **Click the Cells button.**
5. **In the Percentages section, select the Column check box.**
6. **Click Continue.**
7. **Click OK.**

   The cross tabulation table between Degree and Sex is created.

8. **Right-click the cross tabulation table and choose Edit Content ⇨ In Separate Window.**

   You're now in Pivot Table Editor, where you'll be able to edit the table you just created.

   TIP

   You can enter editing mode also by double-clicking a pivot table. By default, however, this won't put you in a new window. The same editing features are available, but editing is done in the main output window.

9. **Choose the Pivot menu ⇨ and then click Pivoting Trays.**

   The Pivoting Trays window visualizes each data element in the table. The Column area at the top of the Pivoting Trays window contains the SEX variable, which is displayed in the column dimension in the table. The Row area to the left of the Pivoting Trays window contains the elements displayed across the rows of the table, in this case the DEGREE variable and the statistics displayed in the table. The Layer area in the Pivoting Trays window is shown in the upper left of the pivoting tray (empty here).

Layers can be useful for large tables with nested categories of information. By creating layers, one simplifies the look of the table, making it easier to read. Layers work best when the table has at least three variables.

10. **Click the Statistics label in the Pivoting Tray and drag it to a new position in the Column area under Respondent Sex.**

11. **Close Pivot Table Editor.**

The chart in Figure 14-2 appears, with counts and percentages in separate columns.

R's highest degree * Respondents sex Crosstabulation

		Respondents sex					
		MALE		FEMALE		Total	
		Count	% within Respondents sex	Count	% within Respondents sex	Count	% within Respondents sex
R's highest degree	LT HIGH SCHOOL	122	11.6%	140	10.8%	262	11.2%
	HIGH SCHOOL	546	51.9%	632	48.8%	1178	50.2%
	JUNIOR COLLEGE	68	6.5%	128	9.9%	196	8.3%
	BACHELOR	205	19.5%	260	20.1%	465	19.8%
	GRADUATE	111	10.6%	136	10.5%	247	10.5%
Total		1052	100.0%	1296	100.0%	2348	100.0%

FIGURE 14-2: Counts and percentages are in separate columns.

See the following for an example of creating a pivot table.

EXAMPLE

Q. Using the GSS2018.sav file, recreate the pivot table in the previous example but change the column labels to N for counts and % for percentages.

A. Create the table as in the previous example. Double-click any of the column labels for Count and change them to N. Then double-click any of the column labels for Percent and change them to %.

R's highest degree * Respondents sex Crosstabulation

		Respondents sex					
		MALE		FEMALE		Total	
		N	%	N	%	N	%
R's highest degree	LT HIGH SCHOOL	122	11.6%	140	10.8%	262	11.2%
	HIGH SCHOOL	546	51.9%	632	48.8%	1178	50.2%
	JUNIOR COLLEGE	68	6.5%	128	9.9%	196	8.3%
	BACHELOR	205	19.5%	260	20.1%	465	19.8%
	GRADUATE	111	10.6%	136	10.5%	247	10.5%
Total		1052	100.0%	1296	100.0%	2348	100.0%

5. Using the GSS2018.sav file, recreate the table as in the previous example with the column name modifications, and then display two decimal places for the percentages.

6. Continuing with the prior example, change the font of the totals so that the values are in bold.

7. Continuing with the prior example, change the font of the data to Times New Roman size 10 and align the data so that it's centered vertically and horizontally in each cell.

8. Continuing with the prior example, adjust the size of the table so that the data better fit the columns.

Answers to Problems in Editing Output

1. Do the following:

   a. Create the chart.

   b. Select all the bars by double-clicking the graph and clicking any one of the bars.

   c. Choose Elements ⇨ Show Data Labels.

   d. Click Apply.

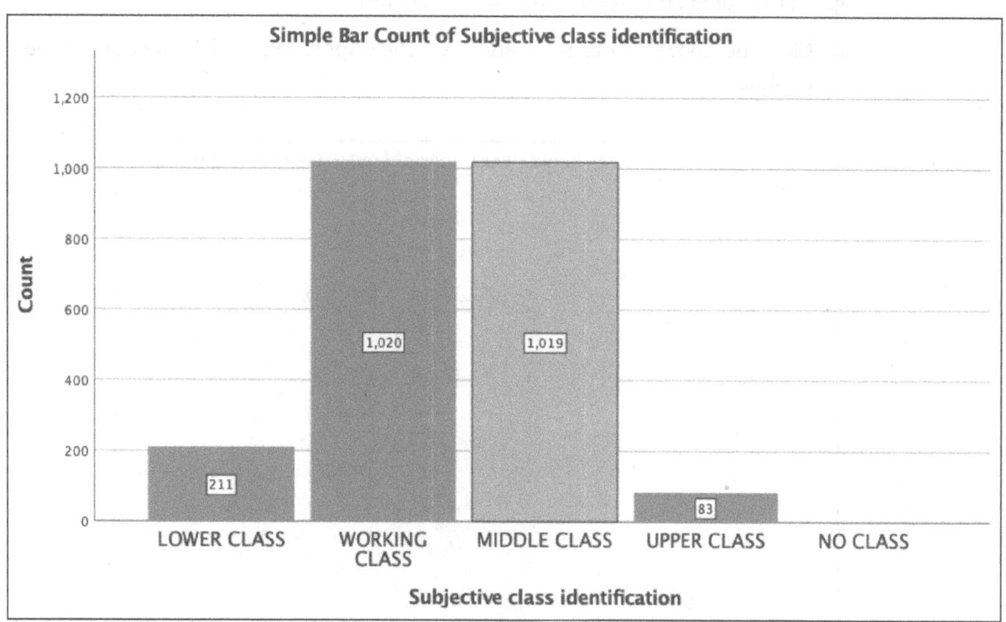

2 Do the following:

a. Create the chart and add the data values.

b. Click one of the data values.

c. Click the Number Format tab in the Properties box.

d. Deselect Display Digit Grouping.

e. Click Apply to remove the commas.

f. Click the Data Value Labels tab in the Properties box.

g. In the Label Position section, select Custom.

h. Click the Above Center icon, and then click Apply to position the data at the top of each bar.

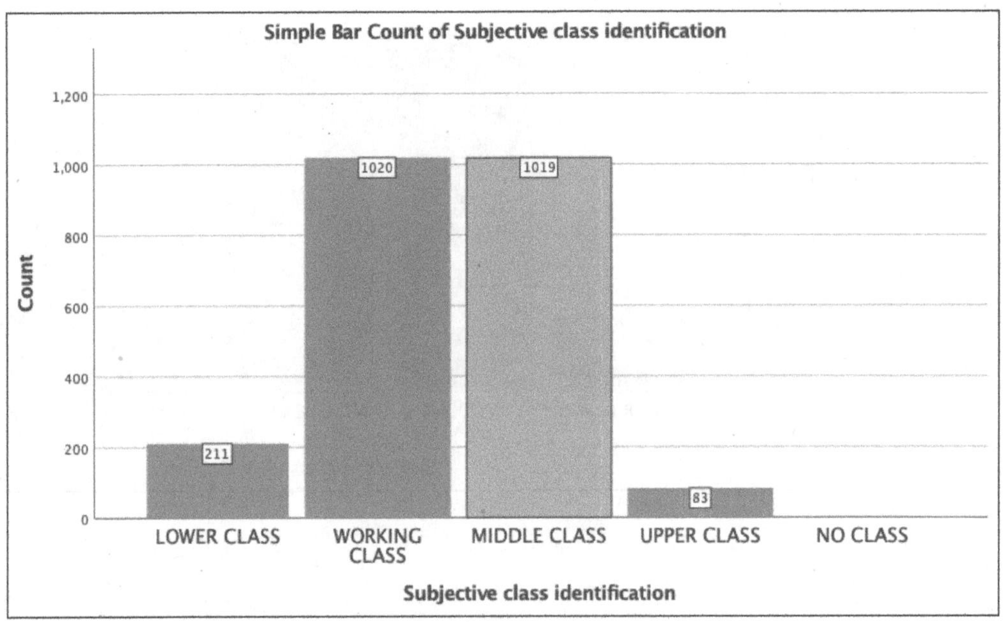

3 Do the following:

a. Create the chart and modify the data values.

b. Click the x-axis label to select it.

c. Click the label again, edit the title, and delete the label.

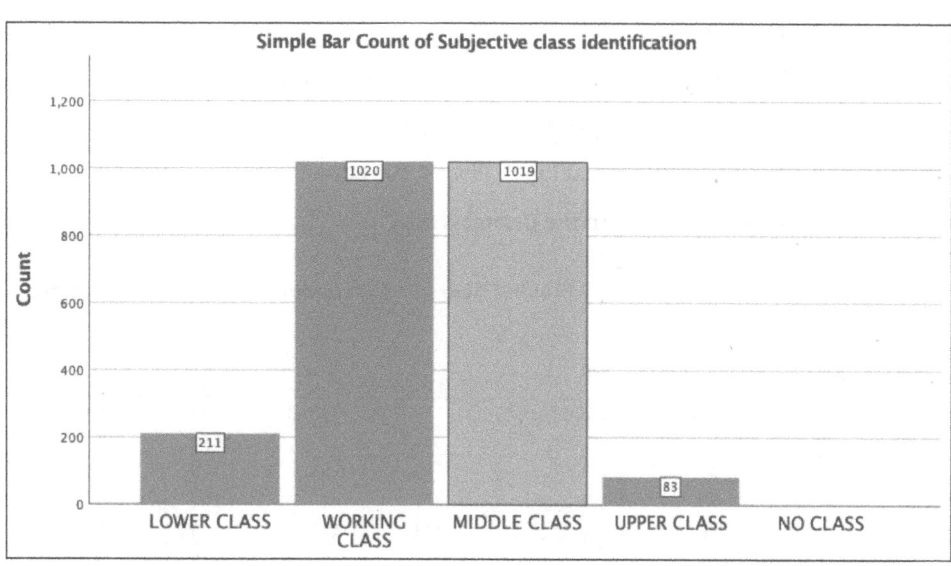

4. Do the following:

   a. Create the chart, modify the data values, and remove the x-axis.

   b. Choose File⇨Save Chart Template.

   c. Click the Data Value Labels section, so that all options in this section are selected.

      This will enable you to save the changes to the data values so you can apply them to new charts. Note that you can't save the change in which you deleted the axis titles.

   d. Name the template and save it.

   e. Recreate the graph and choose File⇨Apply Chart Template.

   f. Navigate to the template you created, and apply it to the newly created graph so that the data value modifications are applied.

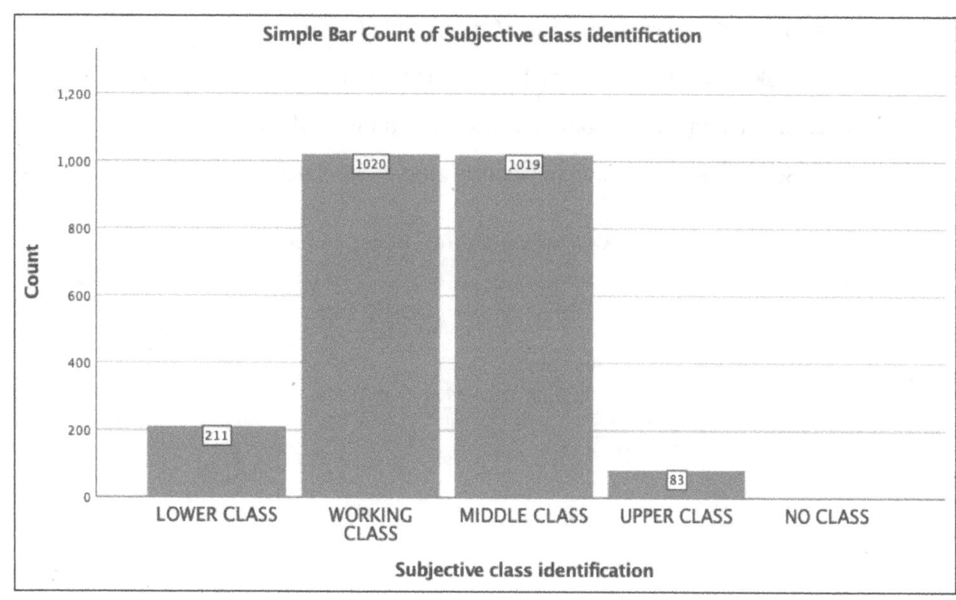

5. Do the following:

   a. Create the table and modify the column names.

   b. Select the values that are percentages.

   c. Select the Cell Format tab in Pivot Table Editor.

   d. Select the value 2 in the Decimals box.

R's highest degree * Respondents sex Crosstabulation

		Respondents sex					
		MALE		FEMALE		Total	
		N	%	N	%	N	%
R's highest degree	LT HIGH SCHOOL	122	11.60%	140	10.80%	262	11.16%
	HIGH SCHOOL	546	51.90%	632	48.77%	1178	50.17%
	JUNIOR COLLEGE	68	6.46%	128	9.88%	196	8.35%
	BACHELOR	205	19.49%	260	20.06%	465	19.80%
	GRADUATE	111	10.55%	136	10.49%	247	10.52%
Total		1052	100.00%	1296	100.00%	2348	100.00%

6. Select all the values for the totals and then click the bold icon in the menu.

R's highest degree * Respondents sex Crosstabulation

		Respondents sex					
		MALE		FEMALE		Total	
		N	%	N	%	N	%
R's highest degree	LT HIGH SCHOOL	122	11.60%	140	10.80%	**262**	**11.16%**
	HIGH SCHOOL	546	51.90%	632	48.77%	**1178**	**50.17%**
	JUNIOR COLLEGE	68	6.46%	128	9.88%	**196**	**8.35%**
	BACHELOR	205	19.49%	260	20.06%	**465**	**19.80%**
	GRADUATE	111	10.55%	136	10.49%	**247**	**10.52%**
Total		**1052**	**100.00%**	**1296**	**100.00%**	**2348**	**100.00%**

7. Do the following:

   a. Select the Area Format tab of Pivot Table Editor.

   b. Make sure Data is selected in the Area section.

   c. Select Times New Roman as the font and 10 as the font size.

   d. Select the vertical and horizontal alignment check boxes.

R's highest degree * Respondents sex Crosstabulation

		Respondents sex					
		MALE		FEMALE		Total	
		N	%	N	%	N	%
R's highest degree	LT HIGH SCHOOL	122	11.60%	140	10.80%	**262**	**11.16%**
	HIGH SCHOOL	546	51.90%	632	48.77%	**1178**	**50.17%**
	JUNIOR COLLEGE	68	6.46%	128	9.88%	**196**	**8.35%**
	BACHELOR	205	19.49%	260	20.06%	**465**	**19.80%**
	GRADUATE	111	10.55%	136	10.49%	**247**	**10.52%**
Total		**1052**	**100.00%**	**1296**	**100.00%**	**2348**	**100.00%**

8 Choose Format ⇨ Autofit.

R's highest degree * Respondents sex Crosstabulation

		Respondents sex					
		MALE		FEMALE		Total	
		N	%	N	%	N	%
R's highest degree	LT HIGH SCHOOL	122	11.60%	140	10.80%	**262**	**11.16%**
	HIGH SCHOOL	546	51.90%	632	48.77%	**1178**	**50.17%**
	JUNIOR COLLEGE	68	6.46%	128	9.88%	**196**	**8.35%**
	BACHELOR	205	19.49%	260	20.06%	**465**	**19.80%**
	GRADUATE	111	10.55%	136	10.49%	**247**	**10.52%**
Total		**1052**	**100.00%**	**1296**	**100.00%**	**2348**	**100.00%**

5 Programming SPSS with Command Syntax

IN THIS PART . . .

Learn how to paste Syntax code.

Use Syntax to create new variables.

IN THIS CHAPTER

» Using the Paste feature to generate Syntax

» Understanding procedures and transformations

» Making variable view specifications with Syntax

Chapter 15

Working with Pasted SPSS Syntax

In this chapter, you use the Paste button to write SPSS Syntax for you. Then you learn a bit about how to work with and modify the code, right in the Syntax window.

With a bit of knowledge and practice, you'll save a lot of time because once you have some code in the Syntax window, it's often much easier to work with it there than going back to the menus and dialogs. Even if Syntax is new to you, dive in and give it a try.

Pasting Procedures

Anything that produces output is a *procedure*. In this section, you start with procedure commands. The most common ones are in the Analyze and Graphs menus:

1. **Choose File ⇨ Open ⇨ Data and load the GSS2018.sav file.**

   You can download the file from the book's companion website at `www.dummies.com/go/spssstatisticsworkbookfd`.

2. **Choose Analyze ⇨ Descriptive Statistics ⇨ Descriptives.**

3. **Select the variables INTECON, INTEDUC, INTENVIR, and INTFARM and place them in the Variable(s) box.**

4. **Click Paste.**

   The Syntax window appears with the resulting code, as shown in Figure 15-1.

5. **Choose Run ⇨ All.**

   You run all the code in the Syntax window. After you start to accumulate more code, you have the option of selecting a portion of code and then choosing Run⇨Selected instead.

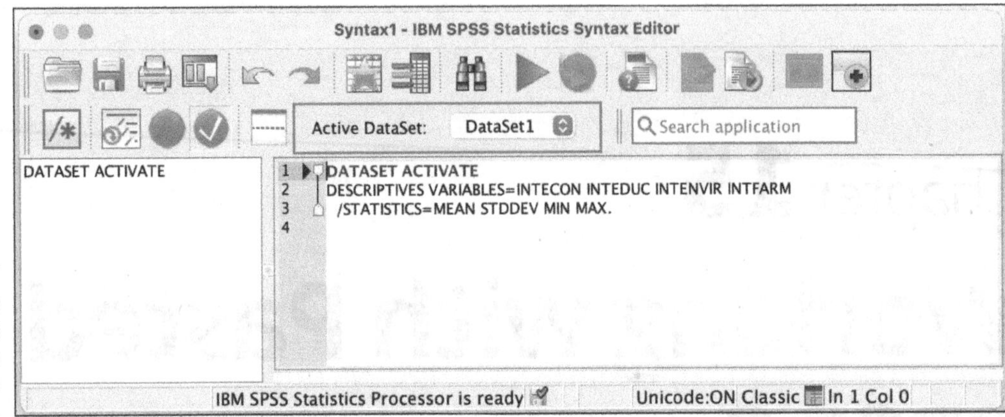

FIGURE 15-1: Pasted Syntax from the Descriptives procedure.

Use the following example to make a simple modification.

Q. Add the INTINTL variable to the command and run it again.

A. Your Syntax should look like the following:

```
DATASET ACTIVATE DataSet1.
DESCRIPTIVES VARIABLES= INTECON INTEDUC INTENVIR INTFARM INTINTL
/STATISTICS=MEAN STDDEV MIN MAX.
```

TIP

The Dataset Activate command is used only to switch back and forth between multiple datasets. You don't need to use the command if you're using only one dataset.

TIP

You don't have to type long lists of variable names manually. In the Utilities menu, a Variables dialog enables you to select variables and then paste the names in the Syntax window. The names will be pasted at the cursor's position, so place it in the correct spot before you click the Paste button.

1. Set up a Frequencies command for INTECON, INTEDUC, and INTSPACE. Change the settings in the Format dialog to Descending Counts. Remove any unnecessary commands.

2. Set up a paired samples t-test to compare INTECON with INTEDUC, and INTECON with INTSPACE. Paste the Syntax. Modify the Syntax to do one more pair, INTECON with INTTECH. (You could paste all three, but try two and then add the third to test your understanding of how Syntax works.)

Pasting Transformations

In this section, you see how transformation commands, generally found in the Transform menu, behave a bit differently when you paste them. For this next example, you revisit a function and some variables from Chapter 5 but click Paste this time:

1. **Choose File ⇨ Open ⇨ Data and load the GSS2018.sav file.**

   Download the file from the book's companion website at `www.dummies.com/go/spssstatisticsworkbookfd`.

2. **Choose Transform ⇨ Compute Variable.**
3. **In the Target Variable box, type** MEAN_ED2 **as the name of your new variable.**
4. **In the Numeric Expression box, type the following equation:** MEAN(EDUC,MAEDUC, PAEDUC)
5. **Click Paste.**

   The resulting code in the Syntax window is

   ```
   COMPUTE MEAN_ED2=MEAN(EDUC, MAEDUC, PAEDUC).
   EXECUTE.
   ```

   The Execute command is necessary only because you don't have a procedure. Transformations wait until they encounter a procedure. If you have a procedure such as the Descriptives procedure in the last section, you don't need an Execute command.

TIP

In SPSS Syntax, you can abbreviate a command to the first four letters.

6. **Use the mouse to highlight the new code.**
7. **Choose Run ⇨ Selection.**

   A new variable is produced.

TIP

You rarely need to use an Execute command. Delete them all. As long as you have a procedure, any procedure, after the transformations in your code, the transformations will execute.

WARNING

Periods are important in SPSS. Watch out for your code turning red when you accidentally delete them.

See the following for an example of using the compute variable transformation.

EXAMPLE

Q. Using the GSS2018.sav file and the MEAN function, calculate the mean interest for the INTECON, INTEDUC, and INTSPACE variables. Then run the Descriptives procedure on INTECON, INTEDUC, INTSPACE, and MEAN_INT. Do it all with Syntax by either typing or pasting, but remove unnecessary commands. Abbreviate Descriptives and see if it runs.

A. Your code should look like the following:

```
COMPUTE MEAN_INT=MEAN(INTECON, INTEDUC, INTSPACE).
DESC VARIABLES= INTECON INTEDUC INTSPACE MEAN_INT.
```

3. Use the GSS2018.sav file. Calculate the mean education (EDUC, MAEDUC, PAEDUC, and SPEDUC) per family by using the MEAN function but only if there are at least three valid values. Name the new variable MEAN_FAM_EDUC. Then add a DESC command for your four original variables and your new variable. This question revisits an example from Chapter 5, but this time you work in the Syntax window. If you use the Paste button to copy the code to Syntax Editor, make sure that you remove any unnecessary lines of code. If you get stuck, check your answer from Chapter 5.

4. Using the GSS2018.sav file and the Count dialog (in the Transform Menu), count the number of missing values in EDUC, MAEDUC, PAEDUC, and SPEDUC. Paste the resulting Syntax. Then add a DESC command to look at all five variables.

Controlling Variable View Specifications with Syntax

In this section, you learn about Syntax Help and how to control the Variable View settings in SPSS by using Syntax. Knowing the commands for Variable Labels, Value Labels, and Missing Values is particularly helpful.

SPSS Syntax isn't just about the Data window and the Syntax window. It can control virtually everything that SPSS can do. The only few items that can't be controlled with Syntax are some settings specific to the menus and dialogs themselves.

If you have an important day-to-day task, or tasks that need to be repeated on occasion, Syntax can control these situations, even if it isn't obvious how to use Paste to write the code.

In the following example you type commands instead of using the Paste button:

1. **Continue to work in the GSS2018.sav file.**
2. **Choose File ⇨ New ⇨ Syntax.**

   In this example, you need an unlabeled variable; we use the MEAN_INT variable from the preceding section.
3. **In the Syntax window, type** VARIABLE, **pausing as you type to look at the screen.**

   After just a few letters, the pop-up window in Figure 15-2 appears.
4. **Select VARIABLE LABELS.**

   For help in completing the command, you can use the Syntax Help button and the Command Syntax Reference document (in the Help menu).

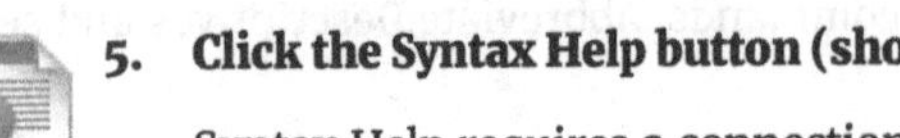

5. **Click the Syntax Help button (shown in the margin).**

   Syntax Help requires a connection to the internet.

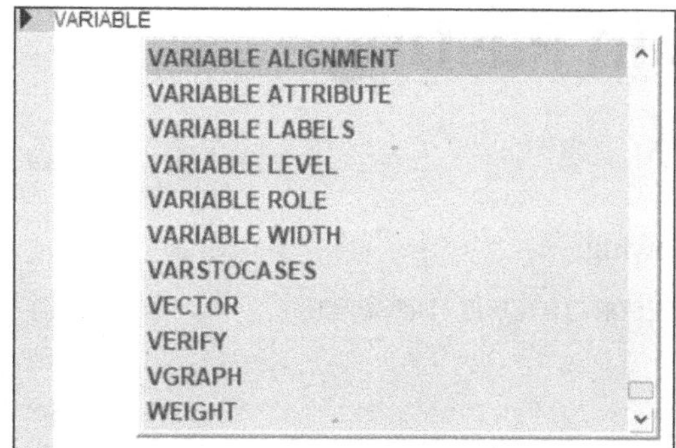

FIGURE 15-2: Variable keywords in Syntax.

You see the following grammar diagram, which you can refer to complete the command. You can use a button on the page to copy the grammar diagram. In the diagram, text in uppercase should be copied exactly, and the text in lowercase should be replaced with your information:

```
VARIABLE LABELS varname 'label' [/varname. . .]
```

6. **Complete the command by typing the following:**

```
VARIABLE LABELS MEAN_INT "Mean Interest".
```

7. **Use the mouse to highlight the new code.**
8. **Choose Run ➪ Selection.**
9. **Click the Variable View tab of Data Editor to verify that the new label appears.**

See the following for an example of using the compute variable transformation.

EXAMPLE

Q. Using the GSS2018.sav file, provide a label for the newly created MEAN_ED2 variable.

A. Your code should look like the following:

```
VARIABLE LABELS MEAN_ED2 "Mean of Respondent and Parental Education".
```

5 Using the GSS2018.sav file, return to the code you used to create MEAN_FAM_EDUC. Add an appropriate variable label. Rerun the code.

6 Using the GSS2018.sav file and the Recode into Same Variables dialog (in the Transform menu), convert the system missing values to 99. Then use Syntax Help to look up the grammar for the Value Labels and Missing Values commands. Declare 99 as missing and provide an appropriate label. Add a FREQ command to force the transformations to run and to check your work. Check the Variable View as well.

Answers to Problems in Pasting and Modifying Syntax

1. The code should look like the following:

```
FREQUENCIES VARIABLES= INTECON INTEDUC INTSPACE
/FORMAT=DFREQ
/ORDER=ANALYSIS.
```

2. The code should look like the following:

```
T-TEST PAIRS= INTECON INTECON INTECON WITH INTEDUC INTSPACE INTTECH
  (PAIRED)
 /ES DISPLAY(TRUE) STANDARDIZER(SD)
 /CRITERIA=CI(.9500)
/MISSING=ANALYSIS.
```

3. The code should look like the following:

```
COMPUTE MEAN_FAM_EDUC =mean.3(educ,speduc,maeduc,paeduc).
DESCRIPTIVES VARIABLES=EDUC SPEDUC MAEDUC PAEDUC MEAN_FAM_EDUC
  /STATISTICS=MEAN STDDEV MIN MAX.
```

Descriptive Statistics

	N	Minimum	Maximum	Mean	Std. Deviation
Highest year of school completed	2345	0	20	13.73	2.974
Highest year school completed, spouse	988	0	20	13.89	3.069
Highest year school completed, mother	2089	0	20	11.88	3.820
Highest year school completed, father	1687	0	20	11.88	4.148
MEAN_FAM_EDUC	1787	.00	20.00	12.8189	2.87584
Valid N (listwise)	729				

4. The code should look like the following:

```
Num_Miss_Educ=EDUC SPEDUC MAEDUC PAEDUC
DESC Num_Miss_Educ EDUC SPEDUC MAEDUC PAEDUC.
```

5 The code should look like the following:

```
COMPUTE MEAN_FAM_EDUC =mean.3(educ,speduc,maeduc,paeduc).
VARIABLE LABELS MEAN_FAM_EDUC "Mean Family Education".
DESCRIPTIVES VARIABLES=EDUC SPEDUC MAEDUC PAEDUC MEAN_FAM_EDUC
  /STATISTICS=MEAN STDDEV MIN MAX.
```

Descriptive Statistics

	N	Minimum	Maximum	Mean	Std. Deviation
Highest year of school completed	2345	0	20	13.73	2.974
Highest year school completed, spouse	988	0	20	13.89	3.069
Highest year school completed, mother	2089	0	20	11.88	3.820
Highest year school completed, father	1687	0	20	11.88	4.148
Mean Family Education	1787	.00	20.00	12.8189	2.87584
Valid N (listwise)	729				

6 The code should look like the following:

```
RECODE MEAN_FAM_EDUC (SYSMIS=99).
MISSING VALUES MEAN_FAM_EDUC (99).
Value labels MEAN_FAM_EDUC (99="Incomplete Information").
FREQ VARIABLES = MEAN_FAM_EDUC.
```

IN THIS CHAPTER

» Writing Syntax that performs a series of steps

» Using DO IF . . . END IF

Chapter 16
Computing Variables with Syntax

In Chapter 15, you were introduced to the Syntax window and computed a few variables. In this chapter, you work with more complex examples. You revisit some of the functions you encountered in Chapter 5, when you created variables by using the Compute Variable dialog, but this time you create the variables by writing code. Once you get the hang of it, you can compute variables faster in the Syntax window than by using Paste from the dialogs.

REMEMBER

If you get stuck looking at an empty Syntax window, simply go back and use the Paste button in the dialogs for help. Then after you've used the commands for a while, switch to modifying.

Calculating New Variable with Syntax

In this section, you repeat the difference calculation from Chapter 5, but you type the formula in the Syntax window rather than use the dialogs:

1. **Choose File ⇨ Open ⇨ Data and load the GSS2018.sav file.**

   Download the file from the book's companion website at `www.dummies.com/go/spssstatisticsworkbookfd`.

2. **Choose File ⇨ New ⇨ Syntax.**

3. **Type** COMPUTE.

   Do not type a period until the entire command is complete. A pop-up menu appears to help you, and the command turns red because it isn't a legal command yet.

REMEMBER

   Typing the commands in uppercase is not a requirement, but it is the convention in SPSS.

4. **On the same line, type the following:**

```
ED_DIFF2 = EDUC - SPEDUC
```

   You use a different name for the variable than the one in Chapter 5 so you can compare the two variables. They should produce the same result.

5. **Add a period at the end of the command.**

   The COMPUTE command turns blue, which means it's grammatically correct. However, a command can turn blue but not produce what you intended.

6. **On the next line, type the following:**

```
DESC EDUC SPEDUC ED_DIFF2.
```

   DESC is the abbreviation for DESCRIPTIVES. This line executes your calculation and gives you a chance to check your work.

7. **Choose Run ⇨ All.**

   The output window looks like Figure 16-1.

Descriptive Statistics

	N	Minimum	Maximum	Mean	Std. Deviation
Highest year of school completed	2345	0	20	13.73	2.974
Highest year school completed, spouse	988	0	20	13.89	3.069
ED_DIFF2	987	-18.00	11.00	.2087	2.75817
Valid N (listwise)	987				

FIGURE 16-1: A DESCRIPTIVES report of EDUC SPEDUC ED_DIFF2.

In the following example, you use some familiar functions to get more practice working in the Syntax window.

EXAMPLE

Q. Revisiting a Chapter 5 example and using the GSS2018.sav file and the same Syntax window, use the COMPUTE transformation command to calculate the ratio of EMAILHR (hours per week using email) to WWWHR (hours per week being online). Call the new variable ONLINE_RATIO. Remember that you're working in the Syntax window. Add the new variable to the DESCRIPTIVES command between ED_DIFF2 and the period.

A. Your Syntax should look like the following:

```
COMPUTE ED_DIFF2 = EDUC - SPEDUC.
COMPUTE ONLINE_RATIO = EMAILHR / WWWHR.
DESC ED_DIFF2 ONLINE_RATIO.
```

The first two questions are easy and repeat two formulas in Chapter 5 so you can get accustomed to the new environment and not worry about the formulas. Remember that you have to write the entire command with the proper punctuation and make sure that the commands turn blue in the Syntax window. If you get stuck on the command, see Chapter 5. Don't forget to include the new variables in a DESCRIPTIVES command, or an alternate procedure command such as FREQUENCIES, to check your work.

Questions 3 – 7 are more challenging and require a different dataset. If you get stuck, refer to Chapter 5, which covers some of the same string functions you need to solve these questions.

1. Using the GSS2018.sav file and the Syntax window, calculate the difference between the CHILDS (number of children) and CHLDIDEL (ideal number of children) variables. Use the absolute value function (ABS) so that all differences are positive.

2. Using the GSS2018.sav file and the Syntax window, create a variable that identifies the maximum years of education for a family. The education variables are EDUC, MAEDUC, PAEDUC, and SPEDUC. Use the MAX function. Add a procedure to check your work.

   The answers for 1 and 2 are shown as one section of code in the answer section of the chapter.

3. Using the Presidents.sav dataset and the CHAR.INDEX function, create a new variable named COMMA that represents the location of the comma in the name. For instance, Washington, George has a comma location of 11.

4. Using the Presidents.sav dataset, COMMA, and the CHAR.SUBSTR function, compute a new variable called LAST_NAME. You don't have to deal with middle names because the code would become too complicated. The last name is everything before the comma.

TIP

Before you can use COMPUTE to create a string variable, you must use a STRING function like this: STRING LAST_NAME (a15). The *a* means alphanumeric, indicating that the variable stores letters, and the 15 means it can fit 15 letters. Size accordingly. Make the width of the variable big enough to fit the longest string that you are likely to encounter.

5. Using the Presidents.sav dataset, COMMA, and the CHAR.SUBSTR function, compute a new variable called FIRST_NAME. The first name is everything after the comma.

6. Using the Presidents.sav dataset and the CONCAT function, reassemble the name so that the first name comes before the last name. You might need to use the RTRIM function and insert a space with CONCAT so that the full name is legible.

7. Assemble all your work into one SPSS Syntax program. This time, eliminate the need for the COMMA variable by nesting the instructions in questions 3 through 6 inside the functions. This process is more advanced, so take your time. You have all the elements you need in your answers for questions 3 through 6.

Using DO IF . . . END IF

Sometimes we want to process a formula a certain way for some cases in our data but a different way for other cases. If you've written code in another computer language, you've probably encountered a way of handling the conditions if, else if, and else. In Syntax, we use the DO IF . . . END IF command.

1. **Choose File ⇨ Open ⇨ Data and load the GSS2018.sav file.**

   Download the file from the book's companion website at `www.dummies.com/go/spssstatisticsworkbookfd`.

2. **Choose File ⇨ New ⇨ Syntax.**
3. **Start to type** DO IF **until the pop-up appears, and then click DO IF in the pop-up.**
4.  **Click the Syntax Help button (shown in the margin).**

   We discussed Syntax Help in Chapter 15. Copy the DO IF grammar chart from Help and paste it where you can refer to it. You could use a text file on your computer or the Syntax window.

   WARNING

   The grammar chart is only for reference. Don't highlight it later when you run your code.

   ```
   DO IF [(]logical expression[)]
   transformation commands
   [ELSE IF [(]logical expression[)]]
   transformation commands
   [ELSE IF [(]logical expression[)]]
   .
   .
   .
   [ELSE]
   transformation commands
   END IF
   ```

   TIP

   If you want to add a comment to a Syntax file, so that a human can read it but SPSS ignores it, just put an asterisk (*) before the comment. It looks like this:

   ```
   * COMMENT.
   ```

   Don't forget the period. The comment should turn gray.

5. **Finish the first line with a condition regarding missing cases:**

   ```
   DO IF MISSING(SPEDUC).
   ```

6. **Referring to the grammar chart, write a separate line to calculate when missing(SPEDUC) is true:**

   ```
   COMPUTE ED_DIFF3=97.
   ```

7. **Referring to the grammar chart again, add a line with ELSE:**

   ```
   ELSE.
   ```

The brackets in the grammar chart mean that you have a choice. A different example might have had an ELSE IF. And you could have more than one ELSE IF. Don't forget the period after ELSE.

8. **Complete the command by writing the following on two lines:**

```
COMPUTE ED_DIFF3 = EDUC - SPEDUC.
END IF.
```

The first line tells SPSS what to do if the DO . . . IF line is false. The END IF terminates the command.

9. **Add a Missing Values command:**

```
MISSING VALUES ED_DIFF3(97).
```

10. **Add the following procedure to force the transformation to run so you can check your work:**

```
FREQ ED_DIFF3.
```

11. **Run the code.**

An advantage of the FREQUENCIES command is that you can see that assigning 97 to missing cases has worked, as shown in Figure 16-2.

Frequencies

Statistics

ED_DIFF3

N	Valid	987
	Missing	1361

ED_DIFF3

		Frequency	Percent	Valid Percent	Cumulative Percent
Valid	-18.00	1	.0	.1	.1
	-11.00	1	.0	.1	.2
	-8.00	4	.2	.4	.6
	-7.00	3	.1	.3	.9
	-6.00	14	.6	1.4	2.3
	-5.00	15	.6	1.5	3.9
	-4.00	47	2.0	4.8	8.6
	-3.00	50	2.1	5.1	13.7
	-2.00	92	3.9	9.3	23.0
	-1.00	68	2.9	6.9	29.9
	.00	323	13.8	32.7	62.6
	1.00	95	4.0	9.6	72.2
	2.00	117	5.0	11.9	84.1
	3.00	49	2.1	5.0	89.1
	4.00	52	2.2	5.3	94.3
	5.00	21	.9	2.1	96.5
	6.00	17	.7	1.7	98.2
	7.00	5	.2	.5	98.7
	8.00	10	.4	1.0	99.7
	10.00	2	.1	.2	99.9
	11.00	1	.0	.1	100.0
	Total	987	42.0	100.0	
Missing	97.00	1360	57.9		
	System	1	.0		
	Total	1361	58.0		
Total		2348	100.0		

FIGURE 16-2: FREQUENCIES report of ED_DIFF3.

See the following for an example of using DO IF . . . END IF.

EXAMPLE

Q. Create a new variable for TV Hours per Week. Use DO IF . . . END IF to create a variable with the following possible values: 1 for TV Hours > WWW Hours, 2 for WWW Hours > TV Hours, and 99 for Insufficient Information. Add the appropriate labels and missing values declarations. Use the Summarize command under Analyze ⇨ Reports ⇨ Case Summaries to force execution and to check your work. This is an elaborate example.

A. Your Syntax should be similar to this:

```
COMPUTE TVHOURS_WEEK=TVHOURS * 7.

DO IF MISSING(TVHOURS) OR MISSING (WWWHR).
COMPUTE MEDIA_TIME = 99.
ELSE IF TVHOURS_WEEK > WWWHR.
COMPUTE MEDIA_TIME = 1.
ELSE IF WWWHR > TVHOURS_WEEK.
COMPUTE MEDIA_TIME = 2.
END IF.

MISSING VALUES MEDIA_TIME (99).
VALUE LABELS MEDIA_TIME 1 'TV > WWW' 2 'WWW > TV' 3 'Insufficient
   Information'.

SUMMARIZE
/TABLES= TVHOURS TVHOURS_WEEK WWWHR MEDIA_TIME
/FORMAT=VALIDLIST NOCASENUM TOTAL LIMIT=100
/TITLE='Case Summaries'
/MISSING=VARIABLE
/CELLS=COUNT.
```

8. This question is the trickiest in this workbook. Use DO IF . . . END IF to figure out the education gap between the respondent and their same-sex parent. Start by figuring out who you can anticipate will be missing and assign them a value of 99. Then ensure that your solution works for both men and women. Use a procedure command to check your work. The Summarize command under Analyze ⇨ Reports ⇨ Case Summaries is a useful choice for this example.

9. Add appropriate labels and missing values declarations to your solution to question 8.

Answers to Problems in Computing Variables with Syntax

1. See answer 2.

2. The answers for questions 1 and 2 are as follows:

```
COMPUTE CHLD_DIFF = ABS(CHILDS - CHLDIDEL).
COMPUTE EDUC_MAX = MAX(EDUC, MAEDUC, PAEDUC, SPEDUC).
DESC CHLD_DIFF EDUC_MAX.
```

3. The code should look like the following:

```
COMPUTE COMMA=CHAR.INDEX(NAME,',').
```

4. The code should look like the following:

```
STRING FIRST_NAME(a25).
COMPUTE FIRST_NAME = CHAR.SUBSTR(NAME,COMMA+1).
```

5. The code should look like the following:

```
STRING LAST_NAME(a35).
COMPUTE LAST_NAME = CHAR.SUBSTR(NAME,1,COMMA-1).
```

6. The code should look like the following:

```
STRING FULL_NAME(a60).
COMPUTE FULL_NAME  = CONCAT (RTRIM(FIRST_NAME),' ',LAST_NAME).
```

7. The following code is one possible solution:

```
STRING FIRST_NAME(a25).
STRING LAST_NAME(a35).
STRING FULL_NAME(a60).
COMPUTE FIRST_NAME = CHAR.SUBSTR(NAME,INDEX(NAME,',')+1).
COMPUTE LAST_NAME = CHAR.SUBSTR(NAME,1,CHAR.INDEX(NAME,',')-1).
COMPUTE FULL_NAME = CONCAT (RTRIM(FIRST_NAME),' ',LAST_NAME).

LIST.
```

LIST is a simple command that we don't encounter much anymore. It simply prints the dataset, including the new variables, to the output window. For this small dataset, using LIST is a good way to check your work.

8. Your code should look similar to the following:

```
DO IF missing(EDUC) or missing(MAEDUC) or missing(PAEDUC).
COMPUTE SAMESEX_GAP = 99.
ELSE IF (SEX=2).
COMPUTE SAMESEX_GAP = EDUC - MAEDUC.
ELSE.
COMPUTE SAMESEX_GAP = EDUC - PAEDUC.
END IF.
```

Here is one possible procedure to use to check your work:

```
SUMMARIZE
/TABLES=EDUC MAEDUC PAEDUC SAMESEX_GAP SEX
/FORMAT=LIST NOCASENUM TOTAL LIMIT=100
/TITLE='Case Summaries'
/MISSING=VARIABLE
/CELLS=COUNT.
```

9. In addition to the code in answer 4, you would need the following three additional lines of code (or something similar):

```
VALUE LABELS SAMESEX_GAP 99 'Unknown'.
MISSING VALUES SAMESEX_GAP (99).
VARIABLE LABELS SAMESEX_GAP "Educ Gap Between Respondent and Same Sex
  Parent".
```

You should put the new lines between the DO IF . . . END IF and the SUMMARIZE COMMAND.

6 The Part of Tens

IN THIS PART . . .

Learn how to overcome difficult situations when preparing and analyzing data.

Get experience taking practice certification questions.

IN THIS CHAPTER

» Using RECODE INTO SAME

» Using COMPUTE with system missing and other missing values

» Merging two or more files

» Creating consistent charts

» Creating consistent tables

» Replicating a multistep process in Syntax

» Producing tables with Custom Tables

» Using the TEMPORARY command

» Explaining p-values to others

» Explaining statistical assumptions to others

Chapter 17

Ten Tricky SPSS Skills to Practice

Mastering SPSS requires two distinct things: knowledge and practice. With *SPSS Statistics For Dummies*, 4th Edition, you gain a lot of knowledge. This workbook gives you the ability to rehearse your skills and check your knowledge.

This chapter provides ten examples of tricky situations that even knowledgeable SPSS users need to practice until they become second nature. They are all discussed in this workbook or *SPSS Statistics For Dummies* or both. If you practice all ten, you'll be on your way to the next level of SPSS mastery.

Using RECODE INTO SAME

We recommend that you master using the RECODE INTO DIFFERENT command and menu before you try RECODE INTO SAME. Although INTO SAME can be faster and easier with practice, there's a risk of overwriting your original variable, so you should never try it for the first time in a high stakes situation or when you're in a rush.

The classic mistake is when you recode a continuous variable such as age or income into a category. Users often think they don't want both versions (continuous and categorical) of the same variable because then they'll have lots of extra columns that take up space and can confuse colleagues. But remember, checking your work is crucial, and having the original variable gives you flexibility to further recode it in other ways.

You can fix a mistake in overwriting a variable as long as you do so before you overwrite the file. After you overwrite the file, you might be in big trouble with an important file. Practice first!

Using COMPUTE with System Missing and Other Missing Values

Whether you compute new variables in the menus or in Syntax, even veteran SPSSers can produce unexpected results when working with missing values. You should always favor user-defined missing values over system missing values. While you might achieve having user-defined missing values with your final dataset, it's virtually impossible to avoid system missing values because it's the SPSS version of a null value. Here's what you should do: Create a small practice dataset based on your work data that has just a couple dozen rows.

Merging Two or More Files

Merging requires practice. If you choose file A to be your active dataset, and then merge with file B, the steps feel a bit different then if you had started with file B. We have watched hundreds of workshop attendees accidentally try to merge a file with itself. Consider using SPSS Syntax to write a program that performs a merge you do regularly. However, you still have to keep track of which file is the active file using Syntax, which itself takes practice.

Finally, although you can't merge more than two files in one step by using the menus, you can do that with Syntax or a more elaborate iterative process in the menus. If you try merges on large complex files without rehearsing first on small practice files, you could get yourself in big trouble.

Creating Consistent Charts

Do you want to know a secret? When we're deciding which topics to include when writing a book and how to explain them, we often find chart templates frustrating. Charting has evolved more than some other SPSS features over the last several version changes. We suggest that you set aside some time to practice using chart templates because they are the best way to make all your charts look consistent. Otherwise, you'll end up manually changing every chart individually, or you'll export all your charts and edit them one by one. Neither is a recommended strategy! Please take the time to become familiar with chart templates.

Creating Consistent Tables

Most people find TableLooks easier to understand than chart templates, especially because a bunch of prebuilt ones are available, giving you a head start whether you want your tables more compact, more bold, or APA compliant. So you have no excuse for fussing around with individual settings one table at a time. To practice using TableLooks, create some output with a few tables and then try switching them all over to another uniform look in as few steps as possible.

Replicating a Multistep Process in Syntax

The first SPSS Syntax skill to develop is the shortcut of using the Paste button, discussed in detail in *SPSS Statistics For Dummies*. If you want to be truly efficient, your SPSS Syntax programs have to be much more comprehensive. Begin with importing data, setting up the data, and then data preparation, analysis, and generating output.

If you're doing something repetitive, such as monthly reports, you should eventually automate every step. If you don't have all the skills to do that now, start with what you know and make a goal of complete automation. Figure out what you already know how to automate, and start from there. Then use the time that you free up to add a little more automation each time the report is due.

Learning these skills takes time, and you should add wiggle room in your schedule for mistakes. Importantly, your mistakes should be practice mistakes — not mistakes that make their way into the final report.

Producing Tables with Custom Tables

In the Part of Tens section of *SPSS Statistics For Dummies*, we list the SPSS add-on modules and what they do. There's a reason why the Custom Tables module has been the most commonly purchased module for years. If you have to produce a report that has more than a couple of pivot tables, especially if you're doing survey analysis, the Custom Tables module probably will be worth the investment. If you're exporting to Excel just to make minor changes, the module almost certainly will help you.

However, you don't want to try to work with the Custom Tables module when you're up against a deadline. Instead, get a free trial of the module at `www.ibm.com/account/reg/us-en/signup?formid=urx-19774` well before your next deadline. Consider running your tables twice. First, use your current step-by-step process. Then try a second time with Custom Tables. Don't be surprised if you can create your ideal table without editing! If you succeed, you can paste the SPSS Syntax and save even more time. Give yourself plenty of time to experiment and get comfortable with the module during the free trial.

Using the TEMPORARY Command

The TEMPORARY command in SPSS Syntax is a powerful command that addresses a frequent need: running an analysis on a specific subset of data. Although you can accomplish the same tasks by using the menus, the process is more efficient in SPSS Syntax. For example, suppose that you want to run an analysis of maternity leave, likelihood of promotion, healthcare coverage, and compensation for only the women in your dataset. But you don't want to accidently delete all of the men. TEMPORARY is just that — it's temporary — and the unselected data returns to active status as soon as the next procedure runs. So why do you need to practice using the TEMPORARY command? In our example, if you made a mistake and then saved the dataset, you would permanently lose the data for the men. To avoid this risk, rehearse the command with practice data.

Explaining P-Values to Others

In 2016, the American Statistical Association (ASA) released a "Statement on Statistical Significance and P-Values" with six principles underlying the proper use and interpretation of the p-value. Even though, the contemporary use of p-values is based on the work of Sir Ronald Fisher nearly 100 years ago and thousands upon thousands of individuals have earned a PhD in the field, the most important professional body in statistics, the ASA, feels the need to publicly address the confusion and misuse of the p-value. So it should not come as a surprise if the customer of your work, whether internal or external, is confused about p-values. And it's more likely that their boss is even more confused about p-values.

When you present findings, any lack of confidence you have in the proper use of p-values will be signaled by hesitation or your body language. Even we take a deep breath and brace ourselves for questions when we present p-values. How can you address this challenge? Practice, practice, practice. Recruit a friend and rehearse how you're going to explain the p-value portion of your report. The person doesn't have to be a statistician — it's actually better if they aren't. When it comes time to explain p-values in a meeting or in front of a large group, you'll be glad that you practiced.

Explaining Statistical Assumptions to Others

Statistical assumptions, like p-values, are a difficult topic to explain. Both topics are technical and rarely understood, but the challenge of explaining statistical assumptions is different than explaining p-values. If you met the statistical assumptions of standard tests, our advice is to not belabor the point. Most audiences will simply want to know that you have competently checked off that box. (However, be prepared in case a statistician is in the room or reading the report). If you failed to meet assumptions and had to use an alternative test such as a non-parametric test, your job is more complicated. We recommend that you report not only the alternate test but also the standard test along with the evidence that you failed to meet assumptions. You're trying to serve two audiences: those who understand the alternate test and those who are familiar only with the standard test. Be sure to rehearse your explanation so that it doesn't cast doubt on your analysis.

IN THIS CHAPTER

» Discovering what's on the exam

» Trying out practice questions

Chapter 18

Ten Practice Certification Questions

The IBM Certified Specialist — SPSS Statistics Level 1 exam can help demonstrate your hard-won competence using SPSS. The certification exam has 55 questions in seven major categories. These categories are listed in Table 18-1, along with the percentage of the exam questions they represent. Also listed are chapters that will help you review. Although not every possible exam question is covered in *SPSS Statistics For Dummies*, 4th edition or in this workbook, together they will give you an excellent head start.

Table 18-1 Breakdown of Chapters in Relation to SPSS Certification Exam Topics

Major Categories	Percent of Exam	Chapters to Review in SPSS Statistics For Dummies	Chapters to Review in This Workbook
Basic inferential statistics	22%	15–19	7, 8, 9, 10, 11
Data management	15%	11	3,
Data transformations	16%	8–10	4, 5,
Data understanding and Descriptives	9%	25	2, 6
Operations and SPSS Syntax	15%	26 and 27	15, 16
Output	7%	12, 13, 24, and 25	1, 12, 13, 14
Reading and defining data	16%	4, 5, and 7	1, 2

Although the following questions are not from the exam, they will give you an indication as to whether you are ready for the official exam.

1 Which of the following could be used to determine the variable in a regression that has the largest effect on the dependent?

a. R^2

b. Adjusted R^2

c. Beta coefficient

d. Standardized beta coefficient

2 Which of the following is a non-parametric test?

a. Independent samples t-test

b. Kruskal-Wallace

c. Paired samples t-test

d. ANOVA

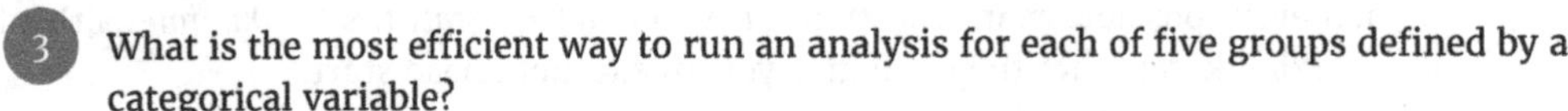

3 What is the most efficient way to run an analysis for each of five groups defined by a categorical variable?

a. Split file

b. Recode

c. Aggregate

d. Split cases

4. Suppose that have a categorical string variable identifying 40 sales offices by city name. What is the easiest way to convert that variable into a numeric integer variable with value labels?

   a. Recode into Same

   b. Auto Recode

   c. LABELS command

   d. STRING command

---

5. How does Optimal Binning determine cut points?

   a. By using equally sized groups to determine cut points

   b. By using equal width intervals to determine cut points

   c. By using an algorithm to determine cut points

   d. By using custom settings to determine cut points

---

6. Which is the definition of a range?

   a. The difference between the mean and the median

   b. The difference between the minimum and the maximum

   c. The difference between the mean and the trimmed mean

   d. The difference between the 75th and the 25th percentile

---

7. Which of the following SPSS Syntax commands would prompt a SELECT IF to cancel after the next procedure?

   a. STOP

   b. UNTIL

   c. TEMPORARY

   d. END IF

8 **What is the most efficient way to make the text in several pivot tables more compact to reduce the number of pages in a report?**

a. Change the font size in each cell.

b. Use the Custom Tables module.

c. Use TableLooks.

d. Use the Output Management System.

9 **What is the appropriate level of measurement for a five-point Likert scale variable ranging from strongly agree to strongly disagree?**

a. Categorical

b. Scale

c. Ordinal

d. String

10 **How do system missing values appear in an SPSS dataset?**

a. As a dot

b. As a dash

c. As a blank

d. As a null

Answers to Practice Questions

1. D. R^2 and adjusted R^2 describe overall fit and not the contribution of an individual variable. Beta coefficients can be misleading in addressing this issue if the compared variables do not all have a uniform range of possible values. See Chapter 11 for more context.
2. B. Only Kruskal-Wallace is a non-parametric test. Non-parametric tests are recommended when you fail to meet distributional assumptions.
3. A. A split file is used to run an analysis more than once for each group in the split variable.
4. B. Auto Recode can do the task in one step, taking a string variable and assigning each string as a label associated with a numeric integer value in a new variable.
5. C. Optimal Binning uses a machine-learning algorithm similar to decision trees.
6. B. The range is the difference between the minimum and the maximum. By the way, the difference between the 75th and 25th percentile is called the *interquartile range.*
7. C. The TEMPORARY command is the answer. STOP and END IF are commands, but neither can accomplish this action. UNTIL is not a keyword in SPSS Syntax.
8. B. The Custom Tables module allows users to modify pivot tables in a multitude of ways, including reducing the size of the tables.
9. C. TableLooks is the answer. Editing the fonts in the tables manually would be very inefficient. Custom Tables and Output Management System are powerful but perform different tasks than the one described here. Custom Tables often produces much more compact tables, but it does so by restructuring the tables and not merely changing the fonts.
10. A. System missing values are indicated with a dot in the data window.

Index

A

B

C

D

E

F

G

H

I

K

L

M

N

O

P

R

S

T

U

V

W

Z

About the Authors

Jesus Salcedo is an independent statistical and data-mining consultant. He is also a senior research analyst at Wiley Publishing, as well as a faculty member for the University of California Irvine Continuing Education Data Science Program. Jesus has a PhD in Psychometrics from Fordham University and has been using SPSS products for over 30 years. He is a former SPSS curriculum team lead and senior education specialist who has written numerous SPSS training courses and trained thousands of users, and he coauthored (with Keith) *SPSS Statistics For Dummies,* 4th Edition, (Wiley) in addition to various other data analytic books. When not working, Jesus enjoys playing racquet sports and fantasy baseball. He also loves to body surf, travel, and eat foods from all cuisines.

Keith McCormick moved to Raleigh-Durham, North Carolina, in the late 90s, planning on graduate school, but a "part-time" SPSS training job turned out to be a career. Keith has been all over the world consulting and conducting training in all things SPSS and statistics. Several thousand students later, he has learned a thing or two about how to get folks started in SPSS. Currently, his focus is on authoring courses on the LinkedIn Learning platform, where he has taught over 700,000 learners about machine learning, data science, and artificial intelligence. Keith is also a sought-after speaker and a consultant currently serving as Pandata's executive data scientist in residence. An avid consumer of data science and machine-learning news, he likes to share his take on his blog at `https://keithmccormick.com/` or in his news feed on LinkedIn. Between assignments, Keith likes to travel to out-of-the-way places, try new food, hike around, and find cool souvenirs to bring home.

Dedication

To Ocean and their love of books

Authors' Acknowledgments

We would like to thank Susan, Levor, and Lindsay for all their hard work and countless hours devoted to this project. Thank you so very much. Jesus also sends a very special THANK YOU to Emily, for putting up with me on those very difficult days when I could not buy an ounce of inspiration — I really appreciate your patience.

Publisher's Acknowledgments

Executive Editor: Lindsay Sandman Lefevere

Project and Copy Editor: Susan Pink

Technical Editor: Levor DaCosta

Production Editor: Saikarthick Kumarasamy

Cover Image: ©Amiak/Getty Images